Local and Regional Development

Local and regional development is an increasingly global issue. For localities and regions, the challenge of enhancing prosperity, improving well-being and increasing living standards has become acute for localities and regions formerly considered discrete parts of the 'developed' and 'developing' worlds. Amid concern over the definitions and sustainability of 'development', a spectre has emerged of deepened unevenness and sharpened inequalities in the development prospects for particular social groups and territories.

Local and Regional Development engages and addresses the key questions: what are the principles and values that shape definitions and strategies of local and regional development? What are the conceptual and theoretical frameworks capable of understanding and interpreting local and regional development? What are the main policy interventions and instruments? How do localities and regions attempt to effect development in practice? What kinds of local and regional development should we be pursuing?

Local and Regional Development addresses the fundamental issues of 'what kind of local and regional development and for whom?', frameworks of understanding, and instruments and policies. *Local and Regional Development* outlines what a holistic, progressive and sustainable local and regional development might constitute before reflecting on its limits and political renewal. With the growing international importance of local and regional development, this book is an essential student purchase, illustrated throughout with maps, figures and case studies from Asia, Europe, and Central and North America.

Andy Pike is Senior Lecturer in the Centre for Urban and Regional Development Studies (CURDS), University of Newcastle, UK.

Andrés Rodríguez-Pose is Professor of Economic Geography in the Department of Geography and Environment, London School of Economics, UK.

John Tomaney is Professor of Regional Governance and Director in the Centre for Urban and Regional Development Studies (CURDS), University of Newcastle, UK.

Local and Regional Development

Andy Pike, Andrés Rodríguez-Pose
and John Tomaney

LONDON AND NEW YORK

First published 2006
by Routledge
2 Park Square, Milton Park, Abingdon, Oxon OX14 4RN

Simultaneously published in the USA and Canada
by Routledge
270 Madison Ave, New York, NY 10016

Routledge is an imprint of the Taylor & Francis Group, an informa business

Typeset in Times New Roman by
Florence Production Ltd, Stoodleigh, Devon
Printed and bound in Great Britain by
TJ International Ltd, Padstow, Cornwall

British Library Cataloguing in Publication Data
A catalogue record for this book is available from the British Library

Library of Congress Cataloging in Publication Data
Pike, Andy 1968–
Local and regional development / Andy Pike, Andrés Rodríguez-Pose and John Tomaney.
p. cm.
Includes bibliographical references and index.
1. Globalisation–Economic aspects. 2. Regional economics.
3. Regional planning. 4. Regionalism–Economic aspects.
5. Economic development. I. Rodríguez-Pose, Andrés.
II. Tomaney, John 1963– III. Title.
HF1359.P545 2006
338.9–dc22 2006005421

ISBN10: 0–415–35717–9 ISBN13: 978–0–415–35717–3 (hbk)
ISBN10: 0–415–35718–7 ISBN13: 978–0–415–35718–7 (pbk)
ISBN10: 0–203–00306–3 ISBN13: 978–0–203–00306–0 (ebk)

For Michelle, Ella, Connell and my parents

For Leticia, Luis and Laura

For Helen and Kate

CONTENTS

PLATES

FIGURES

TABLES

EXAMPLES

ACKNOWLEDGEMENTS

Writing this textbook has been a necessarily and enjoyably collective endeavour. Thanks to Andrew Mould for encouraging us to develop our need for a textbook into the task of writing a textbook, and for supporting us along the way. The insights and ruminations of colleagues have provided inspiration. The Centre for Urban and Regional Development Studies (CURDS), University of Newcastle, continues to provide a supportive context and culture. In particular, we have welcomed the advice and criticism of Young-Chool Choi, Mike Coombes, Stuart Dawley, Andy Gillespie, Sara Gonzalez, Nick Henry, Peter Hetherington, Lynne Humphrey, Neill Marshall, Peter O'Brien, Jane Pollard, Ranald Richardson and Alison Stenning. The participants in the Local and Regional Development Masters programmes at CURDS and the Local Economic Development programme at the London School of Economics have provided a tough proving ground for many of the ideas developed in the book. We thank them for their input. We also acknowledge David Grover's research support for parts of Chapter 5. The final manuscript has benefited from the comments of the anonymous reviewers. We thank them for their efforts in engaging with our work. Thanks to Michelle Wood for the cover art and figures, David Hume for the maps, Michele Allan for help with the images and in particular Sue Robson as well as Amanda Lane for sorting out the manuscript. Andy Pike would like to thank Michelle, Ella and Connell for all their support. Andrés Rodríguez-Pose is grateful to the Royal Society-Wolfson Research Merit Award for financial support during the writing of this book.

We are grateful to those listed for permission to reproduce copyright material:

Figure 2.3 'Average prosperity and equality of distribution by country, 2002' in *Realizing Canada's Prosperity Potential* (2005), reprinted with the permission of the Institute for Competitiveness and Prosperity, Toronto.

Figure 3.10 'Data', information and knowledge' by Burton-Jones, A. (1999) *Knowledge Capitalism*, Oxford: Oxford University Press.

Figure 3.15 'The "economic" in capitalism and non-capitalim' by Ken Byrne in 'Imagining and Enacting Noncapitalist Futures', Community Economies Collective (2001) www.communityeconomies.org, reprinted with permission of Professor Katherine Gibson.

Figure 4.6 'Fragment of an emerging global hierarchy of economic and political relations' by Scott, A. (1998) in *Regions and the World Economy: The Coming Shape of Global Production*, reprinted with the permission of Oxford University Press.

Every effort has been made to contact copyright holders for their permission to reprint material in this book. The publisher would be grateful to hear from any copyright holder who is not here acknowledged and will undertake to rectify any errors or omissions in future editions of this book.

ABBREVIATIONS

ASEAN	Association of South East Asian Nations
CEC	Commission for the European Communities
CED	Community Economic Development
CPE	centrally planned economies
EEC	European Economic Community
EPZ	Export Processing Zones
EU	European Union
FDI	foreign direct investment
FEZ	Free Economic Zone
FTAA	Free Trade Area of the Americas
GATT	General Agreement on Tariffs and Trade
GDP	gross domestic product
GNP	gross national product
GVA	gross value added
ILO	International Labour Organisation
IMF	International Monetary Fund
LETS	Local Exchange Trading Schemes
M&A	merger and acquistition
Mercosur	*Mercado Común del Sur*
MNC	multinational corporation
MW	minimum wage
NAFTA	North American Free Trade Agreement
NESC	National Economic and Social Council
NUTS	*Nomenclature des Unités Territoriales Statistiques* (Nomenclature of Territorial Units for Statistics)
OECD	Organisation for Economic Cooperation and Development
R&D	research and development
RDA	regional development agency
SMEs	small and medium-sized enterprises
TNC	transnational corporation
TUC	Trades Union Congress
UNCTAD	United Nations Conference on Trade and Development
VAT	value added tax
WTO	World Trade Organisation

PART I

Introduction

INTRODUCTION: LOCAL AND REGIONAL DEVELOPMENT

1

Introduction: local and regional development in a global context

Local and regional development has become an increasingly important activity for national as well as local and regional governments across the world since the 1960s and 1970s. In parallel, the context for local and regional development has been dramatically reshaped by deep seated changes in the pattern of economic activity and has become significantly more challenging. First, an alleged qualitative shift has taken place towards a more 'reflexive' capitalism characterised by heightened complexity, uncertainty, risk and rapidity of economic, social, political and cultural change. The economic system has become more internationalised, even 'globalised', knowledge intensive and competitive. While the long-term prospects for the emergent global economy remain uncertain, there is little doubt that its contemporary emergence has raised the spectre of deepened unevenness in the prospects for development for particular social groups and territories and sharpened geographical inequalities in prosperity and well-being. The inclusive and sustainable nature of territorial growth and development has begun to be challenged. Fundamental questions about what constitutes 'success' and 'development' in localities and regions are being posed.

Second, and closely related, structures of government and governance are evolving into multilevel, often devolving systems, working across and between the local, regional, subnational, national and supranational scales. Existing institutions have been reorganised, new institutions have emerged and new relations, often based around 'partnership', have dominated the governance of local and regional development. Third, the reshaped terrain of local and regional development has stimulated new interventions, through instruments and public policies, seeking to harness both internal and external forms of growth and development. Different localities and regions have been able to exercise differing degrees of agency in reshaping existing and developing new approaches and experiments for local and regional development.

Fourth, debates about local and regional development have shifted from a focus on the quantity of development to a concern with its quality. Initially, this involved a focus on the impact of economic development on the natural environment and the constraints this placed on development, but has evolved into a more general concern with questions of the quality of life. This new concern with *sustainable development* has become pervasive, but the term and its implications are highly contested. There are many definitions of sustainable development, but perhaps the best known is that of the World Commission on Environment and Development, or 'Brundtland Commission', which calls for development which 'meets the needs of the present without compromising the ability of future generations to meet their own needs' (World Commission on Environment and Development 1987: 8). Such a definition still leaves considerable scope for dispute (see Williams and Millington 2004). At one end of the continuum are approaches to sustainable development that tend to focus, for instance, on the development of renewable resources and the more efficient exploitation of existing resources, and places faith in technological solutions to ecological problems (sometimes described as 'ecological modernisation'). In this perspective, environmental practices themselves can help stimulate new rounds of economic growth (Gibbs 2000; Murphy 2000a). At the other end of the continuum are versions of sustainability that tend to view the resources of the planet as finite and, consequently, lead to a focus on limiting economic growth, or even reducing it. Certainly, this latter view generally involves a critique of the supposedly wasteful consumption practices of industrial society and, instead, a focus on the promotion of 'well-being' rather than the production and consumption of material goods and an emphasis on greater self-reliance and local development strategies which are respectful of nature (Hines 2000). At whichever point on the continuum, mainstream conceptions of local and regional development look too 'economistic' when measured against the rubric of sustainable development, while traditional measures of *growth* offer, at best, only a partial or intermediate indicator of *development*. Moving from the former to the latter requires new metrics of local and regional development, which focus not just on jobs and incomes, but more broadly on the quality of life (Morgan 2004; see also Nussbaum and Sen 1993; Sen 1999).

An important driver of the changes in the context of local and regional development has been the process of globalisation. Through its increased mobility of capital, workers, goods and services, globalisation is changing the rules by which the economy has been governed during much of the post-war era. Globalisation – which, to a certain extent, may be a political and socially constructed concept by states and neo-liberal economic actors (Peck 1999; Jessop 2002) – is exposing even the most remote spaces to competition and forcing firms, localities and regions to react and adjust to the new economic conditions. Economic and social actors across the world are restructuring their production and consumption habits as a result. This process offers new potentials and challenges. Some firms and places across the world have grasped the opportunities presented by a globalising economy and have established conditions whereby they currently reap the benefits. However, the opening of national economies is also revealing local and regional economic structures with little or no capacity to compete in a globalised environment. The exposure of inefficient, inadequately managed and often low-tech production structures to competition from outside is leading to the restructuring and even demise of local production structures, generating unemployment, and, in others, to a degradation of working conditions (Stiglitz 2002).

Although some claim that this process of globalisation is not really new (Williamson 1997; Hirst and Thompson 1999), the increase in the level of interaction among national economies over recent decades cannot be ignored. Since the late 1980s, trade has expanded significantly and total world foreign direct investment (FDI) flows have increased fivefold (International Monetary Fund (IMF) 2000). Many countries have opened their borders, switching from either *dirigiste* – state-directed – economic planning systems, import substitution industrialisation, or centrally planned economies to varying degrees of liberal free-market structures. The liberalisation of national economies has often been accompanied by macroeconomic stability packages focused upon curbing inflation, reducing fiscal deficits and debt, and low interest rates to encourage long-term investment.

The opening of national economies is in tune with a large body of research in economics, echoing classical theories of comparative advantage discussed in Chapter 3, which both from a theoretical and an empirical perspective has underlined the economic benefits of open economies. The works of Grossman and Helpman (1991), Coe and Helpman (1995), Frankel and Romer (1999) and Fischer (2003) have emphasised the greater capacity of open economies to benefit from trade, capital mobility, technological spillovers and transfers of technology. The restructuring and productivity effects of liberalisation and regional integration have also been highlighted (e.g. Kang and Johansson 2000). Empirically-based research has tended to confirm the supposedly superior economic performance of open economies. Sachs and Warner (1995, 1997), Coe *et al.* (1997) and Fischer (2003) determine the existence of a strong positive relationship between the degree of openness of a country and its economic growth, as a result of their capacity to reap the benefits from an increased mobility of capital and technology. Others, led by Rodrik (2000), Stiglitz (2002) or Wade (2004), have, in contrast, questioned the beneficial effects of trade and the opening of borders for economic growth with evidence of the evolution of social and regional disparities and inequalities. Moreover, talk of globalisation should not lead us to forget the enduring importance of the nation state as a regulator of economic activity, including at the local and regional level.

The combination of economic liberalisation with macroeconomic stability packages has had some positive results. The most spectacular effect has been the reduction of inflation from double or triple digit figures to single digit figures in most countries in the world (Kroszner 2003). There has been a rapid expansion in capital flows to more open countries, export growth has also flourished and economic growth has tended to become less volatile than in the 1980s (Ramey and Ramey 1995; Quinn and Woolley 2001). On the negative side, liberalisation has not been accompanied (with relatively few exceptions where national state interventions have been pivotal such as China, India, or Ireland) by sustained long-term high economic growth or by high employment growth. In a number of countries around the world, recent growth has been lower even than in the 'lost decade' of the 1980s. In addition, economic liberalisation and macroeconomic stability measures are not without risk. The 'Tequila effect' of 1995 in Mexico, the Asian crisis of 1997 or the 2002 Argentine crisis highlight the macroeconomic vulnerability of countries whose fiscal management or currencies were perceived as weak by external investors and whose industries were often unable to cope with rapid restructuring and/or to face competition from either higher technology goods from developed countries or cheaper products from other developing economies.

Globalisation and economic deregulation may also be contributing to the increase of social and territorial inequalities within many of the countries that have liberalised their

Plate 1.1 ***Globalisation and the rise of East Asia: the Singapore waterfront***

Source: Photograph by David Charles

economies (Wade 2004). From a social perspective, there is substantial evidence of increasing social inequality, leading to the exclusion of particular social groups and/or places from mainstream prosperity and well-being (Dowling 1999). Increases in productivity and growth – whenever and wherever they happen – are more related than ever to technological progress. The introduction of new production plants or of new agricultural methods of production is generating greater productivity and efficiency, but frequently at the expense of employment. This frequently jobless economic growth is contributing to the exclusion of large numbers of unskilled workers and to the expansion of the informal economy, both in the 'developed' and in the 'developing' worlds (Schneider and Enste 2000). As a consequence, the economy that seems to be emerging from the process of globalisation is characterised by greater social and often geographical polarisation. The divide between the highly educated and stable wage earners and an increasing group of precarious workers and workers in the informal economy seems to be growing at a greater pace than ever (Esping-Andersen 1999).

From a territorial point of view, only a limited number of localities and regions seem to be reaping the benefits from the new opportunities provided by the process of globalisation. In general, successful regions tend to be those that have something distinctive to offer to markets that expand beyond the traditional realm of the local and regional spheres. The 'winning' regions can be divided into three categories:

- *Large metropolitan regions*: Large urban agglomerations in both the 'developed'and the 'developing' worlds are where many of the high value-added service activities are concentrated. Business, financial, real estate and insurance services are clustering more than ever in large urban regions, as are the headquarters of corporations (Taylor and Walker 2001). The economies of agglomeration derived from such concentration of production factors are leading to the attraction of research and development and design activities to global metropolises of city-regions (Scott and Storper 2003). FDI is also flocking to large metropolitan areas, reinforcing subnational social and economic disparities. For example, Mexico City and its surrounding state have received more than 60 per cent of all FDI in Mexico; Madrid has attracted more than 70 per cent of all FDI flowing into Spain. However, as mentioned earlier, the dynamism of large urban areas does not mean that all its inhabitants have benefited equally. A majority of the large urban agglomerations around the globe suffer from the emergence of a dual economy, in which wealth and high productivity jobs coexist with economic and social deprivation, a growing informal sector, and low paid, precarious jobs in the service sector (Buck *et al.* 2002; Hamnett 2003).
- *Intermediate industrial regions*: The second group of territories that seem to be profiting from the greater mobility of production factors around the world are the intermediate industrial regions. This type of area often combines labour cost advantages with respect to core areas, with human capital and accessibility advantages with respect to peripheral areas, making them attractive locations for new industrial investment. Mountain states and provinces in the United States and Canada are attracting large industrial investments fleeing the old industrial 'rustbelts' of the Eastern and Great Lake areas of North America. Numerous intermediate European

regions in central Italy or southern Germany and France are witnessing a similar trend. From a global point of view, the most advanced regions in the 'developing' world can also be considered as intermediate industrial regions. This is the case of the Mexican states bordering the United States, of São Paulo and the southern states of Brazil, of Karnataka and Maharashtra in India, but, above all, of the coastal provinces of China. The combination of low wages with a relatively skilled and productive labour force and accessibility to markets has made them primary targets for industrial investment. Much of today's mass production is concentrated in these areas.

- *Tourist regions*: Among the regions in the 'developing' world that have managed to find their market niche in a globalised economy are the tourist areas. Places like Cancún in Mexico or Bali in Indonesia have thrived thanks to their capacity to attract large number of tourists from all over the world. Others, without reaching a similar success, have built up a healthy and relatively successful tourist industry.

However, the dynamic areas in a globalised world tend to be the exceptions rather than the rule. More often than not, regions and localities struggle to adapt their economic fabric to the emergent conditions. Globalisation has made economic activity relatively more mobile or footloose. Yet, the ability to invest globally has heightened sensitivity to local and regional differences (Storper 1997). The competitive advantage that certain territories enjoyed in the past as a result of their unique conditions or their proximity to raw materials is becoming less important. Improvements in information technology are contributing to 'delocalise' industrial and agricultural production. A lower degree of delocalisation is occurring in services, often due to their need for face-to-face interaction and trust, although the fate of market services is often linked to the dynamism of economic activity in other sectors. As a consequence, traditional industrial regions, agricultural areas and regions without a clear comparative advantage are finding it difficult to capture new markets and their companies are often losing share in their own traditional markets as a result of the opening of national economies to competition. Basic and mass production industrial companies that had survived and often thrived in conditions of monopoly or oligopoly under fragmented national markets are in many cases crumbling under the market integration and pressures of competition. Traditional agricultural regions have seen their markets invaded by cheaper agricultural produce from more technologically advanced regions, and areas with a strong agricultural potential have to deal with an imperfect and relatively closed world food market (Henson and Loader 2001).

The outcome of recent economic processes is greater economic and social polarisation at the world level (Rodríguez-Pose and Gill 2004). Whereas some national economies such as those in South East Asia, China or Ireland have prospered, albeit with accompanying social and territorial inequalities, under the new conditions, many old industrial and relatively lagging areas in the developed world have struggled, while numerous African, Middle Eastern and Central Asian economies are becoming increasingly detached from world economic circuits. Gross domestic product (GDP) per capita in numerous African countries has stagnated, especially in the 1980s (Bloom and Sachs 1998).

The world's economic polarisation is being reproduced within countries. Different regional capacities to adapt to the new economic context are leading to a greater concentration of economic activity and wealth in a few regions in each country and to increasing economic divergence within countries. As a result, internal economic imbalances are growing in high, low and middle-income countries. Table 1.1 presents one measure of the evolution of regional disparities – in this case, the variance of the natural logarithm of regional GDP per capita – in selected 'developed' and 'developing' countries of the world between 1980 and 2000 or the latest year with available regional information. Several features need to be highlighted from the table. First, the difference in the dimension of internal disparities between 'developed' and 'developing' countries. In 2000, internal economic disparities in Brazil, China, India or Mexico, were twice the size of internal disparities in Spain and three times those in France or the United States. Second, all countries included in the sample, except Brazil, have seen internal economic imbalances grow since 1980. However, whereas greater economic polarisation took place in the United States, France and Germany in the 1980s, the greatest increase in economic imbalances in the lower income countries has taken place during the 1990s. Between 1990 and 2000, the variance of the log of regional GDP per capita has grown by 1.2 per cent in Brazil, 3 per cent in Italy, 11.6 per cent in Spain, 13.6 per cent in Mexico, almost 17 per cent in India and 20 per cent in China.

This internal polarisation has often coincided with the opening of national economies. Whereas in Mexico the 1970s and early 1980s had been characterised by a reduction of internal economic disparities, the opening of the country's borders to trade from 1985 onwards, reinforced by the North American Free Trade Agreement (NAFTA) after 1994,

Table 1.1 ***Variance of the log of regional GDP per capita, 1980–2000***

	Year			*% Change*		
	1980	*1990*	*2000*	*1980–1990*	*1990–2000*	*1980–2000*
'Developing' countries						
China	0.578	0.483	0.581	–16.31	20.20	0.60
India	0.352	0.377	0.441	7.10	16.98	25.28
Mexico	0.388	0.383	0.435	–1.29	13.58	12.11
Brazil	0.588	0.488	0.494	–17.01	1.23	–15.99
'Developed' countries						
United States	0.136	0.152	0.148	11.76	–2.63	8.82
Germany	0.184	0.188	0.186	2.17	–1.06	1.09
Italy	0.265	0.269	0.277	1.51	2.97	4.53
Spain	0.207	0.199	0.222	–3.86	11.56	7.25
France	0.151	0.164	0.163	8.67	–0.29	8.36
Greece	0.156	0.158	0.158	1.21	0.16	1.37
Portugal		0.231	0.236		1.85	
European Union		0.247	0.275		11.24	

Source: Rodríguez-Pose and Gill (2004: 2098)

Notes: Data for Europe: EU 1980–1999; Greece 1981–1999; France 1982–1999. All others as shown. Regional data from EUROSTAT and national statistical offices.

led to an increasing concentration of economic activity in Mexico City and the states along the border with the United States (Sánchez-Reaza and Rodríguez-Pose 2002). Similarly, the increase in internal regional disparities in Brazil has coincided with the progressive opening of the country's economy since the early 1990s (Azzoni 2001). The combination of increasing social and territorial inequality and the greater concentration of high value-added economic activity in core regions and a few peripheral areas that have found a market niche in a globalised economy is leaving numerous – and many of the poorest – areas of the world in a very precarious situation. The challenge of local and regional development has sharpened since the 1970s.

Territorial competition

In the context of globalisation, the contest to attract and retain mobile capital and labour has led to suggestions that localities and regions are now in direct competition with each other. The existence of such territorial competition has focused attention on the 'competitiveness' of local and regional economies as institutions try to provide the conditions that will attract and embed investment. Krugman (1995) has questioned the value of the idea of competitiveness in relation to national economies, labelling it a 'dangerous obsession'. He examines the way it is used to explain national economic performance, arguing that it rests on an inappropriate analogy between the firm and the nation because, unlike firms, countries 'do not go out of business' (Krugman 1995: 31).

Krugman refutes the idea that countries are in competition with each other, arguing that domestic living standards are determined by improvements in domestic productivity. 'Competitiveness', on the other hand, is merely a useful political metaphor which policy-makers deploy to justify policy choices such as supporting particular economic sectors (Krugman 1995, 1996). Krugman (1994) acknowledges, however, that the idea of 'regional competitiveness' may make more sense than 'national competitiveness' because regional economies are more open to trade than national economies and factors of production move more easily in and out of a region than a national economy. Thus, the notion of territorial competition may have some utility when applied to the local or regional level. Camagni (2002) has argued that regions can effectively go out of business, insofar as they are affected by out-migration and abandonment. In policy terms too, as parts of nation states, localities and regions do not have access to the range of policy instruments, such as currency devaluations, which national governments have used traditionally in order to shape levels of economic activity. Localities and regions, then, must compete on the basis of local or regional competitive advantage in order to attract mobile investment. Camagni maintains:

> What really counts nowadays are two orders of factors and process: in an aggregate, macroeconomic approach increasing returns linked to cumulative development processes and the agglomeration of activities; in a microeconomic and microterritorial approach, the specific advantages *strategically* created by the single firms, territorial synergies and co-operation capability *enhanced* by an imaginative and proactive public administration, externalities provided by local

> and national governments and the specificities historically *built* by a territorial culture.
>
> (Camagni 2002: 2405, original emphases)

The implication of Camagni's argument is that localities and regions can and must enter the field of territorial competition. Nevertheless, much of the writing about territorial competition assumes that such strategies are wasteful (see Logan and Molotch 1977, for a classical account). Rodríguez-Pose and Arbix (2001) provide an example of 'pure waste' in the form of the bidding war between Brazilian states aimed at attracting and embedding new investments in the car industry in the 1990s. They show how all states found themselves caught up in the struggle to attract investment justifying it as means of regenerating local and regional economies and generating employment. States frequently exaggerated the alleged benefits of investments to justify the provision of ever larger financial incentives to TNCs, even at the risk of bankrupting their treasuries. As a result, such competition was undermining the potential long-term benefits of FDI. Rodríguez-Pose and Arbix (2001: 150) conclude that 'The bidding wars, presented by state governments as their main – and almost only – development strategy, are a pure waste since they do not lead to a significant increase in welfare at the local, or the national level'.

In the contemporary period, territorial competition involves not just efforts to attract manufacturing firms but also activities which will generate consumption. This form of territorial competition is often pursued by large cities and metropolitan regions. A good example of this type of activity is the competition for international sporting events such as the Olympics, or the award of 'Capital of Culture' status to cities in the European Union (Owen 2002; Shoval 2002; Garcia 2004). Such competition occurs especially between world cities aimed at enhancing their global status in an era of growing inter-urban competition to finance large-scale planned construction projects in those cities. According to Cheshire and Gordon (1998), the incentive to engage in this type of territorial competition is strongest in the economically stronger, leading metropolitan regions and it frequently works against wider spatial equity. This points to the need for effective regulation of competition between localities and regions at the national, or even supranational level. Reese (1992), using evidence from North America, suggests that appropriate forms of regulation can generate positive sum policies. He contrasts the more innovative approach to attracting mobile investment pursued in Ontario, which we discuss in more detail in Chapter 7, where there are federal limitations on financial incentives, with US cities where there are not.

It is probably necessary to distinguish, then, between different types of territorial competition, recognising that some are inherently wasteful, while others may have positive sum effects (Reese 1992; Cheshire and Gordon 1998; Camagni 2002; Malecki 2004). Echoing our discussion in Chapter 2, Malecki contrasts 'imitative "low-road" policies with "high road", knowledge-based policies' (2004: 1103). The Brazilian case described above would fall firmly into the former category. But, other forms of local and regional policy intervention aimed at generating more broadly-based forms of growth or enhancing the networks that underpin local synergies or embedding external firms in local networks to exploit spillovers and increasing returns 'are at the very base

Table 1.2 ***Territorially competitive policies***

Zero sum	*Growth enhancing*	*Network enhancing*
Pure promotion	Training	Internal networking
Capturing mobile investment	Fostering entrepreneurship	External (non-local)
Investment subsidies	Helping new firms	Benchmarking assessments
	Business advice	Airline and air-freight links
Subsidised premises	Uncertainty reduction	Scanning globally for new knowledge
	Coordination	
	Infrastructure investment	

Sources: Adapted from Cheshire and Gordon (1998: 325); Malecki (2004: 113)

of economic development, in its positive-sum, "generative" sense' (Malecki 2004: 1114). Table 1.2 provides examples of territorial competition policies. Cheshire and Gordon conclude:

> The lesson for local policy makers would seem to be: nurture the successful firms already present. Given the evidence presented as to why firms become mobile, one of the easiest and most effective ways of doing this is likely to be by pursuing policies which ensure a ready supply of reasonably cheap premises. Where policy does aim at mobile investment, qualities-based (as distinct from price-based) strategies are more likely to provide gains for local territorial agencies of the communities they represent.
>
> (Cheshire and Gordon 1998: 342)

As we discuss in Chapter 6, the interventions of local and regional development institutions, thus, have a critical role in the attraction and embedding of exogenous resources such as FDI, occupational groups and other internationally mobile activities.

The need for alternative development strategies for localities and regions

What are the options for the people, firms and communities in the localities and regions that are struggling with the new economic situation? What approaches can be taken to address the weaknesses that limit the economic potential of individuals, firms and territories globally? People, firms and societies may need to raise their awareness of – and become more capable to respond to and, perhaps, more able to shape – the challenges presented by the new economic conditions. Within an increasingly dominant and pervasive capitalist global economy, an increased capacity to respond and adjust to global challenges necessarily implies endowing individuals, firms and territories with the factors that will allow them to place their skills, products or services in the global marketplace and to compete with others. Alternatively, it may mean constructing shelters and bulwarks against the harsh forces of global competition by forming assets and resources focused upon local and regional needs and aspirations.

There is, however, no simple and universal way to tackle the challenges posed by globalisation. No unique or universal strategy can be applied to every area or region, regardless of the local context. Past experience has shown that the mere reproduction of development policies in different contexts has more often than not had little or no impact on the generation of sustainable local and regional development and long-term employment (Storper 1997). Traditional top-down policies aimed at achieving economic development have tended to be cut from the same cloth. These have normally consisted of supply-led policies, focused either on infrastructure provision or on the attraction of industries and foreign direct investment. The logic behind this approach was that defective accessibility or the absence of firms that could articulate around them a dynamic industrial tissue and generate technological transfers was at the root of the problems of many lagging areas.

Local and regional development and employment-creation policies have thus, until recently, been usually structured along two axes. The first axis was infrastructural endowment. The supposedly high returns of infrastructural investment identified by some researchers (e.g. Aschauer 1989) fuelled the belief that improving accessibility was the solution for lagging areas. Development and employment policies were thus articulated around the building of motorways, aqueducts, pipelines, telephone lines and other investments in infrastructure. Such investment has unfortunately not always yielded the expected results. One of the most spectacular cases of failure of this sort of top-down and supply-led approach has been the Italian Mezzogiorno, where, despite more than forty years of strong infrastructure investment by the Italian state, the income gap between the North and the South of the country remains at the same level as before the intervention started in the early post-war years (Trigilia 1992). On a wider scale, some studies have also questioned the effectiveness of investment in infrastructure as a sustainable development strategy. Research by Philippe Martin (1999) and Vanhoudt *et al.* (2000) at the European level has unveiled constant or negative economic returns from investment in infrastructure.

The second axis was structured around top-down policies based on industrialisation. The introduction or attraction of large firms to areas with a weak industrial fabric, in combination with other development policies, has been in a few cases – for example in a host of South East Asian countries – a key in the economic take-off of these areas (Storper 1997), often in contexts of strong, state-led national development strategy support. However, these policies have not been particularly successful and the failures outnumber the success stories. Once again the case of the Italian Mezzogiorno is pertinent. Inspired by the 'growth pole' theories of Perroux (1957) discussed in Chapter 3, the establishment during the 1960s and 1970s of shipyards, refineries, car plants and chemical plants in the South of Italy with a relatively weak endogenous industrial fabric did not lead to the desired industrialisation of the South of the country (Viesti 2000). Companies that were lured from the North to the South by the incentive packages offered by the Italian government failed to create around them the industrial linkages and networks that could have delivered sustainable economic growth and employment generation. Inadequate local economic and institutional settings represented a barrier to the creation of networks of local suppliers around the 'imported' large firm, which was the main aim of the policy (Trigilia 1992). As a consequence, most of these large industrial complexes remained detached from their local environments – 'cathedrals in the desert' – whose principal suppliers and customers were located elsewhere rather than locally or in nearby areas (Lipietz 1980). After the demise of the often costly incentive packages that had led to the location of those firms in the South, loss-making firms were left to die *in situ* or moved back to the North.

Similar industrialisation policies in other areas of the world failed also to deliver the expected results. Many of the firms located in less developed cities and regions in France or Spain, again following Perroux's (1957) development pole theory, have not triggered the expected dynamic and innovative effect which was supposed to be at the root of sustainable development (Cuadrado Roura 1994). Similar results have been achieved in most Latin American countries that followed import substitution industrialisation policies until the mid-1980s or the beginning of the 1990s. The protection of national markets in order to foster the emergence of local consumption and, to a lesser extent, intermediate and capital goods industries, led to the creation of a relatively large industrial base in countries such as Mexico, Brazil, Argentina or Chile (see Hernández Laos 1985; Cano 1993). However, the presence of captive markets, direct state subsidies, closeted public procurement, monopolistic and oligopolistic practices and protectionism made most of the industrial base of these countries inefficient relative to world standards. Consumers ended up bearing most of the cost of paying higher prices for products of, in general, lower quality than those available in international markets (Love 1994; Cárdenas 1996). The opening of borders to competition in Latin America has exposed the weaknesses of the industrial base of Latin American countries and led to deindustrialisation and the loss of numerous industrial jobs (Rodríguez-Pose and Tomaney 1999; Dussel Peters 2000).

There are multiple and variable reasons for the failure of traditional local and regional development policies. Some of them are external to the design and implementation of the policies. In some areas, weak or deficient education and skills among people and communities became the main barrier for successful development. In others, weak local economic structures have jeopardised policy efforts towards development. Poorly suited social and institutional contexts have also been highlighted as possible reasons for the poor performance of traditional development policies (North 1990; Rodríguez-Pose 1999). Yet, as important – if not more important – as the external factors are the internal factors related to the design and implementation of the development policies. First among these is the internal imbalance of most traditional development policies. The logic behind most policies was to concentrate on what was perceived to be the most important development bottleneck, with the aim that, once the problem was solved, sustainable development would follow. For example, if the main development bottleneck of an area was perceived to be poor accessibility, heavy investment in transport and communications infrastructure could solve the accessibility problem and, as a consequence, generate internal economic dynamism and bring much needed foreign investment. Sustainable development concerns were not at the forefront of such approaches to local and regional development at this time.

Similarly, the weakness of local industrial tissues could be addressed by luring large firms to the locality or region, which would create direct and indirect jobs, generate technology transfers and spillovers, and trigger entrepreneurship. However, the impact of the implementation of such development policies has generally been disappointing, due to their unbalanced nature. Heavy investment in infrastructure, with little or no emphasis on other development factors such as the support of local firms, the improvement of local human resources, or the diffusion and assimilation of technology, has often created only imperfect accessibility to markets. Where local firms, as a result of their relatively lower levels of competitiveness, have struggled to gain ground in outside markets, more competitive external firms have benefited most from greater accessibility to lagging areas, gaining a greater share of those markets and driving many local firms out of business as a result (see Figure 1.1). The frequent reliance on inward investment has equally not delivered the expected outcomes. Instead of dynamising their environment, and triggering multiplier effects, large industrial complexes brought from other locations have in many cases only been lured by incentives and subsidies and have tended to foster a greater dependency on external economic actors (see Figure 1.1) (Rodríguez-Pose and Arbix 2001; Mytelka 2000).

The second internal factor behind the failure of traditional development policies has been the tendency to replicate standardised policies in different areas of the world, regardless of the local economic, social, political and institutional conditions. Policies that were considered to have succeeded in a specific case have been transferred and implemented almost without changes in different national, regional and local contexts. National planning and development offices, often aided by academics as well as international organisations, were the main culprits behind the universalisation and roll-out

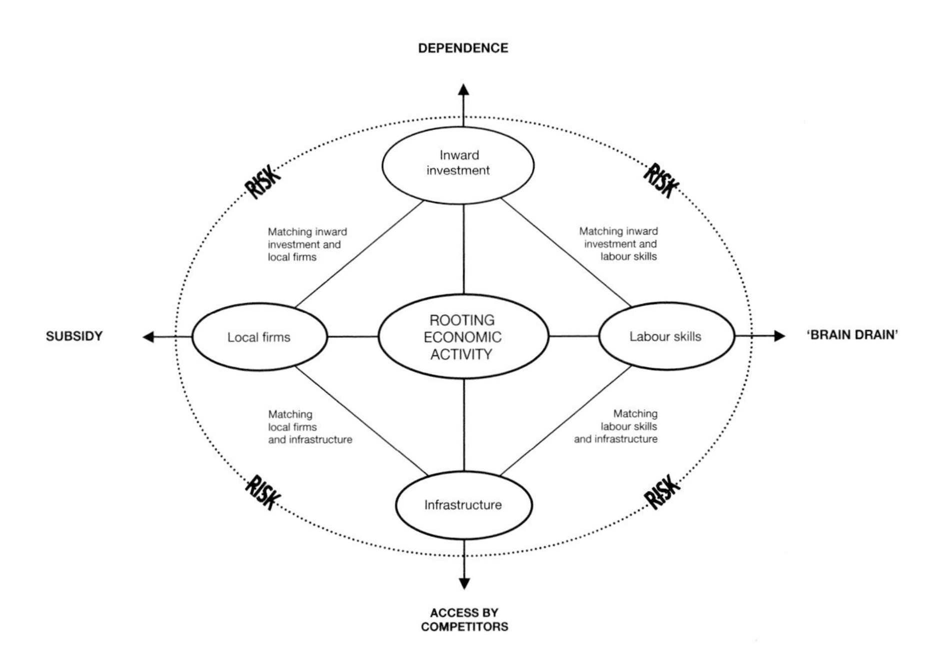

Figure 1.1 ***The bases and the risks of local and regional development strategies***

Source: Adapted from Rodríguez-Pose (2002b: 11)

of such top-down development models and practices. Yet, diverse economic, social and institutional conditions in different local and regional environments mediate the effectiveness of development policies and, in many cases, led to the failure of policies that had proved successful elsewhere. In addition, the reliance on top-down imported policies alienated the local population which, as we discuss in Chapter 4, often had little or no say or participation in building the future economic vision of their territories (Vázquez Barquero 1999, 2003).

The failure of traditional top-down policies, together with the challenges generated by globalisation, has led to a serious rethinking of local and regional development by practitioners and academics. As a result, since 1990 a series of innovative, bottom-up local and regional development policies have emerged (Stöhr 1990; Amin 2000). Although the change from top-down and centralised policies to a bottom-up local and regional development approach neither has been established overnight, nor is based on a single or clearly defined theoretical underpinning, this model of tailor-made approaches to the development of territories has progressively been gaining ground as the foundation for new development strategies (Vázquez Barquero 2003).

The question of what is local and regional development is at the heart of this book and is addressed in Chapter 2. There are many competing theories and models of local and regional development that we review in Chapter 3. There are thus many definitions of local and regional development. Yet, the multitude of theoretical models and the fact that the main sources of inspiration for local and regional development strategies are learning from experience and imitation has prevented the emergence of a widely accepted definition. Therefore, local and regional development strategies tend to resort to outlining the basic features of the approach prior to specifying its particular content. White and Gasser (2001) establish four features that characterise local and regional development strategies: they require participation and social dialogue; they are based on territory; they entail the mobilisation of local resources and competitive advantages; and they are locally owned and managed.

The main differences between local and regional development and traditional top-down approaches are summarised in Table 1.3 and relate to five domains. First, whereas in traditional top-down approaches the decision on where to implement development strategies is typically taken by national central government planners and developers, with little or no involvement of local or regional actors, local and regional development practices favour the promotion of development in *all* territories – not just the underdeveloped or lagging – by mobilising the economic potential and the competitive advantage of every locality and region. The initiative for the development strategy is then taken locally or regionally or with strong local and/or regional support. Second, as a result of where and how the decisions are taken, traditional policies have been generally designed, managed and implemented by national ministries or central government agencies. The involvement of local and regional actors in the delivery and implementation of local and regional development strategies implies, in contrast, a much greater degree of vertical and horizontal coordination of all the actors involved. Vertical coordination entails the synchronisation of local, regional, national and supranational or international institutions. Horizontal coordination comprises local public and private actors concerned with development issues (Table 1.3). This complex multilevel and multi-agent

Table 1.3 ***Top-down and bottom-up local and regional development approaches***

Traditional development policies	*Local and regional development*
1 Top-down approach in which decisions about the areas where intervention is needed are taken in the national centre	1 Promotion of development in all territories with the initiative often coming from below
2 Managed by the national central administration	2 Decentralised, vertical cooperation between different tiers of government and horizontal cooperation between public and private bodies
3 Sectoral approach to development	3 Territorial approach to development (locality, *milieu*)
4 Development of large industrial projects, that will foster other economic activity	4 Use of the development potential of each area, in order to stimulate a progressive adjustment of the local economic system to the changing economic environment
5 Financial support, incentives and subsidies as the main factor of attraction of economic activity	5 Provision of key conditions for the development of economic activity

Source: Authors' own elaboration

context of the government and governance of local and regional development is discussed in Chapter 4.

The third basic difference relates to the type of approach to development. Traditional policies have typically focused upon the promotion of specific industrial sectors that contribute to generate economic dynamism. Local and regional development adopts a territorial approach as a means of achieving economic development. The diagnosis of the economic, social and institutional conditions of every territory and the identification of local economic potential are the foundations upon which such development strategies are built. Closely linked to the sectoral approach of most traditional policies is the development of large industrial projects that were expected to promote additional economic activity and generate the networks and value chains needed in order to achieve sustainable development. The problems of this type of practice were mentioned earlier and have changed the focus and led local and regional development practitioners to identify and use the economic development potential of each area and to stimulate the progressive adjustment of the local and regional socio-economic system to changing economic conditions (Table 1.3).

Finally, the top-down and bottom-up approaches are also set apart by their way of attracting economic activity to localities and regions. While traditional approaches have basically relied upon financial support, incentive packages and subsidies in order to attract and maintain economic activity, local and regional development in general tends to shun such incentive packages and concentrate on the improvement of the basic supply-side conditions for the development and attraction of further economic activity.

According to Vázquez Barquero (1999), local and regional development strategies are usually structured around a threefold scheme that covers the development of economic *hardware*, *software* and '*orgware*'. The development of *hardware* involves many factors common to traditional development policies, such as the provision of basic infrastructure, including the establishment of transport and communication networks, industrial space and the infrastructure for the development of human capital (including education, health and cultural facilities).

The development of *software* implies the design and implementation of comprehensive local development strategies. Based on the diagnosis of the comparative advantages and of the resource bottlenecks of each space, the local interests and organisations or 'stakeholders' – frequently with the participation of outside external experts – define and establish a comprehensive strategy in a bid to fulfil that potential. These strategies are usually articulated around four axes: the improvement of the competitiveness of local firms, the attraction of inward investment, the upgrading of human capital or labour skills, and the building of infrastructure (Figure 1.1). The basic aim is to create a comprehensive and balanced local or regional development strategy that will embed economic activity in a particular territory. From this point of view, the intervention in any of the axes is included in a global strategic framework with the aim of rooting economic activity – in a period when economic activity tends to be both more footloose and sensitive to geographical differences – in a certain area and of fulfilling the economic potential of every territory. This means a comprehensive and balanced strategy in which, for example, any effort put into attracting inward investment is matched by similar and coordinated measures aimed at the improvement of the local economic fabric, of local infrastructure, and of the local labour supply. Similarly, the improvements in labour skills have to be coordinated and synchronised with any effort to boost local firms, to improve infrastructure, and to attract inward resources and so on (Figure 1.1). Such a balanced and integrated approach can be achieved only by the systematic involvement of local economic, social and political actors in the planning and development process and by a careful analysis of the economic potential of any area.

The risks of failing to identify the correct assets, competitive advantages and the structural bottlenecks, or of a poor implementation of the strategy, are high. An excessive emphasis, for example, on the attraction of inward investment is likely to enhance the dependence of an area on external economic actors. Similarly, the improvement of the education and the skills of the population, without a similar improvement of the competitiveness of local industries or without the attraction of foreign resources, may result in a mismatch between the educational and skills supply and labour demand, generating dissatisfaction and possibly 'brain drain' or out-migration of the highly qualified. The upgrading of the competitiveness of local industries not matched by similar progress in labour skills or in the attraction of inward investment may jeopardise in the middle and long run the capacities of local firms to generate and assimilate innovation, and thus undermine their competitiveness (Figure 1.1).

The successful design and implementation of a balanced local or regional development strategy can contribute to generate social, economic and, in many cases, environmentally sustainable development and jobs. And moreover, by making any economic activity taking place in the territory dependent on local conditions and by managing the strategy locally and regionally, it can be inferred that the jobs created are likely to be of a better quality, in the medium and long run, than they would have been if the genesis of employment was exclusively left to local firms with little or no competitive advantage and thus in a very precarious market situation or to outside companies with little or no links in the form of established supply or customer chains to the locality (Rodríguez-Pose 1999).

Finally, any local or regional development strategy also entails what Vázquez Barquero (1999) calls *orgware*, that is, the improvement of the organisational and institutional capacity to design, implement and monitor the whole development strategy. The development of *orgware* goes beyond a mere vertical and horizontal coordination of different levels of government and of local public and private actors and raises important issues of governance that need to be addressed by common institutions (Newman 2000; Hauswirth *et al.* 2003; Leibovitz 2003). The genesis of complex governance systems associated with local and regional development initiatives often involve new forms of cooperation and regional coordination (Brenner 2003: 297). The development of institutions and governance systems also aspires to contribute to the empowerment of the population and to help individuals and communities take charge of their own future. It also fosters the development of civil society and promotes the formation of the networks and partnerships that are fundamental to processes of economic and social progress. It has to be borne in mind however that, although the empowering of local societies is a crucial element in any local and regional development strategy, it cannot be considered as its only goal. It is a means to the end of attaining social, economic and environmentally sustainable development and generating greater economic dynamism and employment.

Overall, there are numerous social and economic advantages related to the adoption of local and regional development strategies in a globalised world in comparison to the resort to traditional development programmes. The social advantages may include the following:

- Local and regional development strategies empower local societies and generate local dialogue. People living in areas of the world that have until recently had little say or control over the economic activity taking place in their territory, by using local and regional development strategies, start to develop a degree of autonomy and adopt a more proactive stance concerning sustainable development and their own economic, social and political futures.
- Local and regional development strategies can help to make local and regional institutions more transparent and accountable and foster the development of the local civil society.

From an economic point of view, the advantages of the approach are as, if not more, significant, and may include the following:

- Local and regional economic development strategies, because of their goal of embedding economic activity in a territory and making any economic activity located in it dependent on the specific economic conditions and comparative advantages of that place, generate sustainable economic growth and employment in firms more capable of withstanding changes in the global economic environment.
- Local and regional economic development strategies, as a result of the involvement of local stakeholders and the rooting of economic activity in a territory, contribute to a general improvement in the quality of jobs.

However, it has to be borne in mind that there are also disadvantages associated with local and regional development policies and that this approach is not without risks. The main drawback of this strategy is that it can be extremely time consuming. The development of local and regional coalitions and the coordination of local and regional stakeholders among themselves and with other institutional actors requires an enormous organisational effort and consumes a considerable amount of time and resources even before the development process proper can be started. And even when the key local and regional institutions are established, there is no guarantee of short-term – or even medium and long-term – success. There is also the risk of not being able to identify, design or implement the most appropriate development strategy. In a rapidly changing and complex context, this is an acutely difficult task. The involvement of local and regional actors may offer numerous advantages, like the empowerment of local societies, greater transparency, accountability and greater closeness to those who have to contribute and will ultimately benefit more from the development process. It may also encourage the local population, in general, and local economic actors, in particular, to take a more active stance in their future and to take more risks. However, the mere involvement of local authors is no guarantee of the selection of appropriate and technically effective strategies for localities and regions. In some cases, it may even result in the adoption of unbalanced development strategies, whose capacity to stimulate medium-term sustainable development is questionable. Vested interest groups, public desire for 'miracle cures' and/or rapid results, often in combination with the presence of populist politicians, may lead to the adoption of short-term, highly visible policies and to the neglect of more balanced strategies, whose long-term impact may have a less immediate impact or appeal to large sections of the public.

The aims and structure of the book

This book aims to provide a clear understanding and explanation of contemporary local and regional development. It flows directly from the changed context of local and regional development and the search for alternative development strategies for localities and regions discussed above. The key questions addressed are:

1 What are the principles and values that shape definitions and strategies of local and regional development?
2 What are the conceptual and theoretical frameworks capable of understanding and interpreting local and regional development?
3 What are the main interventions and instruments of local and regional development policy?
4 How do localities and regions attempt to effect development in practice?
5 And, in normative terms, what kinds of local and regional development should we be pursuing?

This book – *Local and Regional Development* – addresses its central questions in four closely integrated parts. In Part I, Chapter 1 – 'Introduction: local and regional development' – describes the changing context of local and regional development. It emphasises several overlapping dimensions: the growing importance of heightened internationalisation or globalisation; increased inter-territorial competition; more sophisticated or knowledge-intensive forms of economic activity; rapid, uncertain and geographically uneven and unequal economic, social, political, cultural and environmental change; multilayered and devolving systems of government and governance; and new interventions, instruments and policies. Each of these elements are further elaborated throughout the book. The remainder of the Introduction provides this rationale and context for the organisation and structure of the book.

Chapter 2 – 'What kind of local and regional development and for whom?' – examines questions of definition to examine what is meant by local and regional development, establishes its historical context and explains the central importance of the geographical concepts of space, territory, place and scale. The chapter then discusses local and regional development's different varieties, principles and values as well as the socially and geographically uneven nature of who and where benefits and loses from particular forms of local and regional development.

Part II of the book – 'Frameworks of understanding' – examines the concepts and theories of local and regional development and institutions of government and governance. Chapter 3 – 'Concepts and theories of local and regional development' – reviews the main concepts and theories used to understand and explain local and regional development. The approaches reviewed are neo-classical; Keynesian; stages theory; product and profit cycles; long wave theory; Marxism and the spatial division of labour; transition theories; institutionalism and socio-economics; innovation, knowledge and learning; endogenous growth theory; geographical economics; competitive advantage and clusters; sustainable development; and post-developmentalism. For each different school of thought and type of approach, the discussion focuses upon: assumptions and conceptualisation; aims; constituent elements, including causal agents, relationships, mechanisms and processes; linkages to policy and criticisms.

Chapter 4 – 'Institutions: government and governance' – examines the changing nature of the state and its implications for local and regional development. It assesses critically the transition from government to governance and the emergence of a decentralised era of devolution and 'new regionalism' within the framework of multilevel institutional structures ranging from the supranational to the neighbourhood. The relationships between democracy and local and regional development are also addressed.

Part III – 'Interventions: instruments and policies' – examines the practice of local and regional development policy. Chapter 5 – 'Mobilising indigenous potential' – deals with the approach and tools aimed at capitalising upon the indigenous or naturally occurring economic potential and promoting endogenous growth from within localities and regions. Connecting to the different frameworks of understanding in Part II, instruments and policies are addressed for establishing new businesses, growing and sustaining existing businesses and developing and upgrading labour.

Chapter 6 – 'Attracting and embedding exogenous resources' – reviews the approach and policies aimed at implanting and anchoring businesses, investment and people for local and regional development. The discussion addresses the changing roles of transnational corporations (TNCs), global production networks, the role of local and regional institutions, connections between exogenous and indigenous approaches, the securing and retention of occupations and dealing with the consequences of divestment.

Part IV – 'Integrated approaches' – pulls the strands of the book together critically to examine experiences of development in localities and regions internationally and to conclude the book and reflect on the future of local and regional development. Chapter 7 – 'Local and regional development in practice' – draws explicitly on the main themes of the book to assess local and regional development in a range of international case studies. The examples comprise localities and regions coping with economic decline (North East England), trying to effect adjustment (Jalisco, Mexico; Ontario, Canada), sustaining an existing development trajectory (Silicon Valley, California, USA), seeking balanced regional development in a context of regional restructuring (Busan, South Korea), attempting an economic transformation (Seville, Spain) and dealing with the uneven local and regional consequences of transformation (Ireland). Analysis addresses the common and particular ways in which each case has constructed concepts and strategies of local and regional development, their institutions of government and governance, intervention strategies and policies, achievements and issues and future challenges.

Chapter 8 – 'Conclusions' – initially summarises the main themes of the book. In a normative sense and in answer to the question of what kind of local and regional economic development and for whom, it draws upon the core book themes to set out our collective vision of what holistic, progressive and sustainable local and regional development might look like. Detailed practical initiatives are explained. The final section closes the book by reflecting upon the limits and political renewal of local and regional development.

Further reading

For a review on the globalisation debates, see Held, D., McGrew, A., Goldblatt, D. and Perraton, J. (1999) *Global Transformations: Politics, Economics and Culture*. Cambridge: Polity.

On the local and regional development implications of globalisation, see Dicken, P. (2003) *Global Shift: Reshaping the Global Economic Map in the 21st Century* (4th edn). London: Sage; Perrons, D. (2004) *Globalisation and Social Change: People and Places in a Divided World*. London: Routledge.

For a discussion of the relationship between the degree of openness of a country and its economic growth, see Sachs, J.D. and Warner, A (1995) 'Economic reform and the process of global integration', *Brookings Papers on Economic Activity* 1: 1–95.

For a critical engagement with the globalisation debate and the role of international institutions, see Stiglitz, J. (2002) *Globalization and its Discontents*. New York: Norton.

For a discussion of territorial competition, see Malecki, E. (2004) 'Jockeying for position: what it means and why it means to regional development policy when places compete', *Regional Studies* 38(9): 1101–1120.

For the alternative and bottom-up forms of local and regional development, see Stöhr, W.B. (ed.) (1990) *Global Challenge and Local Response: Initiatives for Economic Regeneration in Contemporary Europe*. London: The United Nations University, Mansell; Vázquez Barquero, A. (2003) *Endogenous Development: Networking, Innovation, Institutions and Cities*. London and New York: Routledge.

WHAT KIND OF LOCAL AND REGIONAL DEVELOPMENT AND FOR WHOM?

2

Introduction

> One of the biggest myths is that in order to foster economic development, a community must accept growth. The truth is that growth must be distinguished from development: growth means to get bigger, development means to get better – an increase in quality and diversity.
>
> (Local Government Commission 2004)

The Local Government Commission – a local government association in the United States – has a particular view of the kind of local and regional development it deems appropriate and valuable. Such perspectives may differ from place to place and vary over time. How specific interpretations are determined and how they differ are central to understanding and questioning the nature of local and regional development. To understand what we are dealing with when thinking, writing and doing local and regional development we need to start with first principles. Our basic understandings of what local and regional development is, what it is for and, in a normative sense, what it should be about must be questioned. We can begin by asking the fundamental questions of what kind of local and regional development and for whom? Starting here encourages us to take a critical approach and to consider closely what we are learning and thinking about local and regional development. These basic concerns are addressed in the four sections of this chapter. First, questions of definition are examined to understand what is meant by local and regional development, to establish its historical context and to understand the importance of the geographical concepts of space, territory, place and scale. Second, the nature, character and forms of local and regional development are explored to understand its different varieties and principles and values in different places and time periods. Third, the objects, subjects and social welfare dimensions are addressed to understand the often socially uneven and geographically differentiated distribution of who and where benefits and loses from particular forms of local and regional development. Last, a summary and conclusions are provided. Chapter 3 builds upon the starting points articulated in this chapter and discusses their use in the theories that seek to understand and explain local and regional development.

What is local and regional development?

Questions of definition are a starting point to understand what is meant by local and regional development. However, defining – saying exactly what is meant by – local and regional development is more complex than might be commonly assumed. Definitions are bound up with conceptions of what local and regional development is for and what it is designed to achieve. Referring to conceptions of 'development', Raymond Williams (1983: 103) noted that 'very difficult and contentious political and economic issues have been widely obscured by the apparent simplicity of these terms'. Defining – individually or collectively – what is meant by local and regional development is a critically important and deceptively subtle task if we are to look beyond often unquestioned assumptions and superficial descriptions. In the existing literature, economic dimensions such as growth, wealth creation and jobs have historically been at the forefront of describing what constitutes local and regional development (Armstrong and Taylor 2000). Sometimes, local and regional development is equated with this narrower focus upon local and regional *economic* development. For Storper (1997), the local and regional search for prosperity and well-being is focused upon the sustained increases in employment, income and productivity that remain at the heart of economic development. For Beer *et al.* (2003: 5), there is a 'reasonable consensus about the broad parameters of what is meant by local and regional economic development: it refers to a set of activities aimed at improving the economic well-being of an area'. Such activities may include economic development strategy, research, enterprise, labour market and technology initiatives, political lobbying and so on.

This often dominant economic focus in local and regional development has broadened since the mid-1990s in an attempt to address social, ecological, political and cultural concerns (Geddes and Newman 1999). Reducing social inequality, promoting environmental sustainability, encouraging inclusive government and governance and recognising cultural diversity have been incorporated to varying degrees within definitions of local and regional development (Haughton and Counsell 2004). Moves towards broader notions of quality of life, social cohesion and well-being have been integrated, sometimes uneasily, with continued concerns about economic competitiveness and growth (Geddes and Newman 1999; Morgan 2004). We shall return to the question of the integration, relative balances and differences in emphasis between the economic, social, ecological, political and cultural dimensions of local and regional development below.

In the context of a broader understanding of interrelated dimensions, we can deepen and extend how we think about how to define local and regional development. Such an approach can open up space for innovative thinking about what local and regional development is – in the present. What it can or could be – in terms of visions for the future. And, crucially, what it should be – in the normative sense of people in places making value-based judgements about priorities and what they consider to be appropriate 'development' for their localities and regions. There is no singularly agreed, homogenous understanding of development of or for localities and regions. Particular notions of 'development' are socially determined by particular social groups and/or interests in specific places and time periods. What constitutes 'local and regional development' varies both within and between countries (Reese 1997; Danson *et al.* 2000). In any society, the aspirations for and articulations of 'local and regional development' are fluid and dynamic (Beer *et al.* 2003). They are subject to change over time. Precedents, existing practice and norms are subject to incremental and, sometimes, radical changes – for example in response to the kinds of external shocks outlined in Chapter 1, including currency collapses, political sea changes or environmental catastrophes. The assessment of outcomes and performance can trigger reflection and change. Debate, deliberation and discussion can change the thinking, doing and practising of local and regional development. Models can be imposed and resisted. Political cycles and government agendas can recast public policy for local and regional development. Dissent, struggle and innovation can bring formerly alternative approaches from the margin into the mainstream and vice versa. Local and regional interests do not just define local and regional development in a vacuum, however. At least some consensus exists around common themes, principles and values – introduced in Chapter 1 – to which we will return below. Given this potential for geographical diversity and change over time, reviewing the evolution of definitions of local and regional development can anchor its main themes and dimensions in their historical context.

Definitions of local and regional development: a brief historical context

Given that the definitions and conceptualisations of local and regional development are differentiated geographically and change over time, the historical context and trajectory of their evolution is central to their understanding. Dating back only 250 years to the late eighteenth century, the notion of 'development' as sustained increases in income per capita is a relatively recent phenomenon in human history (Cypher and Dietz 2004). From the nineteenth century, the ascendancy of capitalism as a form of social organisation brought technological change, productivity increases and the dominance of industrial employment, notwithstanding the system's periodic crises and slumps (Harvey 1982; Barratt Brown 1995). From this era, development was focused upon economic dimensions and the relative extent to which capitalism had penetrated the economic and social structures of localities, regions and nations, modernising and replacing pre-capitalist social formations. The late nineteenth century 'Industrial Revolution' laid the foundations of geographical and social inequality in what became known later as the 'developed world' (Pollard 1981). Development was marked by geographical and social unevenness and formed the basis of social and political organisation against its injustices, for example by the labour movement and trade unions (Pollard 1999), in the midst of rapid industrialisation and urbanisation. Trajectories and legacies shaping development were established that, in some old industrial localities and regions, are still being grappled with today (Cooke 1995).

The early part of the twentieth century was marked by international conflicts and the Depression during the 1930s. Uneven local and regional development persisted and the social and political implications typically prompted national state action of the top-down variety described in Chapter 1. Yet, often only the hardest hit areas with localised concentrations of high unemployment received public policy attention. National state interventions and institutional innovations in regional policy emerged – such as President Roosevelt's pioneering Tennessee Valley Authority and its industrial estate model as part of the federal New Deal in the United States during the 1930s – and were mirrored elsewhere, particularly in Western Europe (Hudson and Williams 1994). Some understanding of the nature of the closely connected problems of overdeveloped core regions and underdeveloped peripheral regions within nations was evident but limited (Morgan 2001). The 1940s saw the establishment of international institutions that heralded the post-war era of 'developmentalism' up until the 1970s. Table 2.1 details the characteristics of this broad approach to development across a range of dimensions.

Modernist and progressive notions of 'development' as rational and socialised intervention for the improvement of human existence shaped this epoch (Peet 2002). The development question was largely focused upon so-called 'Third World' countries and the poverty and economic stagnation that afflicted much of Africa, Asia and Latin America at the time (Scott and Storper 2003). Local and regional development was commonly the subject of the top-down national spatial policy discussed in Chapter 1 with an economic and social rationale focused upon growth stimulation and redistribution to lagging localities and regions. Environmental impacts and sustainability were not issues at the time. The post-war growth of the 1950s and 1960s generated optimism about the Keynesian belief in the capacity of the national state as an agent for development and national macroeconomic management, following the successful experiences of the economic reconstruction in Europe, under the Marshall Plan, and in Japan. The deeply embedded structural problems that were to hamper development in developing, transitional and restructuring developed countries were poorly appreciated at the time (Cypher and Dietz 2004). Local and regional questions remained as development concerns for many nation states with 'regional problems' arising from the spatial disparities and inequalities in economic and social conditions within nations (Armstrong and Taylor 2000). In the context of Keynesianism, reducing such geographical inequalities was economically efficient and socially equitable, supporting its position within cohesive national political projects often of a social democratic hue.

Strongly influenced by modernisation theory, developmentalism was typically understood as nations passing through distinct evolutionary stages. Each stage had a progressively more modern character in economic, social and political (democratic) terms (Cypher and Dietz 2004). In the 'Cold War' context, US liberal market democracy described the pinnacle of modernism. For example, Rostow's (1971) 'stages of economic growth' model comprised paths from the traditional society, the preconditions state, the take-off, the drive to maturity and the final high mass consumption stage. Such stages were predictable, linear and constituted model development trajectories for nation states. Formerly 'backward', undeveloped states could progress and modernise through known developmental stages. They became 'developing' – in the course of development – towards the particular western model of capitalist development. The relatively late export-led industrialisation of the 'newly industrialising countries' of Japan, South Korea and Taiwan from the 1960s appeared to reinforce the theoretical relevance and empirical base of this understanding of 'developmentalism' (Storper *et al.* 1998). In particular, it established a precedent for places to jump or leapfrog onto new development

Table 2.1 ***The eras of developmentalism and globalism***

World framework	*Developmentalism (1940s–1970s)*	*Globalism (1970s–)*
Political economy	State regulated markets Keynesian public spending	Self-regulating markets (Monetarism)
Social goals	Social entitlement and welfare Uniform citizenship	Private initiative via free markets Identity politics versus citizenship
Development (model)	Industrial replication National economic management (Brazil, Mexico, India)	Participation in the world market Comparative advantage (Chile, New Zealand, South Korea)
Mobilising tool	Nationalism (post-colonialism)	Efficiency (post-developmentalism) Debt and credit-worthiness
Mechanisms	Import-substitution industrialisation (ISI) Public investment (infrastructure and energy) Education Land reform	Export-oriented industrialisation (EPO) Agro-exporting Privatisation, public and majority-class austerity Entrepreneurialism
Variants	First World (freedom of enterprise) Second World (central planning) Third World (modernisation via developmental alliance)	National structural adjustment (opening economies) Regional free-trade agreements Global economic and environmental management
Local and regional dimension	National spatial policy Economic and social focus Growth redistribution	Supranational and devolved (subnational, regional and local) policy and institutions Economic competitiveness focus Regeneration

Table 2.1 *continued*

Timeline Markers	1940s	1950s	1960s	1970s	1980s	1990s	2000s
	United Nations (1943) Bretton Woods (1944) Marshall Plan (1946) Cold War (1946–)	First Development Decade Korean War (1950–1953) Non-Aligned Movement (1955)	Second Development Decade Vietnam War (1964–1973) Alliance for Progress (1961) UN Conference on Trade and Development (1964)	Oil crises (1973, 1979) New International Economic Order Initiative (1974)	Debt Crisis/ The Lost Decade Debt regime (supervised state/economy restructuring) (mid-1980s) Neo-Liberalism Reaganism, Thatcherism Cold War ends (1989)	Globalisation New World Order begins (early 1990s) Earth Summit (1992) Chiapas revolt (1994)	9/11 (2001) Second Gulf conflict (2004) Growth of China and India
Institutional developments	World Bank and IMF (1944) COMECOM (1947) US$ as world reserve currency		Eurodollar and offshore $ market	Group of Seven (G7) forms (1975)	GATT Uruguay Round (1984) Glasnost and Perestroika in Soviet Union (mid-1980s) IMF and World Bank Structural Adjustment Programmes Single European Market	NAFTA (1994) World Trade Organisation (1995) Euro introduction (1999)	Anti-globalisation protests (Seattle, Davos, Genoa) (early 2000s) European Union enlargement (25 Member States) (2004)

Source: Adapted from McMichael (1996)

trajectories. Geographical differentiation persisted, however, with different national variants as historical and institutional factors hampered some states attempting similar modernisation strategies, for example India up until the 1990s (Chibber 2003) (Table 2.1). The heyday of 'developmentalism' in this era was primarily focused upon economic dimensions and the national level.

In the 1960s, approaches to development explicitly aimed at redistributing opportunities for wealth creation and enhanced economic well-being (Glasmeier 2000). Dissatisfaction grew however concerning the narrow and economistic view of development that equated it solely with increases in income per capita (Peet 2002; Cypher and Dietz 2004). An emerging consensus suggested that there was much more to human social development and well-being than just increasing financial incomes. As Example 2.1 illustrates, attempts were made to broaden the notions of 'development'. Frustration was evident with the limited and/or unequal 'trickle down' of the benefits of economic growth. Radical critiques from the late 1960s, inspired by Marxist thinking, argued that the path of development in the developed North had actively underdeveloped the South through colonialism and the neo-colonialism of its incorporation within an internationalising capitalist system and, as we explore in Chapter 6, the activities of increasingly transnational corporations (Frank 1978; Hymer 1979).

In parallel, greater attention was given to local and regional development questions beneath the level of the nation state (McCrone 1969). Economic and social arguments were marshalled in favour of regional policy as a means of reducing spatial disparities to improve regional *and* national economic efficiency and to contribute to social equity (Armstrong and Taylor 2000). The political claims of 'first wave' regionalism emerged, particularly in western Europe, and articulated its dissatisfaction with their limited relative autonomy and levels of 'development' within often centralised national state structures (Keating 1998). Connections between the colonialism experienced by the former imperial possessions of the Great Powers and their own peripheral subnational regions and localities were even drawn (Hechter 1999). Relationships between the colonial powers and their empires had implications for the regions of the metropolitan powers, shaping their functional specialisations. For example, Cain and Hopkins (1993a, 1993b) have shown how the fate of Britain's industrial regions was crucially linked to the shifting role of the City of London as the dominant financial centre of the British Empire.

The break-up in the mid-1970s of the economic and social settlement of Fordism that had underpinned the relative growth and prosperity of the post-war period, at least for much of the developed North, undermined faith in the power of the national state as an agent of development and regulation (McMichael 1996). Deindustrialisation and transitions towards a service economy unleashed waves of economic and social restructuring. The era of 'developmentalism' gave way – in a highly geographically uneven and contested manner – to an emergent and uncertain era of 'globalism' from the 1970s. Table 2.1 illustrates its markedly different constituent elements from the earlier era of developmentalism.

A counter-revolution set in from the late 1970s against Keynesianism, statism and more radical development theories (Toye 1987; Peet 2002). The ascendancy of monetarism, neo-liberalism and the politics of the New Right sought to roll-back the state and promote the deregulation and liberalisation of markets (Jessop 2002). Transnational

Example 2.1 Broadening the measures of development

Historically, economists typically measured the level of development of the nation, region or locality by using economic growth or income per person. Income levels were considered reasonable approximate measures for levels of development and income per person could be a logical surrogate for gauging social progress. Gross national product (GNP: total value of all income or final product accruing to residents of a country derived from within and outwith that country) and gross domestic product (GDP: total value of all income or final product created within the borders of a country) were typically used, often with adjustments for population size, income distribution, price changes over time, purchasing power parity and the contribution of the informal economy. While recognising that particular forms of development might encompass more than simply rising incomes, economists argued that 'the greatest number of the other dimensions of development that countries might wish to realise are more easily attained at and tend to accompany higher income levels' (Cypher and Dietz 2004: 30).

As dissatisfaction with rates of economic and income growth and geographical and social disparities grew and broader notions of what constitutes 'development' emerged, indicators that go beyond narrow economic measures of growth and income have been sought. Economic growth and the income standard were seen as being too aggregate and unable sufficiently to capture distributional inequalities between people and places. Morbidity and mortality figures too, continued to underline geographical inequalities in social conditions, health, well-being and quality of life. Other measures include the United Nations Development Programme's (UNDP) Human Development Index (HDI). This composite index uses 'longevity, knowledge and decent standard of living' as the indicators of development. HDI uses measures of life expectancy at birth, adult literacy, school enrolment and purchasing power parity GDP per capita. As the UNDP argues:

> Human development is about much more than the rise and fall of national incomes. It is about creating an environment in which people can develop their full potential and lead productive, creative lives in accord with their needs and interests. People are the real wealth of nations.
>
> (UNDP 2001: 9)

Other similarly broad-based composite measures, such as the Gender-related Development Index and Human Poverty Index, have also been developed. In connecting with concerns about sustainability, measures such as the Genuine Progress Indicator (GPI) and environmentally-adjusted net domestic product have been developed too as an attempt to 'green' the national accounts of economic growth. The thinking behind broader measures of development have been explored for localities and regions, for example in the United Kingdom where the focus on gross value added (GVA) has been accompanied by the development of composite 'Quality of Life' and 'Liveability' indices. Going beyond the readily quantifiable and available data sources to assess broader notions of development presents often acute measurement difficulties.

Sources: House of Commons (2003); Cypher and Dietz (2004); Morgan (2004)

blocs emerged around Europe (the European Economic Community (EEC) and later European Union (EU)), North America (NAFTA and the potential Free Trade Area of the Americas (FTAA)) and South East Asia (ASEAN) structuring the emerging global economy and with uneven aims of economic and political integration. Progress and development were now predicated upon nation states and their firms' ability to compete within rapidly internationalising markets. The 1980s marked a high point of market-led restructuring, structural adjustment and uneven and unequal local and regional development in many advanced states. The development problem was seen as a correctable one of market failure. 'Second wave' or 'new' regionalism emerged with an economic focus as regions were encouraged and facilitated by devolution within their national states and prompted by their own social and political aspirations to become responsible agents of their own development (Keating 1998).

In the context of the end of the Cold War and the 'Velvet Revolution' in central and eastern Europe in 1989, a tentative 'Third Way' emerged in the 1990s following dissatisfaction with the highly unequal development outcomes of the market-led 1980s and an unwillingness to return to the national level statism of the post-war age (Giddens 1998). Advanced industrial countries alongside economies undergoing transition from central planning and countries formerly considered as 'developing' have been reincorporated into a much more global development question (Scott and Storper 2003). Combinations of state and market have been sought to cope with the risk, uncertainty and complexity of increasingly 'globalised', rapidly changing and reflexive forms of capitalism (Held *et al.* 1999). However, such experiments coexisted, even overlapping, with the dominant 'Washington Consensus' propelling the turn to neo-liberal approaches to development in 'developed' and 'developing' countries. Free-market capitalism, open economies and conservative macroeconomic policy constituted the political-economic orthodoxy. Constructed by the actions of nation states and supranational institutions (Hirst and Thompson 1999), as we discussed in Chapter 1, 'globalisation' during the 1990s presented formidable local and regional development challenges in a more integrated, interdependent and competitive world (Peck and Yeung 2003). Devolution and the decentralisation of state forms have emerged internationally, often with ambiguous implications for reductions in regional growth disparities (Rodríguez-Pose and Gill 2003). Substantial policy convergence is evident around a neo-liberal agenda among developed and, increasingly, developing nations, focused upon supply-side flexibility, the development of economic potentials and macroeconomic stability (Glasmeier 2000) (Table 2.1). Indeed, as we discussed in Chapter 1, in the context of international fiscal conservatism and trade liberalisation under the World Trade Organisation, uneven development at the local and regional levels may actually have been reinforced by the priority given to increased free trade and low inflation combined with limited government expenditures for economic adjustment (Braun 1991).

By the 2000s, 'development' had broadened further to incorporate sustainability and holistic or integrated approaches to economic, social and environmental concerns (Geddes and Newman 1999; Morgan 2004) (Example 2.1). The government and governance of development had come to the fore at the supranational level in macro-level blocs – the European Union, the Americas and East Asia – with varying degrees of

Plates 2.1–2.2 ***Local and regional development as a global issue: poverty and deprivation in the 'developed' and 'developing' worlds***

Source: Photographs by Michele Allan

integration and within nation states as part of the Organisation for Economic Cooperation and Development (OECD) and World Bank's 'good governance' agenda. Disquiet with 'Western' ideals of 'development' and the imposition of Western models of liberal market democracy have mobilised claims for self-determination and empowerment regarding the definition and means of 'development'. Post-developmentalism – discussed in Chapter 3 – has emphasised the potential role of civil society in 'development', perhaps complementing state and market, as an autonomous entity embedded in localities and regions with better understanding of their social aspirations, needs and potential. A post-'Washington Consensus' has been discerned in development, focused upon a revised neo-liberal model stressing market-friendly state intervention and 'good governance' (Peet 2002).

Currently, local and regional 'development' is a more global issue than hitherto:

> as globalization and international economic integration have moved forward, older conceptions of the broad structure of world economic geography as comprising separate blocs (First, Second and Third Worlds), each with its own developmental dynamic, appear to be giving way to another vision. This alternative perspective seeks to build a common theoretical language about the development of regions and countries in all parts of the world, as well as about the broad architecture of the emerging world system of production and exchange . . . it recognizes that territories are arrayed at different points along a vast spectrum of developmental characteristics.
>
> (Scott and Storper 2003: 582)

This brief historical context of the definitions of local and regional development has identified the evolution of views of 'development'. Each has their own theoretical and ideological basis, definition of 'development', theory of social change, role and agents of development and local and regional development emphasis summarised in Table 2.2. The historical evolution of development has highlighted several central and recurrent themes in discussions of the definition of local and regional development. First, notions of development change over time. Historical evolution, critique and debate are central alongside the interests of those involved in shaping their determination. Changes in existing understandings have emerged from both radical and more moderate, reformist critiques of existing thinking and practice. Second, definitions of 'development' are geographically differentiated. They vary within and between places over time. Third, the historical focus upon economic dimensions has been broadened to include social, ecological, political and cultural concerns. New approaches and measures of development have been sought that are more sensitive to the need for socially determined and sustainable balances between the economic, social, political, ecological and cultural dimensions of development. Last, different emphases on the local and regional level exist in different approaches and definitions of development. The national and, increasingly, supranational focus has evolved to incorporate the local and the regional. The 'where' of development has become important with the recognition that development is not just a national level concern for nation states.

Table 2.2 ***Main views of 'development' and their relations to capitalism***

	Development of capitalism	*Development alongside capitalism*		*Development against capitalism*		*Rejection of development*
	Neo-liberalism	Interventionism		Structuralism	'Alternative' (people-centred) development	'Post-development'
		Intervening to improve 'market efficiency'	Achieving social goals by 'governing the market'			
Vision: desirable 'developed' state	Liberal capitalism (modern industrial society and liberal democracy) (plus achieving basic social/environmental goals)			Modern industrial society (but not capitalist)	All people and groups realise their potential	['development' is not desirable]
Theory of social change	Internal dynamic of capitalism	Need to remove 'barriers' to modernisation	Change can be deliberately directed	Struggle between classes (and other interests)	[not clear]	[not clear]
Role of 'development'	Immanent process within capitalism	To 'ameliorate the disordered faults of [capitalist] progress'		Comprehensive planning and transformation of society	Process of individual and group empowerment	A 'hoax' which strengthened US hegemony
Agents of development	Individual entrepreneurs	Development agencies or 'trustees' of development (states, NGOs, international organisations)		Collective action (generally through the state)	Individuals, social movements	Communities, civil society
Local and regional development	Localities and regions as agents of their own development	Integrated economic, social and environmental approaches to balance competitiveness, social equity and sustainable development		Localities and regions as sites of social revolution, transformation and non-capitalist statist modernisation	Localities and regions as bases for alternative forms of social organisation and grassroots development	Autarchy and autonomous alternative 'development'

Source: Adapted from Thomas (2000)

Where is local and regional development? Space, territory, place and scale

> Territory and its potential endogenous resources is the main 'resource' for development, not solely a mere space.
>
> (Canzanelli 2001: 6)

As Giancarlo Canzanelli suggests, development does not take place in a spatial vacuum devoid of any geographical attachments or context. However it may be understood, development is a profoundly geographical phenomenon. In abstract terms, the social is seen as necessarily spatial. As Manuel Castells (1983: 311) puts it: 'space is not a "reflection of society", it is society'. Any definition of local and regional development requires an appreciation of the fundamentally geographical concepts of space, territory, place and scale. Geography is an integral constituent of economic, social, ecological, political and cultural processes and their geographies condition and shape in profound ways how such processes unfold (Markusen 1985). The 'local' and the 'regional' are not simply containers in which such social processes are played out. For Scott and Storper (2003), spaces – localities and regions – are causal or explanatory factors in economic growth not just receptacles for or manifestations of its outcomes. 'Local' and 'regional' are particular socially constructed spatial scales through which such processes evolve. Social processes are inseparable from their geographies. Any sense of the 'development' – however defined – of such processes or the places in which they are situated and through which they unfold needs to recognise this integral role of space. Put simply, geography matters (Massey and Allen 1984).

Territory refers to the delimited, bordered spatial units under the jurisdiction of an administrative and/or political authority, for instance a nation state, city or region (Anderson 1996). The expression of localities and regions in which different kinds of development may or may not be taking place in specific time periods is often as territorially bounded units with an administrative, political and social identity. For example, the regions of Brazil, Canada or Indonesia face particular local and regional development questions within specific national state structures. Within such territories, states and other quasi- or non-state institutions – such as associations of capital, labour and civil society – engage to differing degrees and in different ways in local and regional development and its government and governance. Chapter 4 addresses this in more detail. Territorial boundaries form defined areas within which particular definitions and kinds of local and regional development may be articulated, determined and pursued. Territory is not a fixed entity in space or time, however (Taylor and Flint 2000). Localities and regions evolve and change over time in ways that affect local and regional development definitions, practice and prospects. For Paasi (1991) 'regions are not, they become'. Localities and regions are seen as evolving economic, social, political, ecological and cultural constructs. Rather than static, unchanging entities, local and regional territories are dynamic, changing over time (Cooke and Morgan 1998). Allen *et al.* (1998) consider localities and regions as 'unbounded', especially where their influence and spatial reach is beyond their territorial boundaries, and relational, in that they are mutually constituted by wider webs of spatialised social relations. Indeed, territorial borders can be changed and 'artificial', newly created regions can acquire attributes and behave like regions too – for example, La Rioja in Spain (Giordano and Roller 2004). Territory gives geographical and institutional shape to the spaces of local and regional development.

The socio-spatial world of local and regional development is not just an homogenous or uniform geographical plane. It is made up of specific and particular places. From Hackney to Honolulu to Hong Kong, each place is particular. Each has its own evolving histories, legacies, institutions and other characteristics that shape their economic assets and trajectories, social outlooks, environmental awareness, politics, culture and so on. Such particularities can be both shared and different and can be materially and symbolically important to defining local and regional development. While each place may be unique, however, different localities and regions may share common histories of development, and face similar challenges and issues. Indeed, each occupies a position within an increasingly integrated and inter-dependent world. The development fortunes of places are increasingly intertwined. Recognition of difference and diversity need not necessarily translate into selfish, parochial and introspective definitions of local and regional development. As Chapter 1 revealed, inter-territorial competition between places in the global context may be inefficient and wasteful of public and private resources (Rodríguez-Pose and Arbix 2001).

Plate 2.3 ***Public demonstrations and political agency: anti-poll tax and anti-warrant sale marches in Scotland during the 1990s***

Source: Photograph by Michele Allan

The geographical diversity of places shapes how and why local and regional development definitions vary both within and between countries and how this changes over time. In this way, local and regional development is context-dependent:

> Economic development is not an objective per se. It is a means for achieving well being, according to the culture and the conditions of certain populations. Nevertheless the well being target is not the same for people living in New York or in Maputo; only who is living in New York or Maputo could fix what they want to achieve in the medium and long term.
>
> (Canzanelli 2001: 24)

The importance of place means local and regional development trajectories are strongly path-dependent (Sunley 2000). Their future development is unavoidably shaped by their own historical evolution (Clark 1990). Phenomena happening in the present 'trail long tails of history' (Allen *et al.* 1998). Such historical legacies can be decisive in understanding and explaining local and regional development – as we shall explore in Chapter 3 and is evident in our case studies in Chapter 7. The particular attributes of places can influence whether or not particular definitions and varieties of local and regional development take root and flourish or fail and wither over time.

In common with space, territory and place, the geographical scales over and through which particular economic, social, political, ecological or cultural processes are manifest are central to local and regional development. Table 2.3 provides examples of scales,

Table 2.3 ***Scales, socio-economic processes and institutional agents***

Scale/level	*Socio-economic process*	*Institutional agents*
Global	Trading regime liberalisation	International Labour Organisation (ILO), International Monetary Fund (IMF), World Trade Organisation (WTO), nation states
Macro-regional	Information and communication technology network expansion	European Union, Member States, regulatory bodies, private sector providers
National	House price inflation	Central banks, building societies, borrowers
Sub-national	Transport infrastructure expansion	Public transport bodies, private companies, financial institutions
Regional	University graduate labour market retention	Universities, regional development agencies, employers, training providers
Subregional	Labour market contraction	Employment services, trade unions, business associations, employers, employees
Local	Local currency experimentation	Local Exchange Trading Systems, households
Neighbourhood	Social exclusion	Local authorities, regeneration partnerships, voluntary groups
Community	Adult literacy extension	Education and training institutions, households, families

Source: Authors' own research

socio-economic processes and institutional agents. It illustrates the ways in which different processes potentially constitutive of local and regional development can work across and between different scales through the actions of particular agents. While the 'local' and the 'regional' are the specific spatial scales for the processes of development that are the focus of this book, what goes on at the 'local' and the 'regional' scales cannot be divorced from their relations with processes unfolding at other levels and scales (Perrons 2004). Each scale and level is mutually constitutive – they each make up the other. As Jones *et al.* (2004: 103) suggest: 'localities cannot be understood as neatly bounded administrative territories, and places are intrinsically multi-scalar, constituted by social relations that range from the parochial to the global'. Phenomena and processes that may somehow be thought of as 'external' or outside, perhaps since they appear to be beyond the control of particular localities and regions, can have profound impacts. Example 2.2 discusses the local instability caused by a global trade dispute. Scales are often contested and socially determined, for instance by states and associations of capital, labour and civil society (Swyngedouw 1997). Taking into account the

Example 2.2 Global trade disputes and local economic instability

The connections between the different scales of local and regional development were demonstrated in the so-called 'Banana Wars' in the late 1990s. As part of a trade dispute between the United States and the European Union through the World Trade Organisation (WTO) regarding the preferential treatment of Caribbean banana imports, the United States retaliated by listing a range of EU exports subject to a punitive import tariff of 100 per cent, effectively pricing them out of the lucrative US market. These specialised high-value exports, chosen politically to apply pressure upon EU negotiators and Member States, included French handbags, German coffee makers and Italian cheese. Such products were typically produced by highly localised industrial clusters. One targeted sector, for example, encompassed cashmere producers in the Anglo-Scottish Borders in the United Kingdom. Here, the local cashmere sector employed over 1000 people and was concentrated in the 'knitwear capital' of Hawick. Jim Thompson of the Hawick Cashmere Company said the sanctions would cause major problems for the industry in Scotland: 'The Americans are unaware how polarised the cashmere industry is in the Borders. It will have catastrophic effects – if this actually goes through we are looking at most definitely a thousand jobs in the Borders'. As the result of a global trade dispute, this particular local and regional economy was destabilised by a prolonged period of damaging uncertainty regarding a key export market. The episode was finally resolved following over two years of political lobbying by respective local interests at the national and European levels. Initially, only cashmere was removed from the list and a negotiated settlement brokered at the WTO. From 1 July 2001, the United States agreed to suspend the sanctions imposed against the remaining EU products.

Sources: Pike (2002a); 'US claim banana trade war victory', *Guardian* 7 April 1999

broader context of the scale or level of economic, social and political processes is central to defining local and regional development. The 'where' of local and regional development is a geographical concern. Together, the concepts of space, territory, place and scale are central to definitions of local and regional development.

What kind of local and regional development?

In common with the preceding discussion about definitions, there is no easily accepted and singular meaning given to the different kinds of local and regional development determined by different people and groups in different places at different times. Its nature, character and form can evolve in geographically uneven ways. Thinking about the possible kinds of local and regional development encourages us to consider its different varieties and the principles and values utilised in its determination. What local and regional development is for and what it is trying to do in its aims and objectives are framed and shaped by its definitions, varieties, principles and values.

Varieties of local and regional development

Different kinds or types of local and regional development exist. Building upon the issues of defining what is meant by local and regional development, we can draw distinctions about its different sorts and nature. Examples are offered in Table 2.4, although this list is not exhaustive. Other dimensions might be apparent or receive priority in different localities and regions. Emphasis given to some concerns may differ and change over time. The distinctions do not read down each column vertically in a linked fashion. The table should be read across each row. Given the complexity and geographical unevenness of the social world, such distinctions may be a question of degree or extent. While absolute development might mean an aspiration for geographically even

Table 2.4 ***Distinctions in local and regional development***

Dimension	*Distinction*	
Approach	Absolute	Relative
Autonomy	Local, regional	National, supranational
Direction	Top-down	Bottom-up
Emphasis	Strong	Weak
Focus	Exogenous	Indigenous
Institutional lead	State	Market
Inter-territorial relations	Competitive	Cooperative
Measures	'Hard'	'Soft'
Objects	People	Places
Rate	Fast	Slow
Scale	Large	Small
Spatial focus	Local	Regional
Sustainability	Strong	Weak

Source: Authors' own research

development within and across localities, regions and social groups; relative development suggests uneven development. Whether by default or design, relative development privileges the development of particular localities, regions and/or social groups, often exacerbating rather than reducing disparities and inequalities between them. There is substantive difference between absolute development *of* or relative development *in* a locality or region (Morgan and Sayer 1988). Such geographical and social welfare concerns link to the discussion later in this chapter of local and regional development for whom? Autonomy describes where the power and resources for local and regional development reside, for example the traditional top-down or more recent bottom-up approaches introduced in Chapter 1. Emphases may be 'strong', high priority and/or radical in their intent, or 'weak', low priority and/or conservative and reformist.

As we discussed in Chapter 1, the direction of local and regional development may be top-down, bottom-up or combine elements of both approaches. The focus may emphasise exogenous (growing or originating from the outside and subject to external factors) and indigenous (native or inherent) and endogenous (from within) forms of growth to varying degrees. Institutional leads may encompass both state and market, 'Third Way'-style, or even civil society. Inter-territorial relations may be wedded to differing degrees to competition and/or cooperation. As discussed in Chapter 1, measures may include interventions focused upon 'hard' infrastructure and capital projects and/or 'soft' training and technology support. The rate of development may seek to balance 'fast' development to address pressing social need with a 'slow', perhaps more sustainable, outlook. Large- and/or small-scale projects may be combined. The spatial focus may distinguish the particular geographical scale of development efforts. Views of sustainability may be 'strong' or 'weak'. The objects of local and regional development may be people and/or places. The subjects can be the themes upon which 'development' is based.

Echoing the broadened notions of development to incorporate economic, social, political, environmental and cultural concerns, a resonant distinction in the kinds of local and regional development is between its quantitative level or extent and its qualitative character or nature. The quantitative dimension of local and regional development may relate to a numeric measure, for example a per capita growth rate of GDP, a number of jobs created or safeguarded, new investment projects secured or new firms established. Putting aside for a moment issues of data availability and reliability, a quantitative approach focuses objectively on the numbers: how much of a particular something. The focus can be on absolute or relative change over specific time periods between and within localities and regions.

The qualitative dimension is concerned with the nature and character of local and regional development, for example the economic, social and ecological sustainability and form of growth, the type and 'quality' of jobs, the embeddedness and sustainability of investments, and the growth potential and sectors of new firms. The qualitative approach focuses upon more subjective concerns that connect with specific principles and values of local and regional development socially determined within particular localities and regions at specific times. For example, the 'quality' of jobs might be judged by their terms and conditions of employment, relative wage levels, opportunities for career progression, trade union recognition and so on. The sustainability of a development may be assessed by its ecological impact or 'footprint'.

The quantitative and qualitative dimensions of local and regional development can be integrated but are not always or necessarily coincidental. Localities and regions can experience 'development' in quantitative terms but with a problematic qualitative dimension, for example through increased employment levels in low 'quality' jobs in unsustainable inward investors and/or short-lived start-up firms. Conversely, localities and regions can witness development in qualitative terms that is problematic in quantitative terms, for example insufficient (although potentially good quality) jobs, too few new investments and new firms.

The qualitative dimension has become increasingly important in recent years in tandem with broader understandings of local and regional development and following concerns about the potentially damaging effects of weak and unsustainable forms of local and regional development (Morgan 2004). Some studies have concentrated on high-productivity, high-cohesion forms of growth, while leaving other less desirable, but widespread, types of growth under-researched (Sunley 2000). At global lending institution the IMF in the mid-1990s, for example, the focus was shifted towards a particular kind of 'high quality' economic growth:

> that is sustainable, brings lasting gains in employment and living standards and reduces poverty. High quality growth should promote greater equity and

Plate 2.4 ***Waterfront urban regeneration: the Quayside in Newcastle upon Tyne in North East England***

Source: Photograph by Michelle Wood

> equality of opportunity. It should respect human freedom and protect the environment . . . Achieving high quality growth depends, therefore, not only on pursuing sound economic policies, but also on implementing a broad range of social policies.
>
> (IMF 1995: 286, cited in Cypher and Dietz 2004: 30)

However, amid much critical commentary (McMichael 1996; Stiglitz 2002) and in the wake of the Asian financial crises in the late 1990s, the extent to which such principles have been practised by such international institutions remains open to question.

In grappling with the qualitative dimensions of growth, 'high' and 'low' roads to local and regional development have been identified (Cooke 1995; Luria 1997). As Example 2.3 explains, the 'high' road equates with qualitatively better, more sustainable and appropriate forms of local and regional development. However, what is considered 'better' and 'appropriate' is shaped by principles and values that – as we suggested above – are socially determined in different places and time periods. What constitutes 'successful' or 'failed' local and regional development in this context will vary across space and time.

Plate 2.5 ***Maritime engineering in old industrial regions: shipyards on the River Tyne in North East England***

Source: Photograph by Michelle Wood

Example 2.3 The 'high' and 'low' roads of local and regional development

The different kinds of local and regional development have been articulated in conceptions and metaphors of 'high' and 'low' roads. The distinction between these related understandings and potential routes to 'development' focuses upon the qualitative dimension of the nature of local and regional development. Across interrelated dimensions – productivity, wages, skills, value-added and so on – localities and regions can pursue more or less 'high road' strategies based upon high productivity, high wages, high skills, high value-added or the converse 'low road' strategies based upon low productivity, low wages, low skills, low value-added. Faced with competition from nation states such as China and India with lower wages and weaker regulatory regimes but comparable skills and productivity, many localities and regions in the developed world perceive 'low road' competition as a 'race to the bottom', through deregulation and the weakening of social protection, incompatible with maintaining or improving living standards and social and economic well-being and local and regional development. For example, like other German Länder with high manufacturing densities, North Rhine-Westphalia (NRW) has pursued a 'high road' strategy through substantial investment in technology-led regional industrial policy to network research and development (R&D) centres, technology transfer, innovative small and medium-sized enterprises (SMEs) and technology support institutions locally and throughout the region. Close to 100,000 new jobs were created between 1984 and 1994 by this strategy, many in environmental engineering. For NRW, the 'low road' was not perceived as a sustainable option, although recent high unemployment across Germany has raised questions about its economic and social model. Elsewhere, discussions about the offshore relocation of low value, price sensitive activities to lower cost regions in Europe and beyond has prompted concerns about how localities and regions might move from 'low' to 'high' road development paths. However, in recognising the persistence of basic skills deficiencies among the workforce in Wales, the Wales Trades Union Congress (TUC) has argued that:

> The conventional response to this problem [of offshore relocation] from the WDA [Welsh Development Agency] and the [National] Assembly is to say that Wales needs to 'move up market' into the 'knowledge-driven economy'. But the big question is how Wales gets from here to there when one in four of the Welsh population is functionally illiterate and two in five non-numerate?
>
> (Wales TUC n.d.: 4.1)

While offering a way of thinking about the issues and evident in international debates over strategic local and regional development policy, the 'high' and 'low' road distinction is relatively simplistic. It may be better seen as a continuum, differing across and between different activities in different places and something which changes over time (high to low, low to high) with different expressions in different localities. Distinguishing between 'high' and 'low' roads is also problematic for developing countries and peripheral localities and regions whose low wages and weak social protection may be perceived as competitive advantages within a globalising economy.

Sources: Cooke (1995); Wales TUC (n.d.)

Principles and values

Principles and values serve to shape how specific social groups and interests in particular places define, interpret, understand and articulate what is defined and meant by local and regional development. The fundamental or primary elements of local and regional development may be collectively held unanimously, shared with a degree of consensus or subject to contest and differing interpretations by different social groups and interests within and between places over time. The worth, desirability and appropriateness of the different varieties of local and regional development may be similarly articulated as the objects of cohesion or division. Principles and values of local and regional development raise normative questions concerned with values, ethics and opinions of what ought to or should be rather than what is. Value judgements are implied in thinking about principles and values. What could or should local and regional development mean? What sorts of local and regional development does a locality or region need or want? What kinds of development are deemed appropriate and, as a consequence, inappropriate? What constitutes the 'success' or 'failure' of a specific kind of development for a locality or region?

Principles and values of local and regional development are socially determined within localities and regions. Principles might reflect universal beliefs held independently of a country's levels of development such as democracy, equity, fairness, liberty and solidarity. They often reflect the relations and balances of power between the agency of the state, market, civil society and public. Political systems and the government and governance of local and regional development – discussed in Chapter 4 – are central to how such questions are framed, deliberated and resolved. The agency of state, market, civil society and public is not wholly autonomous or independent to act and decide its own course of development, however. Each is often circumscribed by the structural context in which it is embedded and the constraints this creates in any consideration of what 'development' is, could or should be about.

For reasons we shall discuss in Chapter 3, people and institutions within localities and regions are rarely free to choose their development paths and trajectories. Their development aspirations and strategies do not start with blank sheets of paper. The social determination of the principles and values of local and regional development is a geographically uneven and historical process. Particular constructions or notions of 'development' condition the social use of resources with potentially quite different implications in economic, social, ecological, political and cultural terms, for example whether a locality chooses an internal focus upon social needs or an emphasis upon external markets (Williams 1983).

Different interest groups will often seek to influence the principles and values to their advantage, often claiming potential contributions to particular notions of 'development' in or of localities and regions. For instance, organised labour may lobby for greater social protection to enhance the well-being of the employed workforce while business associations may demand more flexible labour markets to foster wealth creation. Charities and environmental organisations may lobby for higher environmental standards in trade regulations to encourage the upgrading of ecologically damaging economic activities. The state often has to balance and arbitrate between such competing interests. Individuals and institutions with social power and influence can seek to impose their specific visions of local and regional development but these may be contested and resisted (Harvey 2000). It is, then, a critical starting point to ask whose principles and values are being pursued in local and regional development.

Linking to the discussion of geographical concepts above, the social determination of principles and values has a space, territory, place and scale. Particular individuals and/or institutions may act in the interests of their particular social class, for example capital or labour, or their territory, for example their locality, region or nation. Class can work across spaces and territory works within spaces to provide the bases for the determination and articulation of local and regional development principles and values (Beynon and Hudson 1993). The principles and values that shape social aspirations concerning the desired kind of local and regional development deemed valuable and appropriate by particular localities and regions are geographically differentiated and change over time. They may reflect perceived economic, social and political injustices, for example regarding the allocation of public expenditure, the actions of local or transnational firms, ecological damage or the relative degree of political autonomy. 'Development' may then mean a 'fairer' allocation of public funding, greater regulatory control over firms, enhanced environmental standards and protection and further political powers and responsibilities.

The attributes and characteristics of places influence the collectively held and articulated principles and values reflected – to greater or lesser degrees – in local and regional development. Social aspirations for development and what can, can't or could be achieved are geographically rooted and conditioned by past experience and assessments of local assets and networks. Geographically embedded principles and values have material influences upon the kind of local and regional development considered desirable, feasible or possible in particular localities and regions. Example 2.4 explores the particular principles and values of associative or cooperative rather than individualistic entrepreneurialism in Wales. Put simply, place matters for the principles and values of local and regional development. Indeed, as we discussed in Chapter 1, the need for more context-sensitive policy that acknowledges the importance of principles and values in place has been recognised (Storper 1997). This theme will be explored in more detail in Part III, 'Interventions: instruments and policies' and Part IV, 'Integrated approaches'.

Although it has been the subject of the conceptual and theoretical debate discussed further in Chapter 3, sustainability has come to the fore as a highly significant recent influence upon the definitions, geographies, varieties, principles and values of local and regional development (Geddes and Newman 1999; Haughton and Counsell 2004; Morgan 2004). Sustainability questions the fundamental aims and purposes of local and regional development, particularly its focus upon economic growth, and its durability, longevity and longer-term implications. Sustainability has economic, social, ecological, political and cultural dimensions. In contrast to an earlier environment-led era, recent versions of sustainable development have sought holistic approaches to integrate rather than trade-off these specific facets. Internationally, as discussed in Example 2.5, 'smart growth' has been promoted in North America and Australia as a means of simultaneously achieving economic, social and environmental aims. In the European Union too,

Example 2.4 Principles and values of associative entrepreneurialism in Wales

The former coal-mining and heavy industrial heartland of south Wales has undergone prolonged restructuring throughout the twentieth century. Economic and social regeneration have been recurrent and central tasks for local and regional development. The emphasis upon enterprise and new firm formation has proved a tangible failure, however. Wales still performs poorly relative to the other nations and regions in the United Kingdom. In 2002, Wales' value added tax (VAT) registration rate for new businesses was the second lowest nationally at 26 per 10,000 resident adult population compared to 37 for the United Kingdom as a whole (Office for National Statistics 2004). Molly Scott Cato argues that the particular form of an individualistic and self-serving model of enterprise and entrepreneurialism jars with the principles and values of Welsh society. The particular set of Welsh values are based upon community, mutual aid and solidarity. These values sit uneasily with the heroic individualism and competitive entrepreneurialism that are claimed to provide the dynamism and flexibility of the Anglo-American economic model. In particular, the historically embedded antipathy towards the neo-colonialism of the mainly English coal owners during the dominance of traditional industry in Welsh economy and society has made private enterprise synonymous with public exploitation for many in Wales. For the Welsh people:

> Rather than seeing both private and public employment as equally legitimate domains there has been a popular tendency to equate esteem and worth with service to the community. . .This may be described as the result of the predominance of an other-regarding rather than a self-regarding value system.
>
> (Casson *et al.* 1994: 15, quoted in Cato 2004: 228)

The efforts of various public agencies in seeking to promote Welsh enterprise using the Anglo-American model are seen as inappropriate and doomed to failure. Instead, Molly Scott Cato argues that a particular form of associative or cooperative entrepreneurship goes with the grain of Welsh principles and values and is more likely to succeed, citing the case of the employee-owned cooperative Tower Colliery in Hirwaun, Glamorgan, in the South Wales Valleys. This particular form of local and regional development is interpreted as a more appropriate fit with Welsh principles and values. Whether this innovative approach can stimulate higher levels of new business start-ups in Wales remains to be seen.

Source: Cato (2004)

the European Spatial Development Perspective attempts to promote a more integrated spatial development framework for the whole of the European territory and to balance and reconcile the sometimes contradictory interests of economy, society and environment (Figure 2.1). Questions of social justice and equalities in local and regional development have been addressed too, for example in gender-sensitive approaches (Rees 2000).

Example 2.5 'Smart growth' and local and regional development

The quest for sustainable and 'liveable' communities has stimulated 'smart growth' approaches to local and regional development, especially in North America and Australia. Current patterns of urban and suburban development are interpreted as detrimental to quality of life. Urban sprawl and inner city population flight have fostered concerns about the geography of public service provision, particularly how inner and outer city services can be funded and supported in the context of suburbanisation and shifting tax bases, and the growing separation between where people work from where people live. Problems identified include congestion and pollution from automobile dependence, loss of open space, pressure upon public infrastructure and services, inequitable distribution of economic resources and the loss of the sense of community.

Local and regional development has a role in making communities 'more successfully serve the needs of those who live and work within them' (Local Government Commission, Ahwahnee Principles, p. 1). The Local Government Commission in America has developed its 'Smart Growth: Economic Development for the 21st Century' agenda based upon a set of principles informed by sustainability. The argument is that:

> We can no longer afford to waste our resources, whether financial, natural or human. Prosperity in the 21st Century will be based on creating and maintaining a sustainable standard of living and a high quality of life for all. To meet this challenge, a comprehensive new model is emerging for smart growth which recognizes the economic value of natural and human capital. Embracing economic, social and environmental responsibility, this approach focuses on the most critical building blocks for success, the community and the region. It emphasizes community-wide and regional collaboration for building prosperous and livable places. While each community and region has unique challenges and opportunities, the following common principles should guide an integrated approach by all sectors to promoting economic vitality within their communities, and in partnership with their neighbours in the larger region: integrated approach, vision and inclusion, poverty reduction, local focus, industry clusters, wired communities, long-term investment, human investment, environmental responsibility, corporate responsibility, compact development, livable communities, center focus, distinctive communities and regional collaboration.

Elements of this agenda have been adapted by particular states and cities in North America, including North Carolina's 'Smart Growth Alliance', Greater Boston, Baltimore and Portland, and in Australia, including Queensland.

Sources: Local Government Commission, Smart Growth: Economic Development for the 21st Century, www.lgc.org/economic/localecon.html; Haughton and Counsell (2004)

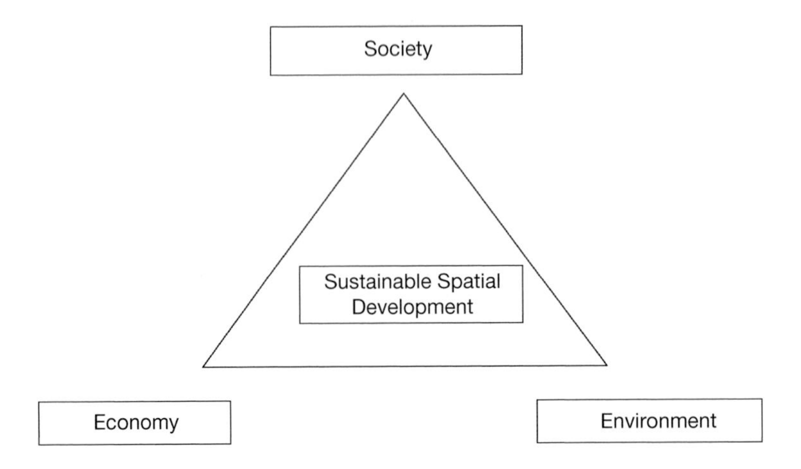

Figure 2.1 ***European Spatial Development Perspective***

Source: Adapted from European Commission (1999)

Different kinds of local and regional development are closely connected to socially determined principles and values that can differ from place to place and change over time. Distinctions can be made between varieties of local and regional development, including 'high' and 'low' road versions and its quantitative extent and qualitative character. Principles and values shape how specific social groups and interests in particular places define and articulate what is meant by local and regional development. They shape the normative questions about the perceived worth and desirability of its different varieties. Different degrees of commitment to sustainability, social justice and equalities in more holistic and integrated forms of local and regional development may emerge in different places and change over time. The kinds, principles and values of local and regional development can have very different implications – economically, socially, environmentally, politically and culturally – for different social groups and places in localities and regions.

Local and regional development for whom?

Definitions and kinds of local and regional development are closely connected to questions of local and regional development for whom? Answers to which question concern, first, the objects and subjects of local and regional development and, second, the social welfare dimensions of the often uneven and geographically differentiated distribution of who and where benefits and loses from particular varieties of local and regional development. Specific social groups and/or institutional interests may be advantaged by particular forms of local and regional development. 'High' or 'low' road strategies may favour or disadvantage specific social groups, occupations, firms, sectors, institutions and localities and regions. For example, property-led approaches may benefit property development companies and real estate speculators at the expense of first-time house buyers and local communities. Understanding and explaining the objects, subjects and social welfare outcomes are central to local and regional development.

The objects of local and regional development refer to the material things to which 'development' action is directed and the subjects are the themes and topics upon which 'development' is based. Understanding the objects and subjects is an important starting point in thinking through the effects and implications of specific definitions, principles and values of local and regional development processes and policies. Table 2.5 provides examples of the different and sometimes overlapping levels and scales for the objects and subjects of local and regional development.

A range of policy instruments can be designed to intervene and shape the extent and nature of local and regional development. The different forms and possible examples are detailed in Table 2.6. Policy instruments can be coordinated and integrated within comprehensive development programmes for localities and regions. Particular policy instruments may have specific aims and both intended and unintended consequences. The interrelations and spillovers between policy areas can cause either negative or positive consequences, and knock-on effects. They require careful deliberation in the problem definition, policy design and delivery stages of the policy cycle. The translation of the objects and subjects of local and regional development into policy interventions can usefully distinguish between those with an explicitly spatial focus and those without but with spatial impacts. For example, area-based regeneration partnerships are explicitly spatial. Their objects are specific types of places and their subject is economic and social well-being in particular kinds of neighbourhood or community. This spatial policy intervention seeks a spatial outcome. Whereas individuals or households may be the objects of policy changes in the tax and benefits system whose subject might be welfare reform and public expenditure efficiency. However, this policy may influence disposable incomes and expenditure patterns in the local economy. A non-spatial policy

Table 2.5 ***The objects and subjects of local and regional development***

Level/scale	*Objects*	*Subjects*
People	Individuals	Education
	Households	Homecare services
	Families	Childcare services
Spaces, places and territories	Neighbourhoods	Neighbourhood renewal
	Communities	Community regeneration
	Villages	Rural diversification
	Localities	Strategic partnerships
	Towns	Market town revival
	Cities	Growth strategies
	City-regions	Local authority collaboration
	Sub-regions	Spatial strategies
	Regions	Regional economic strategies
	Sub-nations	Economic development strategies
	Nations	Regional development
	Macro-regions	Economic and social cohesion
	International	Aid distribution
	Global	Trade liberalisation

Source: Authors' own research

Table 2.6 ***Policy instruments for local and regional development***

Forms	*Examples*
Direct intervention	
Direct provision of services	Education, health
Commissioning of services from public, private and/or voluntary sectors	Employment zones, subsidised public transport services
Economic instruments	
Taxes	Fuel duty, VAT, development or roof tax
Charges	Congestion charges, road pricing
Subsidies, tax credits and vouchers	R&D tax credits, pre-school education vouchers
Benefits and grants	Social transfer payments, education maintenance grants
Tradeable permits and quotas	Carbon emissions trading scheme
Award and auctioning of franchises and licenses	Mobile phones, airport landing slots, broadcasting
Government loans, loan guarantees and insurance	Student loans, Social Fund, export credit guarantee
Regulation and other legislation	
Price and market structure regulation	Competition legislation, price regulation, privatised utilities
Production and consumption regulation	Planning rules, public service obligations on privatised utilities, renewable energy obligations, licensing
Standards setting regulation	Accreditation for education and training qualifications, trading standards
Prescription and prohibition legislation	Criminal justice
Rights and representation legislation or regulation	Human rights, freedom of information
Information, education and advice	
Provision of information	On-line services, leaflets, multi-language information, information access for disabled
Public education campaigns	Health, education
Reporting and disclosure requirements	Financial services, public appointments
Labelling	Food and drink ingredients, household products
Advisory services	Careers services, micro- and small business advice institutions
Representation services	Ombudsmen, area forums
Self-regulation	
Voluntary agreements	Advertising standards, corporate social responsibility initiatives
Codes of practice	Banking Code
Co-regulation	Industrial relations and dispute resolution institutions

Source: Adapted from Prime Minister's Strategy Unit (2004)

can, then, have distinct geographical effects. From time to time, specific scales of policy intervention can emerge as the focus of local and regional development given particular interpretations of the problems of specific types of area, for instance Community Economic Development (CED) (Example 2.6) and the more recent focus upon neighbourhoods.

The social welfare distribution of who and where benefits and loses from particular varieties of local and regional development is geographically differentiated and changes over time. In abstract terms, spatial disparities in economic and social conditions are inherent in the combined and uneven development within capitalism as a socio-economic system (Glasmeier 2000). Inequalities exist between the impacts and experiences of socio-economic processes by particular social groups, often depending upon their class, ethnicity, gender and social identity. The distribution of social power and resources within society shapes who and where gains from local and regional development (Harvey 1982). Within capitalism, a recurrent and normative issue for local and regional development concerns the social welfare implications of the relationship between economic efficiency and growth and social equity (Bluestone and Harrison 2000; Scott and Storper 2003). Are they contradictory or complementary and to what degree is each considered desirable and/or appropriate for local and regional development?

Historically, Kuznets' (1960) nationally focused work argued that further economic growth tended to create more inequality at low income levels. Richardson (1979) concurred and argued that regional inequalities may only be a problem in the early stages

Example 2.6 Community Economic Development

In response to the inadequacies of the kinds of 'top-down', state-led approaches to local and regional development outlined in Chapter 1 during the 1960s and 1970s, community-focused economic development emerged where neither the private nor public sectors had managed to ameliorate persistent deprivation or had struggled to provide services to 'hard to reach' social groups and localities. Community Economic Development is a bottom-up approach characterised by community-led and determined regeneration. Civil society through the voluntary and community sector is seen as the key deliverer of community services as a not-for-profit non-market and non-state, especially local authority, organisation. So-called 'Third' sector initiatives beyond state and market constitute a 'social economy' and increasingly significant components of local and regional development and policy. They may include cooperatives, social enterprise, credit unions, intermediate labour markets, Local Exchange Trading Schemes (LETS) and support for informal activities such as volunteering. CED is potentially beneficial for disadvantaged localities since it seeks to use and develop the skills of local people, recirculate local resources through the local economy through local ownership and foster self-determination in local communities. The potential for CED while significant may remain limited by its context of internal and external constraints, for example low levels of local disposable income, weak education and skills locally and barriers to existing market entry.

Sources: Geddes and Newman (1999); Haughton (1999); Amin *et al.* (2002)

of a nation's growth. As income levels per capita increased, a critical threshold of income is reached and further economic growth and higher average per capita income tended to reduce a nation's overall income inequality. Figure 2.2 depicts the relationship in Kuznets' inverted-U hypothesis. Richardson (1979) saw compatibility between some regional efficiency and equity objectives, potentially reinforced through strong and redistributive regional policies. Recent debates encapsulate the 'knife-edge' dilemma between growth and equity:

> some analysts hold that development policy is best focused on productivity improvements in dynamic agglomerations (thereby maximising national growth rates but increasing social tensions), while other analysts suggest that limiting inequality through appropriate forms of income distribution (social and/or inter-regional) can lead to more viable long-run development programmes.
>
> (Scott and Storper 2003: 588)

Growth and equity considerations remain a central issue and in constant tension for local and regional development.

Globally, a comparison of the relative levels of prosperity (measured in GDP per capita) and income inequality (measured by the GINI coefficient – the ratio of income of the richest 20 per cent of income earners to the income of the poorest 20 per cent – varies between 0 and 1, values closer to 1 mean greater income inequality, values closer to 0 mean less income inequality) reveals a markedly uneven picture (Figure 2.3). At the national level, many higher-income countries are grouped around relatively similar income distributions. Japan – more equal – and the United States – more unequal – stand out. Lower-income countries vary across the range from the former centrally planned economies (CPE) countries in Europe (e.g. Poland, Romania and Ukraine) with comparable and higher levels of income equality to lower income countries with highly unequal income distributions (e.g. Brazil, Colombia). Similarly, as an indicators of living

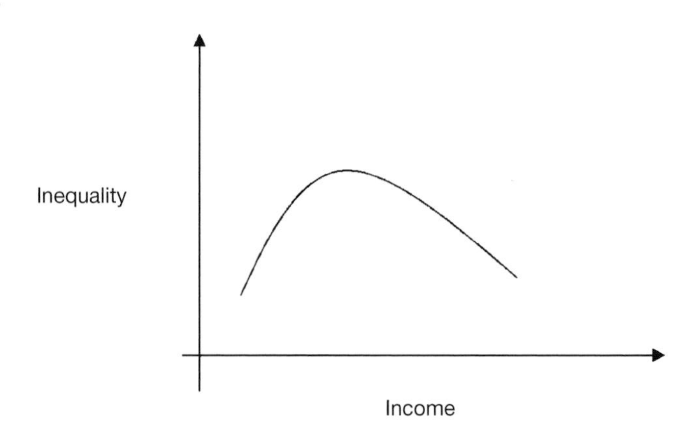

Figure 2.2 ***The Kuznets inverted-U hypothesis***

Source: Adapted from Cypher and Dietz (2004: 54)

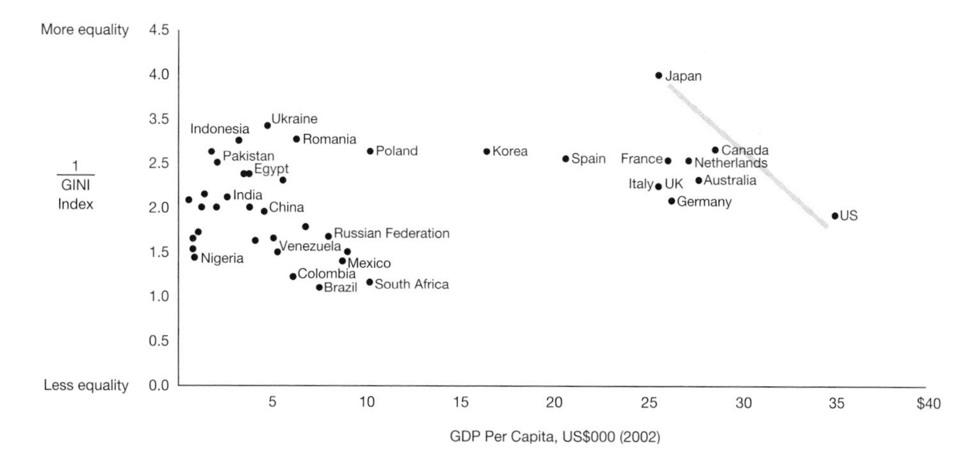

Figure 2.3 ***Average prosperity and equality of distribution by country, 2002***

Source: Institute for Competitiveness and Prosperity (2005: 10)

standards and well-being, comparisons of age-adjusted mortality rates exhibit large regional differences within countries, particularly in the United States, Australia and Canada, compared to the much lower variations in Japan, the Netherlands and Portugal (Figure 2.4).

In the European Union, discussion has addressed the trade-offs between economic efficiency, social rights and territorial cohesion. Cohesion is seen as a 'dynamic' and 'subjective' concept defined as 'the political tolerability of the levels of economic and social disparity that exist and are expected in the European Union and of the measures that are in place to deal with them' (Mayes 1995: 1). The issue has focused upon whether progress towards greater economic efficiency, for example through the Single European

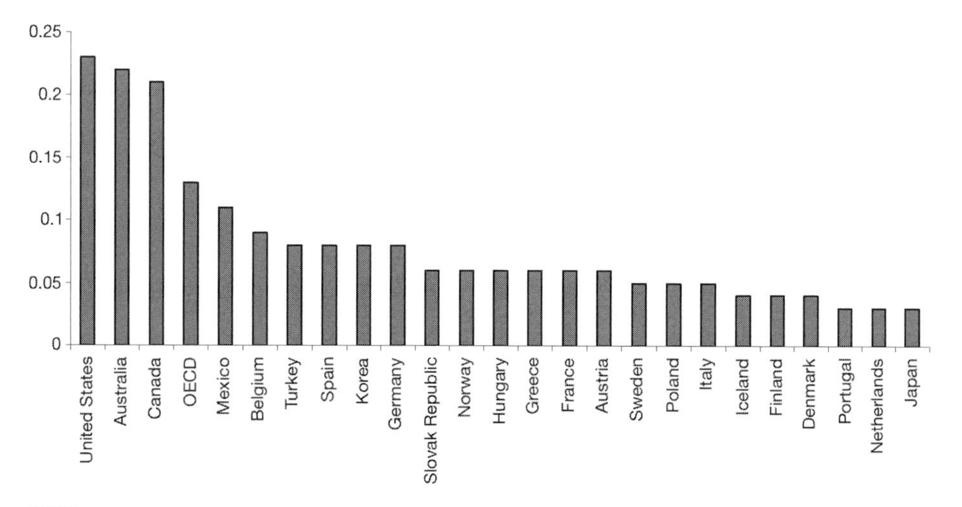

Figure 2.4 ***Coefficient of regional variation of age-adjusted mortality rates by country, 2000***

Source: Adapted from OECD (2005b: 150)

Market and the single currency the euro, and the extension of social rights, for example through the working time regulations, will be made or traded-off at the expense of economic and social cohesion across the regions of the European Union. Figure 2.5 illustrates the trade-offs involved. The extent to which social cohesion is a result or cause of economic growth remains debatable (Perrons 2004).

In parallel with the broadened definitions of local and regional development, social welfare analysis has addressed questions of equality. Recent research has focused upon the utilisation of the experience and participation of women and how this could make the underlying definitions, principles and varieties of local and regional development more gender-sensitive (Example 2.7). Similarly, the contribution of Black and Minority Ethnic communities in shaping approaches to local and regional development has been critical in tackling discrimination, promoting positive role models, raising educational aspirations and increasing economic participation (Ram and Smallbone 2003). The ways in which concepts and theories of local and regional development seek to understand and explain such dilemmas and broader claims for recognition, the roles of markets and states and public policy interventions to shape who and where gains and loses are addressed in Chapters 3 and 4.

We have suggested in this chapter that the social definition and the geographies of space, territory, place and scale matter to who and where is advantaged or disadvantaged in particular forms of local and regional development. The variety of local and regional development pursued and its underlying principles and values condition its extent and nature. The objects, subjects and social welfare aspirations of local and regional development result from answers to the question of local and regional development for whom? The answers are geographically differentiated and change over time.

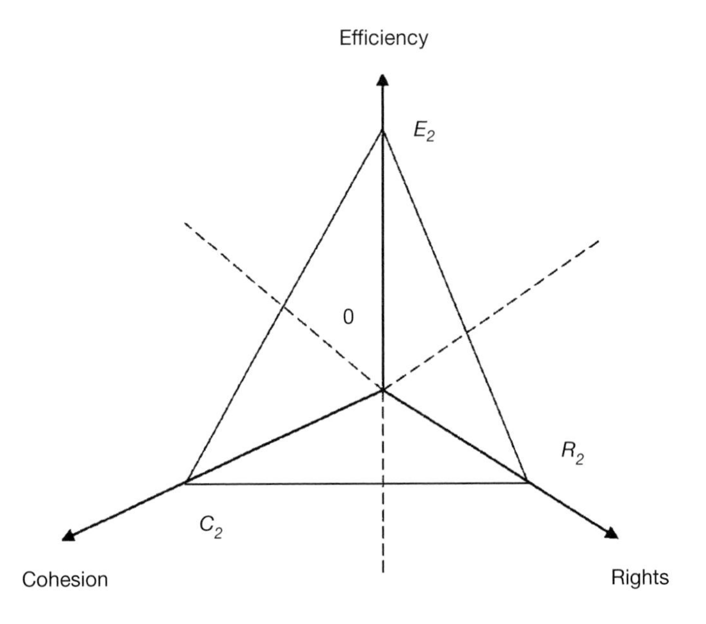

Figure 2.5 *Trade-offs between cohesion, efficiency and rights*

Source: Adapted from Mayes (1995: 2–3)

For some localities and regions, the socially determined answer may be purely meritocratic and the result of individual and institutional exploitation of the 'equality of opportunity' provided by competitive markets in the global context with minimal state and

Example 2.7 Gender-sensitive regional development

'Women should be identified as a potential for regional development and as capable actors in the regional policy process' (Aufhauser *et al.* 2003: 17).

Traditional forms of regional policy have often been blind to the particular interests and participation of women. Despite the marked gender effects of regional economic and social restructuring and the increasing relevance of what were formerly regarded as women's issues (e.g. childcare, work/life balance) in local and regional development, the potential contribution of women remains undervalued. Women remain under-represented in public bodies, especially in senior decision-making roles. Regional policies have tended to utilise highly simplified images of women that fail: 'to do justice to the actual and increasing variety of female lives' (p. 6). The development potential of women's participation in economic, social, environmental, political and cultural terms and the relevance, efficiency and effectiveness of regional policy interventions are inhibited as a result.

As part of raising awareness and fostering 'gender competence', Aufhauser *et al.* (2003) have developed fundamental principles of 'gender-sensitive regional development'. They seek to adapt and broaden regional policy interventions to ensure heightened gender awareness. For Aufhauser *et al.*:

> 'Gender-sensitive regional development' may be understood as a concept for the design of spatial development processes aimed at bringing about the co-existence of women and men on an equal footing and in particular at contributing towards the improvement of the possibilities of female self-determination and participation.
>
> (Aufhauser *et al.* 2003: 3)

The principles include: enabling women and men to choose self-determined life concepts and lifestyles, challenging gender stereotypes, taking account of gender-based regional inequality structures in policy design, developing integrated approaches with heightened gender awareness and promoting women as opinion formers in the regional development process.

Practical measures to implement this agenda include 'mainstreaming' gender issues across regional public policy, encouraging regional policy design and development 'by, for and with women' (for example, by building upon the EU's EQUAL partnerships with a gender focus) and facilitating women's participation in the relevant institutions. Future developments seek to make progress on this agenda, although implementation of gender-sensitive measures and instruments are typically still not regarded as priorities by the predominantly male actors in local and regional policy.

Source: Aufhauser *et al.* (2003)

institutional regulation – irrespective of the geographical and social unevenness of development. For others, towards the opposite end of the economic, political-ideological spectrum, the answers may be about state and institutional support and internationalism to overcome inequality and disadvantage for people and places in a quest for more geographically and socially even forms of local and regional development. We return to the questions of what kind of local and regional development and for whom in the conclusions in Chapter 8.

Conclusion

Understanding local and regional development requires an engagement with its most basic nature. What it is, what it is for and, in a normative sense, what it should be are critical starting points. This chapter has addressed the fundamental questions of what kind of local and regional development and for whom? Questions of definition were reviewed to examine what is meant by local and regional development, its historical context and the importance of its geographies of space, territory, place and scale. Definitions of local and regional development have broadened to include economic and social, environmental, political and cultural concerns. Definitions are socially determined in the context of historically enduring themes, principles and values, incorporating geographical diversity and changes over time. The historical evolution of 'development' in the post-war period emphasises its changing meanings, geographical differentiation, broadened focus and approaches to local and regional development. Geography matters as a causal factor in local and regional development. Territories evolve as defined areas in which particular definitions of local and regional development are constructed and pursued. Places shape the geographical diversity and unevenness of local and regional development. Economic, social, political, environmental and cultural processes influence local and regional development across, between and through different scales.

The varieties and principles and values in different places and time periods were then explored. Different kinds of local and regional development connect to socially determined and normative principles and values. They may differ geographically and change over time. Distinguishing the objects, subjects and social welfare aspects of local and regional development helps understand the often socially and geographically uneven distribution of who and where benefits or loses from particular forms of local and regional development. The next chapter builds upon the discussion of what kind of local and regional development and for whom, and engages with the concepts and theories that seek to explain local and regional development.

Further reading

For the more international understanding of local and regional development questions, see Scott, A.J. and Storper, M. (2003) 'Regions, globalization, development', *Regional Studies* 37(6–7): 579–593; Cypher, J.M. and Dietz, J.L. (2004) *The Process of Economic Development*. London: Routledge; Beer, A., Haughton, G. and Maude, A. (2003)

Developing Locally: An International Comparison of Local and Regional Economic Development. Bristol: Policy Press.

For the evolution of local and regional development thinking, see Geddes, M. and Newman, I. (1999) 'Evolution and conflict in local economic development', *Local Economy* 13(5): 12–25.

On the new metrics and broader, more sustainable understanding of local and regional development, see Morgan, K. (2004) 'Sustainable regions: governance, innovation and scale', *European Planning Studies* 12(6): 871–889.

On the local and regional foundations of economic growth, see Sunley, P. (2000) 'Urban and regional growth', in T.J. Barnes and E. Sheppard (eds) *A Companion to Economic Geography*. Oxford: Blackwell.

PART II

Frameworks of understanding

CONCEPTS AND THEORIES OF LOCAL AND REGIONAL DEVELOPMENT

3

Introduction

> Despite ever increasing integration of local economies into global flows of trade and capital, . . . local economic differentiation remains endemic to capitalism, and may even be intensifying as transport and communications costs fall. Despite the numerous glossy predictions of the death of distance and the end of geography, local and regional differences in growth may be intensifying across the industrialized world . . . the search for simple trends in urban and regional disparities has been confounded by the new complexity and unpredictability of local economic changes. In the developing world too, regional and urban inequalities have reached unprecedented scales. Thus, it seems more important than ever to understand the processes causing local economic growth.
>
> (Sunley 2000: 187)

Connecting with the growing importance and profoundly changing context of local and regional development introduced in Chapter 1, this chapter addresses this challenge for concepts and theories: to provide frameworks to understand local and regional development. Concepts and theories are developed to help us interpret and make sense of how and why things work out in the ways that they do. They should provide us with usable definitions of concepts, an understanding of the main causal agents and relationships and how these may be articulated in mechanisms and processes. Concepts and theories are developed to help us understand and explain local and regional development across space in place and over time. The kinds of concepts and theories we might use for interpretation are closely linked to our answers to the questions of what kind of local and regional development and for whom discussed in Chapter 2. This chapter reviews the most important and influential concepts and theories of local and regional development. It provides an accessible and critical discussion of the main frameworks of understanding and explanation. Each approach is reviewed, its limitations discussed and connections are made to the substantial literature on theories of local and regional development for further reading and reflection.

The chapter is organised around the different schools of thought and types of approach to understanding and explaining local and regional development. Different theoretical traditions take different starting points and make a variety of assumptions. Their epistemology – theory of the method or grounds of knowledge – and ontology – essence of things or being in the abstract – often differ. New approaches can develop on the basis of criticism and the rejection of existing frameworks of understanding. Theories evolve over time in response to critique and their own conceptual development, often in the light of ongoing empirical research and changing political circumstances. Concepts and theories are not set in stone; they are constantly evolving in parallel with the world they seek to understand and explain.

Each section in this chapter examines how particular theories address common questions. These comprise, first, what are the conceptual building blocks used by the theories and how are they defined? How do they conceptualise localities and regions and their 'development'? Second, what is the purpose and focus of the theories? What are they seeking to understand and explain? For some, this may include fundamental questions of regional growth: why do some regions grow more rapidly than others? What are the dynamics of regional convergence and divergence? Why are local and regional disparities in social welfare persistent over time? Third, what are the constituent elements – causal agents, relationships, mechanisms and processes – of the theories? Fourth, what kinds of explanations do the theories provide? Fifth, how do the concepts and theories relate to local and regional development policy? Last, what are their criticisms and limitations? We begin our review by focusing upon one of the earliest and most influential: the neo-classical approach to local and regional development.

The neo-classical growth theory of local and regional convergence

In the tradition of the classical economics of David Ricardo, John Stuart Mill and Adam Smith, neo-classical economics is characterised by microeconomic theory developed to examine static rather than dynamic equilibrium within economic systems. Disparities in regional growth are a traditional concern of neo-classical approaches to local and regional development (Borts and Stein 1964; Williamson 1965). In this approach, regional growth determines regional income and economic and social welfare. Local and regional 'development' within this theory is focused upon the long-run reduction of geographical disparities in income per capita and output. The causal mechanisms in the theory predict that such spatial disparities will reduce and move towards or converge upon an economically optimal equilibrium in the long run (Martin and Sunley 1998). The theory seeks to explain where and why such convergence does not occur and why disparities continue to grow or diverge between regions. 'Regions' are understood as subnational territorial units and have been the main geographical focus of the theory.

Conceptually, measures of regional growth are several in neo-classical theory (Armstrong and Taylor 2000). Output growth refers to the expansion of productive capacity within a region and illustrates the extent to which the region is attracting the key factors of production capital and labour. Output growth per worker is a measure of productivity and reveals how efficiently resources are being used within a regional economy. This measure relates directly to the relative competitiveness of specific regions in comparison with other regions. Output growth per capita relates growth to the population of a region and illustrates the relative level of economic and social welfare in the region.

In the neo-classical model, regional output growth is dependent upon the growth of three factors of production: capital stock, labour force and technology. Figure 3.1 illustrates these determinants of regional output growth. Technological progress is seen as a key contributor to growth due to its influence upon productivity growth rates in the long run (Armstrong and Taylor 2000). Innovation and technology have the potential to increase output growth per worker. In this basic version of the neo-classical theory,

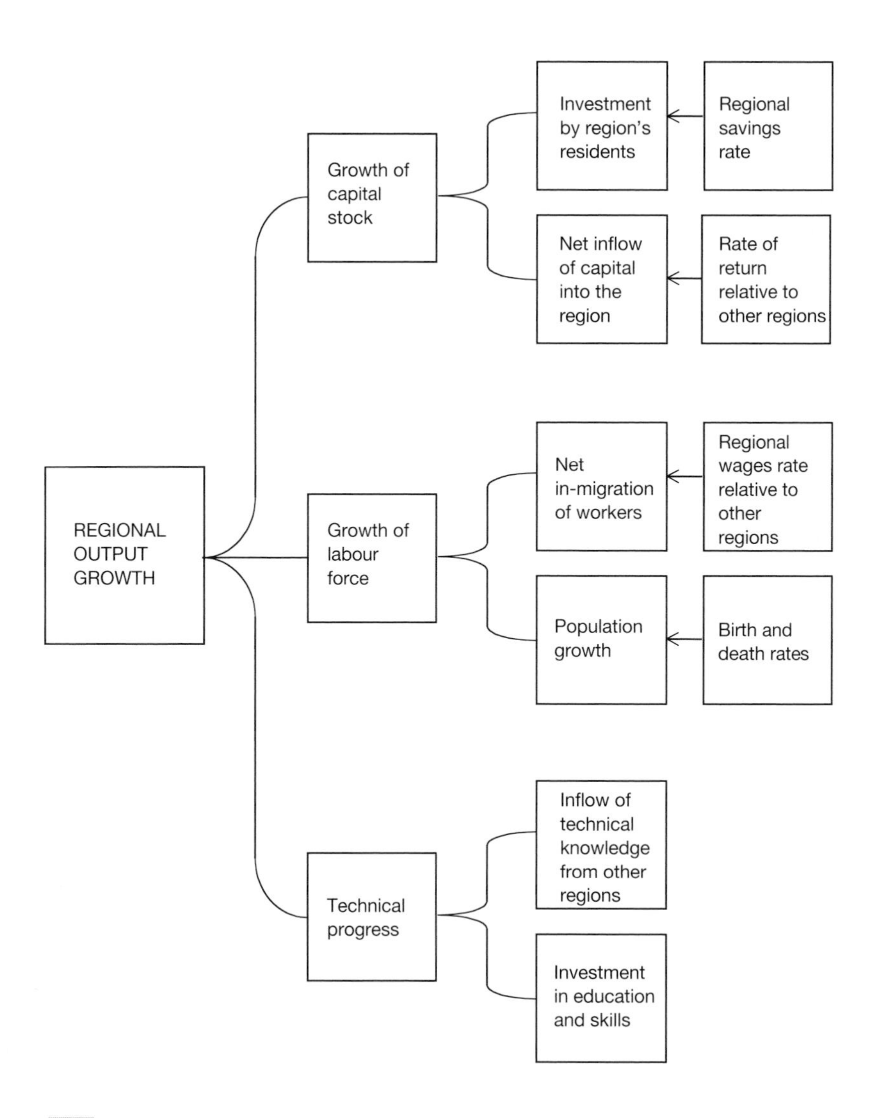

Figure 3.1 ***The determinants of regional output growth***

Source: Adapted from Armstrong and Taylor (2000: 72)

technological change as well as other important determinants such as human capital, savings and population growth rates are 'disembodied' or treated independently of capital and labour inputs. Hence, this theory is often referred to as exogenous growth theory. Regional growth disparities are explained in the neo-classical approach by variations in the growth of the main factors of production: the rate of technological progress and the relationship between capital and labour – the capital/labour ratio. Productivity – output per worker – will increase only if capital per worker increases (Figure 3.2). This is a positive relationship, often referred to as 'capital deepening' (Clark *et al.* 1986). However, this productivity increase occurs at a falling rate due to diminishing marginal returns. This is a central concept in neo-classical economics: beyond a specific level of input further input will result only in decreases in the additional marginal output of the product per unit input. When the additional or marginal product of labour reaches zero, then an equilibrium position is achieved. At this point, there is no incentive to increase the capital/labour ratio.

Neo-classical growth theory has evolved to understand changes over time. It focuses upon the supply of factors of production and assumes their perfect mobility across and between regions (Barro and Sala-i-Martin 1995). The theory assumes perfect knowledge about factor prices and the economically rational and efficient choices of buyers and suppliers in response to market signals. Economic returns to the increasing scale of economic activities are assumed to be constant. Perfectly functioning markets are seen as capable of ameliorating or reducing rather than exacerbating or increasing geograph-

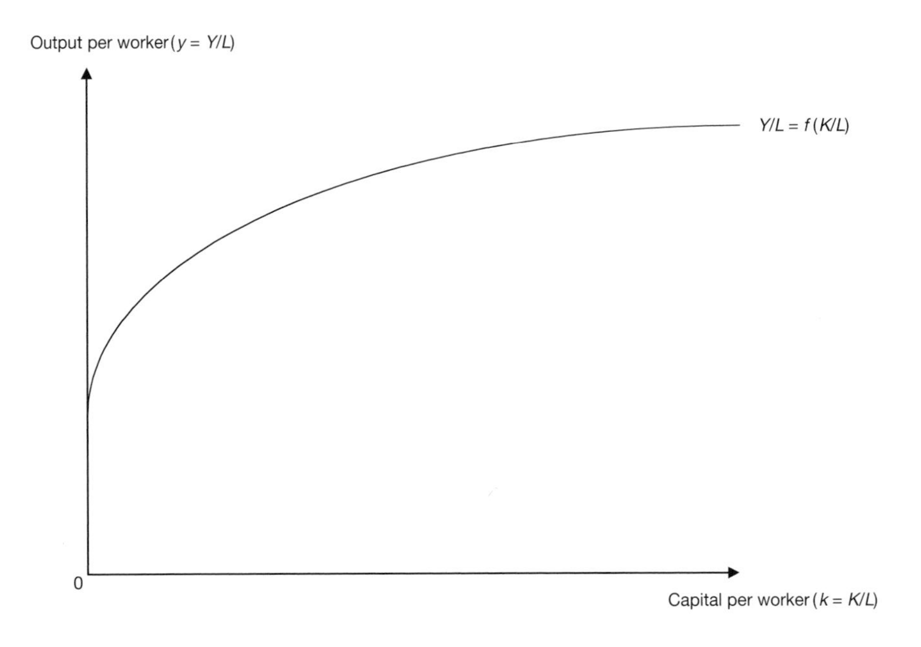

Figure 3.2 ***The capital/labour ratio***

Source: Adapted from Armstrong and Taylor (2000: 68)

ical disparities in economic and social conditions. Under the strict economic rationality and market-based conceptualisations of the neo-classical model, the perfect mobility of factors of production of capital and labour move to regions offering the highest relative rates of return. Firms look for the most profitable locations and labour seeks the highest wages. The adjustment mechanism works because regions with high capital/labour ratios have high wages and a low return or yield on investment. Capital and labour therefore move in opposite directions. High wage regions lose capital and attract labour. Conversely, regions with low capital/labour ratios have low wages and high returns on investment. Low wage regions lose labour and attract capital. This market adjustment mechanism works over the long run to reduce regional disparities in the capital/labour ratio and regional growth. Regions with less capital per unit of labour tend to have higher relative rates of return and higher initial growth rates than regions with higher levels of capital per worker (Barro and Sala-i-Martin 1995).

In the neo-classical theory, regional disparities are only ever temporary since spatial inequalities set in motion the self-correcting movements in prices, wages, capital and labour to underpin the eventual convergence of economic and social conditions between regions (Martin and Sunley 1998). In concert, technology diffuses across regions to allow 'catch-up' and geographical equalisation in levels of technological progress (Malecki 1997). In theory, convergence in output growth between regions occurs and an equilibrium position is achieved.

The neo-classical approach describes different types of regional convergence. Conditional convergence refers to movement towards a steady state growth rate resulting in constant per capita incomes, consumption levels and capital/labour ratios between regions. It is conditional because the savings rates, depreciation rates and population growth rates that influence regional growth but are treated as external to the neo-classical growth model can differ across countries. Conditional convergence does not necessarily result in equal per capita income levels across countries. Absolute convergence results when the growth model parameters are equal. Richer countries will tend to grow slower than poorer countries which start from a lower level of development. For absolute convergence, the neo-classical model suggests that per capita incomes will become equalised across countries over time. The model has different measures of spatial convergence between regions. Beta (β-convergence) measures the speed of convergence. It occurs when poor regions grow faster than richer regions. Over the long term, per capita incomes equalise across economies. In this measure there is a negative relationship between the growth of per capita income and the level of per capita income at the start of the period. Sigma (σ-convergence) is a measure of income inequality and is the dispersion or spread of per capita income between regions at a given point in time. Convergence occurs when the dispersion of per capita income between regions, although not necessarily between people within regions, falls over time; β-convergence can occur without σ-convergence.

Another important neo-classical approach that addresses inter-regional convergence is the theory of comparative advantage (Armstrong and Taylor 2000). In this approach, nations and regions specialise in economic activities in which they hold a comparative advantage, principally in industries that utilise their abundant factors of production. This can include labour, land, capital and natural resources. Trade between nations and regions is based upon differences in such factor endowments. Trade is a positive sum game in which trading partners benefit. In a static rather than dynamic framework, specialisation and trade promote efficient resource allocation and inter-regional convergence.

The evidence

Empirical examinations of neo-classical growth theory suggest regional convergence is often a slow and discontinuous process (Barro and Sala-i-Martin 1991; Martin and Sunley 1998; Armstrong and Taylor 2000). Considerable variations exist in the speed and extent of convergence over different time periods in different places (Armstrong and Vickerman 1995; Scott and Storper 2003). Convergence often fluctuates with the economic cycle, increasing faster during the upswing of growth than in the downswing periods of recession, although convergence in general appears to have slowed considerably since the early 1980s (Dunford 1993). While interpretations differ, Example 3.1 shows how the European Union appears to be experiencing national level convergence between Member States and regional level stability or divergence. Club convergence is often evident where the growth performance of countries with similar structural characteristics and initial conditions converge. The relatively more prosperous and developed OECD countries, developing and underdeveloped countries form three distinct and separate convergence clubs without any necessary convergence in economic growth between them (Martin and Sunley 1998). At the regional level, geographical clustering of growth rates is evident in the United States and Europe with spatially proximate grouping of fast and slower growth regions (Armstrong and Vickerman 1995).

Example 3.1 Regional convergence and divergence in the European Union

Despite using a common neo-classical approach to regional growth, different interpretations exist of the changes in regional disparities over time in the European Union. Often the analyses differ because of the use of a particular technique, data set or time period used. Research using econometric models has concluded that the regional convergence evident across Europe will continue (Mur 1996). Other analyses suggest that, despite increasing expenditures on regional policy, regional disparities in Europe have not narrowed substantially. Indeed, some measures suggest such inequalities have widened (Puga 2002) and the impact of regional policy has been at best marginal for growth in the periphery (Rodríguez-Pose and Fratesi 2004). At the national level, income disparities between Member States have fallen. However, income and unemployment inequalities have risen between regions within Member States. International convergence has been accompanied by regional divergence (Rodríguez-Pose 2002a).

As Table 3.1 illustrates, the European Commission (2004) interprets some convergence in GDP per head across the European Union. Using one of the key measures discusse above, Beta (β-convergence) – the inverse relationship between growth and initial GDP per head – is evident for each of the time periods from 1980. Regions with the lowest initial levels of GDP per head have, on average, the highest growth rates. In particular, the Objective 1 regions – with growth under 75 per cent of the EU15 average – experienced strong growth between 1988 and 1994, driven by high growth rates in the new Länder following the unification of the former East and West Germany. The role of regional policy in this convergence process through the Structural Funds is recognised and promoted by the European Commission. On the other measure, Sigma (σ-convergence) – the dispersion of per capita income between regions at a given point in time – is also evident in the European Union between 1980 and 2001. Regional disparities remain an important issue for the European Union, however, particularly in the context of the enlargement of the European Union and the accession of ten new Member States in 2004. Cyprus, Czech Republic, Estonia, Hungary, Latvia, Lithuania, Malta, Poland, Slovakia and Slovenia each has lower levels of GDP per head than most existing Member States and, in some cases, similarly marked regional disparities (Rodríguez-Pose 2002a). The European Commission estimates that even if growth in the Accession Countries (plus applicant countries Bulgaria and Romania) can be sustained at 1.5 per cent above that in the rest of the European Union – for example growth of 4 per cent per year relative to 2.5 per cent in the EU15 – average GDP per head in the twelve countries would still remain below 60 per cent of the enlarged EU27 average until 2017. Regional convergence in the European Union, then, is a long-term issue.

Table 3.1 ***Regional growth disparities in the European Union, 1980–2001***

	No. of regions	*GDP per head (% growth rate)*	*Beta convergence rate per year (%)*
1980–1988			
All EU15 regions	197	2.0	0.5
Objective 1 regions	55	1.9	0.4
Other regions	142	2.0	2.1
1988–1994			
All EU15 regions	197	1.3	0.7
Objective 1 regions	55	1.4	3.1
Other regions	142	1.2	0.8
1994–2001			
All EU15 regions	197	2.3	0.9
Objective 1 regions	55	2.6	1.6
Other regions	142	2.1	0.0

Source: European Commission (2004: 146)

The neo-classical approach to regional policy

Neo-classical theory and its explanation of regional growth disparities are highly influential for regional policy. Detailed in Table 3.2, regional policy underpinned by neo-classical growth theory has been described as a 'free-market' approach. This view claims that convergence will happen regardless of intervention due to the causal mechanisms of the growth model that move regions towards equilibrium or that intervention will either hinder or increase the speed of convergence. In the European Union, for example, the output measure of GDP per capita is used in the geographical analysis of the eligibility of regions for regional policy (European Commission 2004). The focus has been the determinants of regional per capita income levels and how low-income regions can converge or 'catch-up' with relatively higher-income regions. Identifying the interventions to correct market failures and 'speed-up' convergence has been central to regional policy. Local and regional development policy is discussed in more detail in Part III.

The critique of the neo-classical approach

Criticisms of the neo-classical growth model have focused on several issues. First, its main assumptions are interpreted as unrealistic (Martin and Sunley 1998). Factor mobility is less than perfect (Armstrong and Taylor 2000). The access to and availability of capital is markedly uneven geographically (Mason and Harrison 1999). While capital is relatively mobile, labour's economic position, for instance in the housing market, and ties of social reproduction, for instance through family and the education of children, form attachments to places that can often militate against geographical mobility. Indeed, neo-classical approaches have focused upon such issues in explaining persistent regional

Table 3.2 ***Neo-classical regional policy: the 'free-market' approach***

Dimensions	*Characteristics*
Theoretical approach	Neo-classical economics Supply-side flexibility Correcting market failure
Causes of regional economic disparities	Market failures Inefficiency problem in regions due to labour market rigidities Lack of entrepreneurial 'culture' Excessive state intervention
Political ideology	New Right/neo-liberal Popular capitalism Deregulation/liberalisation Privatisation Small state sector Enterprise culture
Approach to reviving disadvantaged regions	Correction of market failures Deregulation of regional labour markets Tax incentives to promote efficiency
Regional policy	Minimal expenditure Selective assistance

Sources: Adapted from Martin (1989); Armstrong and Taylor (2000)

unemployment disparities (Armstrong and Taylor 2000). Perfect information is questionable. Investors and workers are not perfectly informed and able to respond rationally to price signals. Competition is often imperfect too with many markets for goods and services not reflecting the ideal of many buyers and sellers each without significant market power (Robinson 1964). The limitations of the comparative advantage theory comprise its static framework based on inherited factor endowments and its neo-classical assumptions of diminishing returns and technological equivalence between regions and nations (Kitson *et al.* 2004). Howes and Markusen (1993: 35, cited in Martin and Sunley 1996: 274) go further in challenging the Ricardian model of comparative advantage and the social welfare implications of the free trade model: 'there is some danger that the unfettered pursuit of free trade will actually depress wages and employment and lower world living standards'.

Second, the external or exogenous treatment of technology and labour weakens the model. Technological progress is profoundly uneven geographically and technology diffusion exhibits strong distance-decay effects (Malecki 1997). Shifts in the technological frontier have questioned the assumption of constant returns to scale and the productivity relationship described by the capital/labour ratio. Such issues have stimulated the development of endogenous growth theory – discussed below – which seeks to embody or internalise technology and human capital. Indeed, Armstrong and Taylor (2000) suggest the long-run persistence of disparities in regional growth rates may be due to the differential ability of regions to generate their own technology and adapt technology from elsewhere. Linking to the notion of stages of development unfolding over time, the likelihood of inter-regional convergence has been linked to the later stages of national development (Williamson 1965; Richardson 1980). This convergence is explained by the eventual equalisation of labour migration rates, capital market development, reduction of public policy bias towards core regions and the growth of inter-regional linkages.

Third, evidence suggests the neo-classical adjustment mechanism typically fails to work or operates only in the very long run and/or in specific time periods. Fingleton and McCombie (1997) even suggest that observed convergence is consistent with explanations other than that provided by neo-classical growth theory, in particular technological diffusion and regional policy. In explaining the dynamics of regional disparities, Armstrong and Taylor (2000: 85) suggest that: 'The neo-classical adjustment mechanism may play a relatively minor role'. Fundamentally, the very determinants of neo-classical growth theory – capital stock, labour force and technology – are inherently geographically variable (Martin and Sunley 1998). Yet neo-classical theory still predicts conditional convergence even given labour and capital's heterogeneity across space (Barro and Sala-i-Martin 1995). Despite these criticisms, the neo-classical approach is still highly influential in local and regional development as we explore throughout the remainder of the book.

Keynesian theories of local and regional divergence

Keynesian economics takes its name from the eminent economist John Maynard Keynes whose distinctive approach focused upon the under-employment of resources, the demand-side of the economy and the role of the state in managing aggregate demand. Although his work focused on national economies, his ideas have been taken up by regional economists. Keynesian theories focus upon the reduction of regional growth disparities in their approach to local and regional development. Building upon the critique of neo-classical approaches, the emphasis is upon understanding and explaining regional divergence: the reasons why regional growth disparities persist and are reproduced over time. Similar to the neo-classical approach, 'development' is equated with the reduction of regional disparities and 'regions' are the geographical focus. In contrast, the theories emphasise the medium rather than the long run. The adjustment mechanism in the Keynesian model focuses upon the role of demand rather than factor supply. Markets are seen as potentially exacerbating or increasing rather than ameliorating or reducing disparities in economic and social conditions:

> because market forces, if left to their own devices, are spatially disequilibriating. Economies of scale and agglomeration lead to the cumulative concentration of capital, labour, and output in certain regions at the expense of others: uneven regional development is self-reinforcing rather than self-correcting.
>
> (Martin and Sunley 1998: 201)

Drawing upon the ideas of John Maynard Keynes (1936), Keynesian theories use the approach and language of neo-classical economics to reach contrary conclusions.

Export base theory

Export base theory typifies the Keynesian emphasis upon demand. Differences in regional growth are explained by regional differences in the growth of the region's exports – the goods and services that are sold outside the region. External demand for the region's output determines the region's growth rate. In contrast to the endogenous approaches discussed below, regions are seen to develop from 'without' rather than from 'within' (Armstrong and Taylor 2000). Initially focused upon the exploitation of natural resources and the integration of resource-based regions into international trade (Innis 1920; North 1955), the export base approach developed theories of regional specialisation and adaptation for continued growth as well as decline. Regional specialisation in specific export commodities was explained using the neo-classical comparative advantage theory discussed above. Regions specialise in the production and export of commodities that use their relatively abundant factors intensively – whether they are raw materials, labour, capital and/or technology (Armstrong and Taylor 2000). As illustrated in Figure 3.3, the region's response to external demand stimulates growth in the basic or export sector and in the subservient 'residentiary' or non-basic sector. Multiplier effects are triggered as income and expenditure chains are stimulated within local and regional economies. Multipliers can be positive or negative.

While an oversimplification and hamstrung by the assumption of the relative immobility of factors of production, export base theory established the importance of

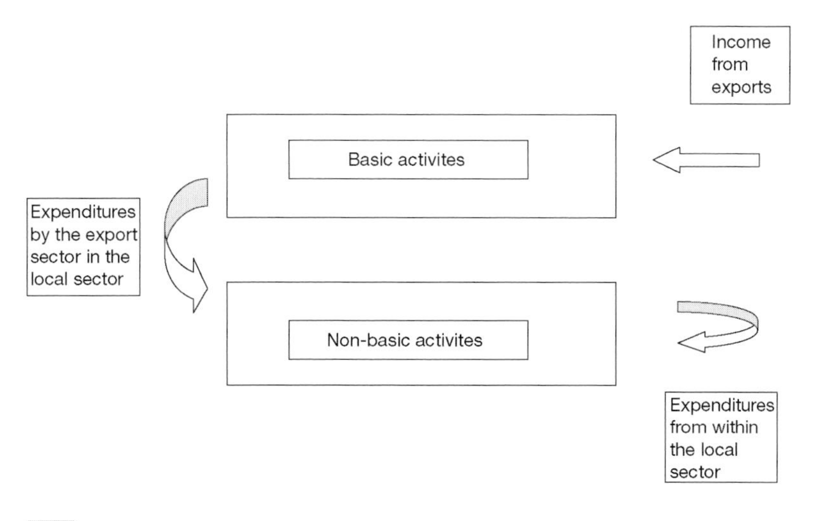

Figure 3.3 ***Export base theory***

Source: Authors' own research

specialisation and the impact of external demand for a region's products upon its growth. Demand is determined by the price of the region's exports, the income levels of other regions and the price of substitutes in external markets. The international competitiveness of the region's export sector relative to those in other regions determines its growth. Product quality and after-sales service influence demand too. On the supply-side, production cost factors, including wages, capital, raw materials, intermediate inputs and technology, influence the region's export competitiveness.

The sensitivity or elasticity of demand for the region's exports to changes in price and income is critical. Demand for inputs from other regions is also important. With favourable demand and supply, the region's export sector grows, demand for factor inputs bids up their prices relative to other regions and induces inflow of capital and labour. Regional disparities result. The duration of any such growth differential depends upon factor shortages, subsequent inflationary pressures and competition from alternative suppliers in other regions. Adaptation may require improved competitiveness through cost reduction and/or productivity increases and the development of new export markets, depending upon the degree of factor mobility between regions. The export-led growth process can be cumulative with positive multiplier effects upon regional income, an induced accelerator effect on investment, increased labour inflow and demand for local goods and services and the growth of subsidiary industries and external economies (Armstrong and Taylor 2000). A cumulative reversal of the process and relationships may also set in, however, for example through shifts in the demand for exports, technological change and competition. Example 3.2 examines how an export-based approach has been used to explain the pronounced disparities in growth and income in the regions of China.

Example 3.2 Regional disparities in China

The dramatic economic growth in recent decades in China has been accompanied by pronounced regional disparities. As part of economic reforms to open up the economy to international trade, China's national development policy has prioritised investment in export-oriented processing activities to exploit its comparative advantage of abundant and relatively cheap labour. As Table 3.3 illustrates, during the 1990s, marked differences have opened up in growth and income between the coastal and inland regions. Comparative growth rates between the coastal and inland regions were similar during the 1980s. Accompanying government-sponsored liberalisation, during the 1990s the real GDP per capita increased 95 per cent in the inland regions and 144 per cent in the coastal regions. China's coastal regions are catching up and even surpassing the other South East Asian economies of Malaysia, the Philippines, Indonesia and Thailand. Taking a Keynesian perspective, recent research has emphasised the role of exports and foreign direct investment (FDI) in explaining these regional inequalities. Fu (2004) argues that exports exerted a significant positive impact upon growth in the coastal regions. FDI-based and labour-intensive, processing-type exports have induced substantial growth in the coastal regions, attracting the mobile resources of labour from the inland regions. In contrast, the inland regions have not benefited from any significant linkages or spillovers from the growth of the coastal regions. Migration of labour from the inland regions has increased regional income inequalities, although the growing urbanisation of the interior is acting as a counterbalance. For the Chinese government, further labour migration to the capital-rich, coastal regions will exacerbate regional disparities. Regional policy is therefore focused upon encouraging the dispersal of domestic and foreign-owned capital to the labour surplus inland regions.

Table 3.3 ***GDP, FDI and exports by coastal and inland regions in China, 1999***

Regions	*Real GDP (PPC) (yuan at 1990 constant prices)*	*GDP % change 1978–1999*	*GDP % of national total*	*FDI % of national total*	*Exports % of national total*
Coastal					
Beijing	9,960	255	2.7	4.13	3.2
Tianjin	8,017	218	1.8	3.94	3.3
Shanghai	15,459	184	4.9	8.19	9.4
Liaoning	5,062	242	5.1	4.16	4.2
Hebei	3,479	339	5.6	1.99	1.4
Jiangsu	5,352	472	9.4	12.13	9.5
Zhejiang	6,041	739	6.5	3.11	7.0
Fujian	5,418	812	4.3	9.78	5.4
Shandong	4,353	533	9.4	5.9	6.3
Guangdong	5,886	637	10.3	28.25	40.4
Guangxi	2,082	325	2.4	2.09	0.6
*Average or sum**	5,204	411	62.4*	83.7*	90.7*

Inland					
Shanxi	2,372	199	1.8	0.42	0.8
Inner Mongolia	2,685	289	1.5	0.17	0.4
Jilin	3,182	284	2.0	0.84	0.6
Heilongjiang	3,844	213	3.5	1.09	0.8
Anhui	2,362	345	3.6	0.88	0.8
Jiangxi	2,339	289	2.4	0.81	0.5
Henan	2,456	387	5.6	1.22	0.6
Hubei	3,269	353	4.7	1.78	0.8
Hunan	2,562	312	4.1	1.48	0.7
Sichuan	2,234	306	4.5	1.54	0.6
Guizhou	1,242	226	1.1	0.13	0.2
Yunnan	2,234	354	2.3	0.27	0.5
Shaanxi	2,058	222	1.8	0.9	0.5
Gansu	1,851	144	1.1	0.13	0.2
Qinghai	2,340	151	0.3	0.01	0.1
Ningxia	2,245	179	0.3	0.04	0.1
Xinjiang	3,247	377	1.4	0.11	0.5
*Average or sum**	2,497	292	40.0*	11.8*	8.7*
National average	3,631	358	—		

Note: * Sum of column cells.

Source: Fu (2004)

Increasing returns and cumulative causation

Explicitly rejecting the neo-classical approach, Kaldor (1970, 1981) explained regional growth per capita by a region's ability to specialise and exploit scale economies. Sectoral structure was important too. Manufacturing was interpreted as a 'flywheel of growth' capable of fostering innovation and generating significant productivity benefits and faster growth for manufacturing specialised regions compared to resource-based regions. Kaldor emphasised increasing returns – rather than the neo-classical model's constant or diminishing returns – whereby increases in inputs generate disproportionately larger increases in quantities of outputs. Growth processes founded upon increasing returns are cumulative as fast growing regions steal a march on other regions and further reinforce their regional specialisation (Armstrong and Taylor 2000). Such increasing returns are central to the extended neo-classical theories discussed below.

The way in which the growth process tends to feed on itself in a circular and cumulative way and generate unbalanced regional growth is central to Gunnar Myrdal's (1957) theory of cumulative causation. Following the Kaldorian and Keynesian approach, this theory emphasises increasing rather than constant or diminishing returns to scale, agglomeration or external economies and the positive growth implications for localities and regions that were first to industrialise. The cumulative growth process is outlined in Figure 3.4. Beneficial effects between factors of production further advantage and propel growth in developed regions, often at the expense of lagging regions. Growth in developed regions may benefit lagging regions through 'spread' effects or what Hirschman (1958) called 'trickle down', including technological diffusion and export markets for their products. However, although relatively underdeveloped or peripheral regions could offer low-wage labour, this may be offset by more powerful agglomeration economies and the centripetal forces they generate in attracting factors of production in the developed or core regions. 'Backwash' effects could further reinforce disparities through encouraging capital and labour flows from lagging to developed regions. Rational responses to market price signals therefore reinforce rather than reduce regional inequalities. Liberalised trade further intensifies this polarised development between core and peripheral regions by catalysing growth in developed regions at the expense of lagging regions. Kaldor's (1970) elaboration of cumulative causation emphasised how increasing returns gave early industrialising regions advantages in international trade:

> Actual monetary wages may be the same in all regions, but efficiency wages, defined as monetary wages divided by a measure of labour productivity, tend to be lower in industrialized regions due to scale economies. Since regions with lower efficiency wages can produce more output, which in turn leads to further reductions in the efficiency wage (and so on), growth may build on itself without bound.
>
> (Dawkins 2003: 139)

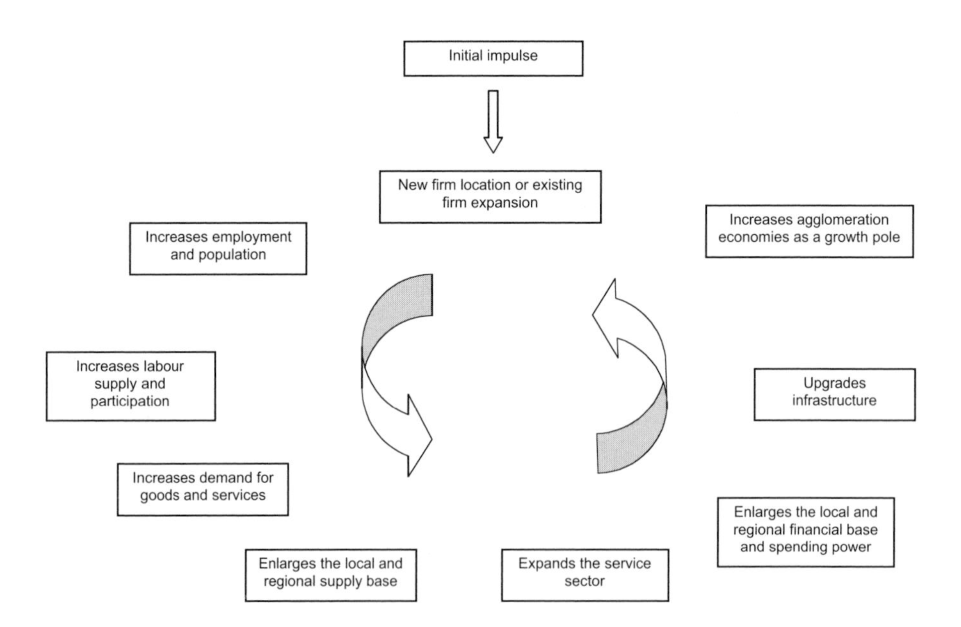

Figure 3.4 ***Cumulative regional growth***

Source: Adapted from Chisholm (1990: 66)

Through such feedbacks, cumulative causation can work in a positive direction and create virtuous circles of growth and development locally and regionally. Conversely too, negative relationships can reverse the process and create vicious circles of decline perhaps resulting from a loss in the competitiveness of the region's exports or external shocks such as price rises in factor inputs.

The Verdoorn effect and growth pole theory

Unbalanced regional growth and divergence are central to Keynesian theories of local and regional development. As Figure 3.5 outlines, Dixon and Thirlwall's (1975) explanation emphasises the feedback effect of the region's growth upon the export sector's competitiveness, the knock-on effect upon output, and further beneficial effects for the export sector's productivity and competitiveness. Dixon and Thirlwall (1975) emphasise the operation of the Verdoorn effect whereby the growth in labour productivity is partly dependent upon the growth of output. Positive and strong growth in labour productivity and output become mutually reinforcing. Growth pole theory draws upon cumulative causation too, in particular the potential linkages between propulsive firms capable of generating induced growth through inter-industry linkages – both backwards and forwards through supply chains (Hirschman 1958) – and localised industrial growth (Perroux 1950). Growth centres or poles may emerge, generated by agglomeration economies, to propel local and regional growth and development. In common with export base theory, Friedmann's centre-periphery model emphasises the potential for the external inducement of growth, the powerful external economies of core regions and the

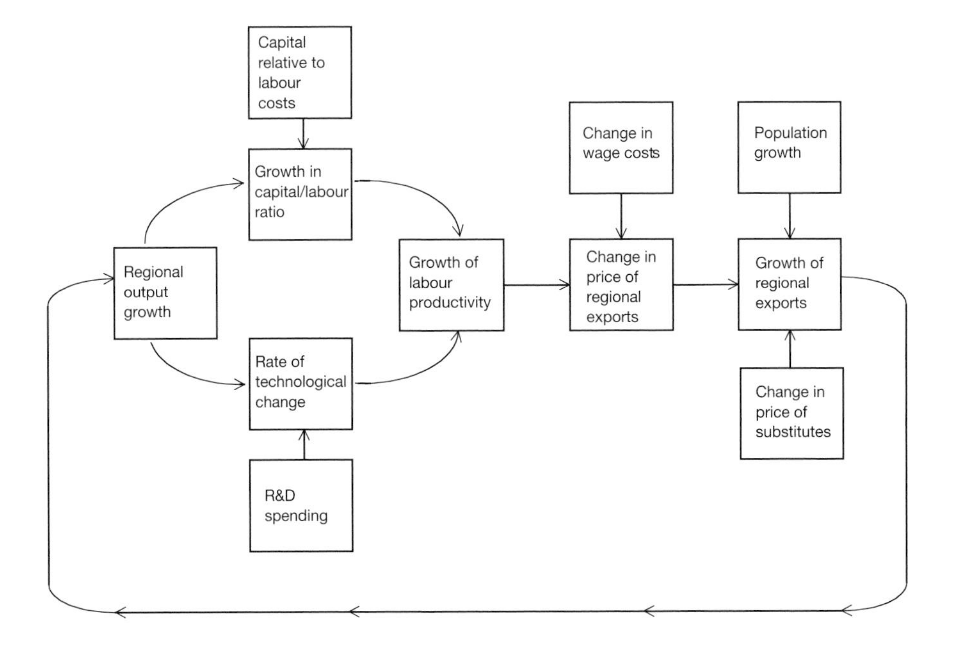

Figure 3.5 ***The Dixon–Thirlwall model of regional growth***

Source: Adapted from Armstrong and Taylor (2000: 95)

role of political and economic leadership and entrepreneurship in translating export demand into growth in the non-basic residentiary sector (Friedmann 1972). In a closely interdependent way, non-core regions are defined by their relationship with the core and their relative degree of autonomy, for example as 'resource frontiers' or 'downward transitional' areas.

The Keynesian approach to regional policy

In common with neo-classical theories, Keynesian theories of regional divergence have strongly influenced regional policy. The potential of markets to reinforce rather than reduce regional disparities has established a role for the state, especially at the national level, and public policy intervention (McCrone 1969; Kaldor 1970). Contrasting with the neo-classical and its free-market regional policy, a long history of Keynesian interventionist regional policy is evident (Table 3.4). Government-directed growth was considered feasible due to its potential to stimulate increasing returns to scale (Rosenstein-Rodan 1943). Balanced and geographically even growth may be attainable through government intervention to establish virtuous circles of high savings and high growth (Nurske 1961; Singer 1975).

Inspired by the Keynesian approach, dedicated regional development agencies and industrial estates were pioneered under President Roosevelt's New Deal in the 1930s by the Tennessee Valley Authority in the United States. As we discussed in Chapter 1, during the era of 'developmentalism', growth pole experiments were evident in the 1960s with then propulsive firms and industries of the day, such as chemicals and automobiles, implanted to stimulate new economic growth in lagging regions (Rodríguez-Pose

Table 3.4 ***Neo-Keynesian regional policy: the 'interventionist' approach***

Dimensions	*Characteristics*
Theoretical approach	Reconstructed Keynesianism Demand-side stimulation Supply-side support for industry and services
Causes of regional economic disparities	Market deregulation and liberalisation Structural weaknesses Low investment Drain of capital to developed regions Inadequate and insufficient government participation in regional development
Political ideology	State intervention Social democratic National territorial cohesion and solidarity
Approach to reviving disadvantaged regions	Proactive policies at the local and regional level Public investment in infrastructure
Regional policy	Extensive regional aid Automatic assistance Decentralisation of regional regeneration powers to local and regional agencies and authorities

Sources: Adapted from Martin (1989); Armstrong and Taylor (2000: 211)

1994). Criticism of the Keynesian approach peaked in the neo-classical and free-market dominated 1980s due to its uneven performance record, 'deadweight' effects subsidising activities that would have happened irrespective of public support, uncertainty regarding how to maintain growth and some dramatic failures costly in public expenditure terms (Taylor and Wren 1997). Despite such criticism, the experience of recent decades has questioned the strength of Hirschman's trickle-down effects, particularly in the most disadvantaged and lagging regions:

> Most contemporary policy, rhetoric aside, relies upon a model of development diffusion based on the principle of benefits trickling down to troubled targets. Conventional policy has yet to find the key that unlocks the fates of truly troubled locales.
>
> (Glasmeier 2000: 568)

Further discussion of the Keynesian-inspired approaches to local and regional development policy will be addressed in Part III.

The critique of the Keynesian approach

Although it attempts to integrate a consideration of the demand and the supply-side, export base theory has been criticised as oversimplistic, ignoring significant factors within regions (e.g. entrepreneurialism, public policy) and not providing a systematic explanation of the determinants of demand for a region's exports (Armstrong and Taylor 2000). Dixon and Thirlwall's (1975) model has been criticised too for failing to specify the type of exports in which a region may specialise, assuming the export sector is the only source of regional growth and generating controversial empirical evidence. In addition, problems have been identified in the model's failure to clarify the complexities of the Verdoorn effect and exactly how the division and specialisation of labour and technical change fosters output growth and productivity gains (Armstrong and Taylor 2000). More generally, Hirschman (1958) argued that polarised or dualistic development between developed cores and underdeveloped peripheries can benefit both growing regions and their hinterlands through 'trickling-down' effects that create demand for the products and labour of lagging regions. Although the polarisation effects identified by cumulative causation theory can be strong stimuli to regional divergence, Hirschman (1958) argues that they are countered by such trickle-down processes, especially when supported by interventionist regional policy. Deliberate state-led decentralisation of propulsive industries may reverse geographical polarisation (Townroe and Keen 1984). Whether such countervailing forces are sufficient only to keep regional divergence in check rather than to promote regional convergence is open to empirical question. Evidence from the European Union suggests absolute regional differences in unemployment in EU countries rates tend to vary counter-cyclically, widening during recessions and narrowing during booms (Baddeley *et al.* 1998; Boldrin and Canova 2001). Despite such criticisms, as we shall see in Chapters 5 and 6 and in the case studies in Chapter 7, the Keynesian approach remains highly influential in understanding and explaining local and regional development policy.

Theories of structural and temporal change

In contrast to the neo-classical and Keynesian emphases upon regional convergence or divergence, theories of structural and temporal change focus upon local and regional development as historical and evolutionary processes, sometimes incorporating periods of structural or systemic change. Theories have used metaphors of stages, cycles and waves to conceptualise the geographically uneven character of local and regional development. In contrast to the neo-classical focus upon exchange and factor prices, theories of temporal and structural change have taken a broader view, encompassing production, technology, consumption and institutions of government and governance.

Stages theory

Stages theories of economic growth have tended to focus upon the national and regional levels and sectoral change (Perloff *et al.* 1960). As Figure 3.6 illustrates, through time, regions and nations are interpreted as moving through progressively more advanced stages of economic growth and development, from agriculture to manufacture to services to quaternary or knowledge-based forms of development (Clark 1939; Fisher 1939). A 'ratchet effect' is evident whereby growth patterns get locked in to place and guard against future contraction, for example through densely localised linkages, specialised public infrastructures, localised demand and labour markets and innovation potential (Thompson 1968). The adaptability this may afford localities and regions is developed in the recent approaches to innovation, knowledge and learning discussed below. Scale diseconomies from congestion and bureaucracy may counter this effect.

Periods of rapid transformation are possible as a critical mass of investment and activity may generate a 'take-off' to underpin sustained periods of growth and 'development' (Gerschenkron 1962; Rostow 1971). This model was a hallmark of the developmentalism and liberal-market democracy promoted by the United States during the post-war Cold War with the Soviet Union. Indeed, some argue that 'development' requires such structural change as 'leaps' and 'transformations' from existing states to new, more developmental states (Cypher and Dietz 2004). Echoing the neo-classical theory of comparative advantage, over time, specialisation and trade replace self-sufficiency. Diminishing returns and changes in the internal division of labour propel the transition between stages. 'Development' equates with growth and industrialisation and constitutes transitions through each ever more advanced stage of economic activity. Regional convergence is considered more likely in the latter stages of this development

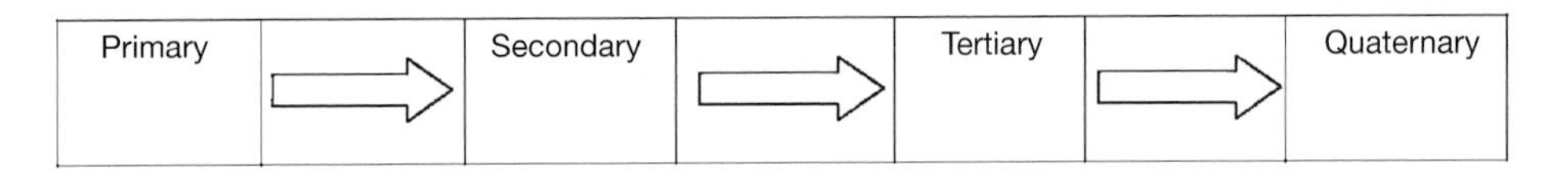

Figure 3.6 ***Stages theory***

Source: Adapted from Fisher (1939)

model (Williamson 1965). Criticism has focused upon the linear and programmatic logic of such models: all nations and regions were predicted to follow the same developmental path. Underdevelopment theorists challenged the idea of linear stages and concluded that there were similar national patterns of development notwithstanding persistent differences (Kuznets 1966). Mirroring the development trajectory of advanced industrial nations and regions, this particular form and theory of 'development' was closely associated with the national focus of post-war developmentalism summarised in Chapter 2 (Figure 2.1). With its emphasis upon the conditions and requirements for structural change, stages theory remains an important, if not readily acknowledged, influence upon local and regional development policy.

Cycle theories

Cycle theories focus upon the temporal evolution of local and regional industrial structures and their relation to local and regional development. Geographical variations in spatial factor costs are linked to the differential stages of product and industry life cycles through the product cycle model (Storper and Walker 1989). Building upon its initial micro-level focus upon the locational behaviour of US multinational corporations (MNCs) (Vernon 1979), the product cycle sought to link regional development to the export-oriented evolution of regional industrial structures (Norton and Rees 1979; Storper 1985; Sternberg 1996). As Table 3.5 describes, initially, innovating firms introducing new products retain locational proximity to key suppliers and R&D functions. Low elasticity or sensitivity of demand to price changes for new products renders initial regional cost differences relatively less significant. Large urban markets provide sizeable and sophisticated markets to prove immature products. With maturity and standardisation, economies of scale become relatively more important than flexibility. Decentralisation to exploit relatively cheaper labour in underdeveloped regions and nations occurs with potential export back into the core regions that, by this stage in the cycle, have already developed new products to restart the process (Weinstein *et al.* 1985).

Criticisms of the product cycle model concern the model's narrow focus upon individual products rather than industries and markets, its relevance to historically specific time periods, its emphasis upon labour as the primary cost consideration, its reliance upon a set of essential causal relationships and linkages, its emphasis upon the determining role of technology and its limited conceptualisation of innovation (Sayer 1985; Taylor 1986; Schoenberger 1989). Product cycle theory has had a limited impact upon local and regional development policy apart from highlighting the potential significance of cycles of industrial development and the need for localities and regions to attract and retain different kinds of industrial functions and occupations.

Building upon notions of Schumpeterian innovation and Marxian uneven development discussed below, Markusen's (1985) profit cycle theory counters many of the problems of the product cycle model. The theory moves beyond simplistic factor cost explanations of local and regional development by emphasising market power and corporate strategies in the closely intertwined evolution of industries and regions (Gertler 1984). The theory focuses upon the meso-level of industries, beneath the macro-level

Table 3.5 ***The product life cycle***

	Introduction	*Growth: market enlarges*	*Maturity: mass production*	*Saturation*	*Decline*
Sales volume	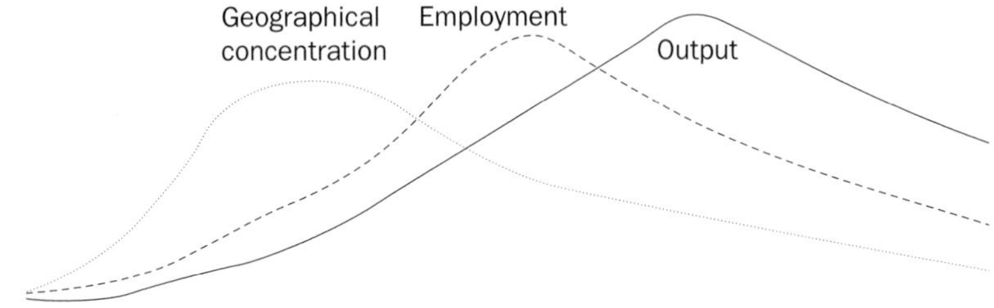	Geographical concentration	Employment	Output	
Demand conditions	Very few buyers	Growing number of buyers	Peak demand	Declining demand	Steep fall-off in demand
Technology	Short production runs Rapidly changing techniques	Introduction of mass-production methods Some variation in techniques but less rapid change	Long-run production runs and stable technology Few innovations		
Capital intensity	Low	High because of high rate of obsolescence		High because of large quantity of specialised equipment	
Industry structure	Entry is determined by 'know-how' Few competitors	Growing number of competing firms Increasing vertical integration	Financial resources critical for entry Number of firms starts to decline	General stability at first, followed by exit of some firms	
Critical production factors	Scientific and engineering skills External economies (access to specialist firms) Agglomeration economies	Management Capital	Semi-skilled and unskilled labour Capital		
Employment	Employment grows along with output		Productivity drives down employment		
Geography	Location either random (i.e. home of inventor) or in core region close to R&D and headquarter functions	Initial plants close to R&D in core regions	Relocation of production to lower cost peripheries is permitted by standardisation of product and production process and impelled by increased price competition Relocation is either to less developed countries or lower cost peripheral regions within core countries		
Regional development implications	Highly innovative firms High rates of R&D Skilled scientific and engineering employment Some local agglomerations	Shift to mass production Modern plants with new capital Requirements for management and engineering skills as well as semi-skilled production	Branch plant economy based upon low skilled and low wage production Potential for rationalisation and plant closure		

Source: Adapted from Dawley (2003); Storper and Walker (1989); Markusen (1985)

Table 3.6 ***Profit cycles and regional development***

Stage	*Profit stage*	*Locational behaviour: spatial succession*
I	Zero profit: the initial birth and design stage of an industry.	Concentration: often arbitrary locations related to location of invention.
II	Super profit: the era of excess profit from temporary monopoly and innovative edge.	Agglomeration: proliferation and growth in size of the innovating firms and their tendency to draw linked sectors, and a labour force, to them at the initial site.
III	Normal profit: the stage of open entry, movement towards market saturation, and absence of substantial market power.	Dispersion: firms grow in size and decline in numbers. Attempts to expand and locate in new markets. Sites of lower factors costs, such as labour, become increasingly attractive as oligopolies break down and competition increases. Increased automation in the production process allows lower skilled labour to be located and utilised. These locations are relatively remote from the 'core'.
IV	Normal-plus or normal-minus profit: the post saturation stage, where either successful oligopolisation boosts profits again or predatory and excessive competition squeezes profit.	Relocation: certain sectors may have been spatially retarded by the centripetal forces of agglomeration. However, during the onset of a decline in profits these sectors may relocate at an accelerated rate in the latter stages. If this dispersion occurs in conjunction with cuts or new plant formation, then relocation will occur.
V	Negative profit: the obsolescence stage of the sector.	Abandonment: production retirement as quickly as possible either through plant closure or relocation to cheaper sites.

Sources: Markusen (1985); Dawley (2003), adapted from Storper and Walker (1989)

of the economy and above the micro-level of individuals and firms, and attempts to construct a historically dynamic approach. Table 3.6 outlines five sequential stages of profitability and competitive structure through which an industry will evolve with generalisable patterns of employment, locational behaviour and local and regional development implications (Markusen 1985).

Technologically dynamic regions start with a fairly competitive stage in which externalities beyond the firm and industry are important then progress to an oligopolistic stage dominated by large firms as products mature and technology diffuses to other areas. Innovators earn monopoly rents from the sole supply of new goods and services in the initial period of 'super-profits'. Firm or innovation location is often the result of historical accident such as the initial base of the founder. Co-location occurs to benefit from the externalities of technological spillovers and labour pooling. New entrant competition erodes super profits and creates normal profits and may be drawn to the site of the initial industry's innovation or regions favourable to the industry. Firm size growth, concentration and consumer market orientation underpin the geographical concentration of oligopolistic firms that exert market and political power.

Eventual market saturation and destabilisation underpins the emergence of oligopolistic organisational forms driven to search for additional profits, including decentralisation in search of relatively lower labour cost inputs or more flexible, non-union labour. 'Negative profit' results from decline in the face of substitute products and/or services and imports associated with the abandonment of location-specific facilities. Despite its relative theoretical flexibility to cope with complex processes of industrial change compared to product cycle theory (Schoenberger 2000), criticism of profit cycle theory has focused upon its reliance upon a set of essential causal relationships, its abstraction or generalisation of causal process from particular temporal and spatial empirical circumstances and its limited contextualisation (Storper 1985). While profit cycle theory revealed the potential stages for intervention, its influence upon local and regional policy has been limited.

Wave theories

Originating in the late nineteenth century, macro-technological long-wave theories of capitalist development based upon macro-level technological shifts revived following the structural changes during the late 1960s and 1970s. Long-wave theory retains a focus upon internal change within regions as the explanation for local and regional development (Marshall 1987). Drawing upon Kondratiev's description of fifty-year long waves in commodity price cycles (Barnett 1997), Schumpeter's (1994) theory of long waves provides the theoretical basis. Each long wave is underpinned by a progressively more advanced 'techno-economic' paradigm. As Figure 3.7 illustrates, each has its own distinctive geography of local and regional development – the current being the fifth Kondratiev based upon microelectronics. Transitions between long waves occur through a process of what Schumpeter called 'creative destruction': downswings cause a 'bunching' of innovations and stimulate entrepreneurial activity to lay the foundations of structural change and a successive 'techno-economic' paradigm (Sternberg 1996). The initially narrow focus of long-wave theory was broadened to incorporate the social, political and institutional context (Freeman and Perez 1988; Hall and Preston 1988).

Long-wave theory echoes Markusen's (1985) profit cycle with initially competitive markets giving way to oligopoly when the returns from innovation eventually diminish as the technological-economic paradigm matures. Crafts (1996) has explored a possible link between Schumpeterian long-wave theories and the endogenous growth models discussed below with transition between long waves as periods of local and regional divergence. In the context of understanding local and regional development, long-wave theory has been criticised for the determining role given to technology and its limited and functionalist views of socio-institutional processes (Hirst and Zeitlin 1991; Malecki 1997). The theoretical focus upon macro-level generalisation and abstraction as well as the causal power attributed to the mechanistic waves neglect local and regional complexity and differentiation. Indeed, most of the long-wave theories are aspatial and offer a limited ability to explain specific outcomes in particular times and places (Sternberg 1996; Dawley 2003). Given its macro-level and historical focus, local and regional development policy has drawn little from long-wave theory except attempts to promote the conditions for adaptation to emergent techno-economic paradigms and to encourage the creative destruction of innovation and entrepreneurialism.

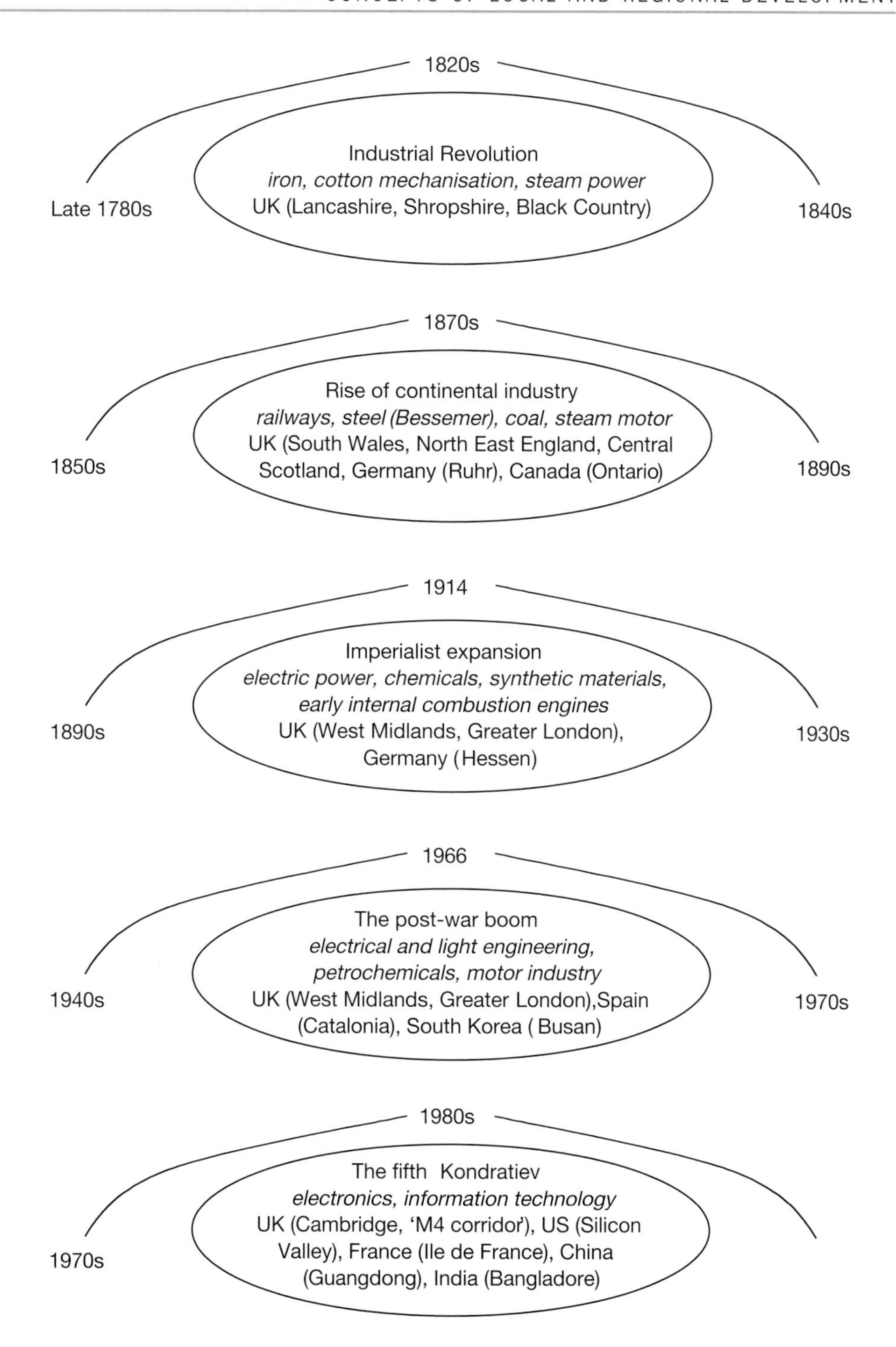

Figure 3.7 ***Long waves of economic growth***

Source: Adapted from Dicken (2003: 88)

Marxism and radical political economy: the historical geography of uneven development

From the late 1960s and into the 1970s and 1980s, structural changes in the nature of capitalism and its local and regional development implications prompted much interest in Marxist or radical approaches. Deindustrialisation, the shift to services, heightened international mobility of factors of production and growing local and regional inequalities in economic, social, gender and ethnic terms prompted radical critiques of prevailing approaches to local and regional development (Bluestone and Harrison 1982; Harvey 1982). Such issues had hitherto been dominated by objective, positivist and, often quantitative, regional science. Marxist approaches changed the focus of the local and regional development question towards understanding and explaining periodic industrial restructuring and the changing 'spatial divisions of labour' – the geographically constituted organisation of the social relations between capital, labour and the state (Lovering 1989; Massey 1995). This approach argues that aggregate growth figures at the local and regional level concealed hierarchical spatial structures of interrelations with implications for job quality and regional functional specialisation (Sunley 2000). Through the geographical division of labour within organisations illustrated in Figure 3.8, places were becoming specialised in particular functions, such as headquarters, R&D and assembly, that underpinned the hierarchical relations between places. The historical regional industrial specialisation in which all functions were geographically concentrated had been fragmented and spatially extended over time. For Marxian analysis, periodic accumulation crises inherent in capitalist development fostered new spatial, technological and social 'fixes' that underpinned further equally unstable configurations of local and regional growth and decline (Harvey 1982; Storper and Walker 1989).

Building upon a political economy critique of neo-classical economics, Marxist theory interpreted regional growth as episodic and capable of historical periods of both convergence and divergence (Martin and Sunley 1998). The uneven geographical fragmentation of regional industrial specialisation fostered a geographical division of the ranges of corporate functions and their associated jobs and occupations between core and peripheral localities and regions (Massey 1995). 'Development' constituted the upgrading of regional functional specialisation to incorporate higher-level activities, such as headquarters and R&D, and better quality and higher-paid jobs with more positive implications for local and regional development. Transitions in local and regional development were explained by the changing position and role of localities and regions within the spatial division of labour. For example, the United States experienced the shift of industry from the north eastern 'rustbelt' to the southern and western 'sunbelt' (Sawers and Tabb 1984). Capital accumulation and the social forces of class conflict are integral to Marxist political economy and emphasise the critical role of external forces in shaping economic and social change in localities and regions (Dunford and Perrons 1994; Perrons 2004). Space and place are the geographical focus (Beynon and Hudson 1993). At least one attempt was made to link Keynesian cumulative causation models with this Marxist approach (Holland 1976). Policy implications emphasised the critical role of the state and regional policy in progressive action to ameliorate local and regional inequality in democratically accountable ways.

Locationally concentrated spatial structure – without intra-firm hierarchies

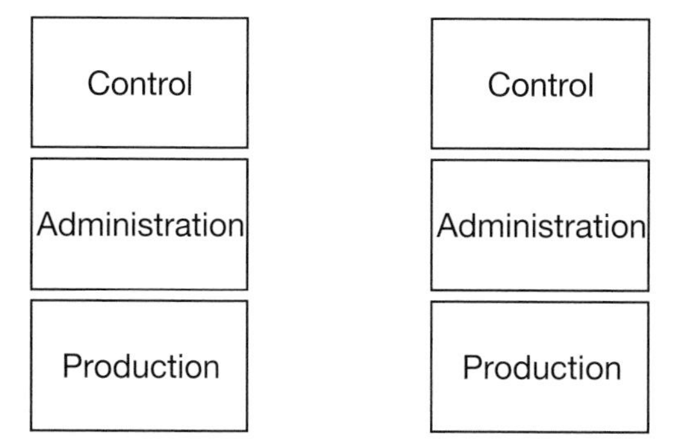

Cloning branch plant spatial structure – hierarchy of relations of ownership only

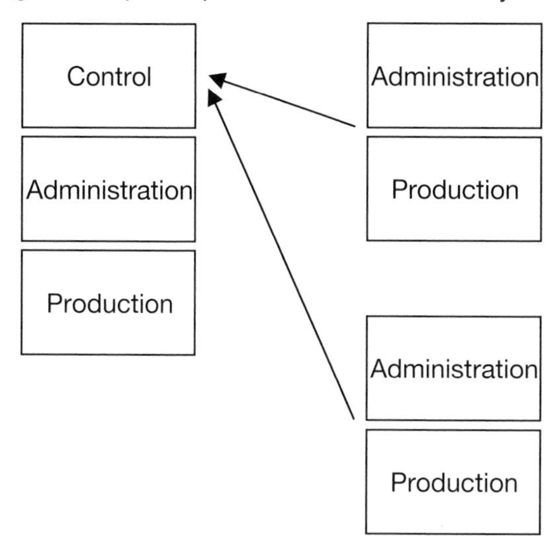

Part-process spatial structure – plants distinguished and connected by relations of ownership and in technical division of labour

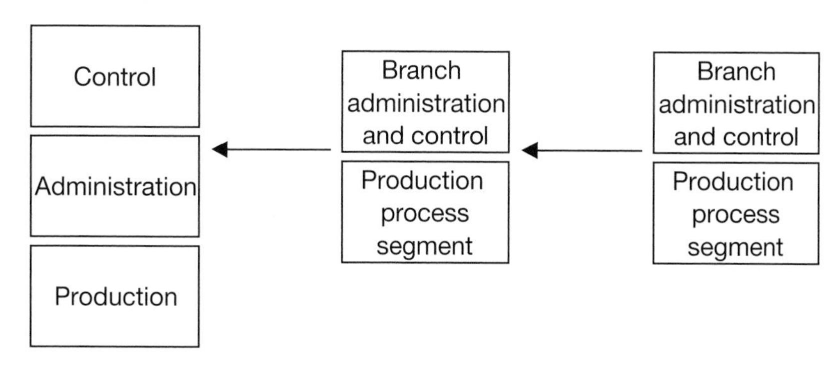

Figure 3.8 ***Spatial divisions of labour***

Source: Adapted from Massey (1995: 75)

Amid concerns about the dominant explanatory power given to abstract social structures and economic logics (Sayer 1985), criticism focused upon the spatial division of labour's attempt to balance the determining roles of structure with social agency, its national intra-regional focus and demand-led view of local and regional change, its limited conception of local labour market regulation and reproduction and its narrow state-centred conception of institutions (Warde 1985; Sunley 1996; Peet 1998; Dawley 2003). Building upon Massey's (1995) pioneering early work, geographical political economy remains a highly influential approach in economic geography and local and regional development (Castree 1999; Pike 2005; MacKinnon and Cumbers 2007).

Transition theories: the resurgence of local and regional economies

Debate followed the failure of neo-classical growth theory to explain whether the slow down and even reversal in regional convergence since the mid- to late-1970s was due to cyclical change, *ad hoc* exceptional events or a more fundamental systemic transition in local and regional development (Dunford and Perrons 1994). In the mid-1980s, the focus shifted away from long-term evolutions in regional growth and decline (Martin and Sunley 1998). In addition, the focus on production and technology of stage, cycle and wave theories broadened. Various theories of structural change emerged that sought to explain substantive transitions in the nature of capitalism and their implications for local and regional development. Central to the emergent themes were specific forms of local and especially regional economies whose particular social, technological and institutional foundations had underpinned relatively faster growth performance (Scott 1986; Becattini 1990). 'Development' became an issue of the extent to which localities and regions could ape the characteristics and relative economic success of exemplar types of 'industrial districts' – whether craft based (Third Italy; Hollywood, Los Angeles), high-tech (Silicon Valley, California; Rhônes-Alpes, France) or financial centres (City of London; Wall Street, New York). 'Resurgent' regions became the focus of local and regional development theory and policy (Storper 1995; Scott 1998).

Flexible specialisation

An early and influential institutionalist transition theory focused upon the idea of 'industrial divides' – systemic discontinuities in the social organisation and regulation of production – between the pre-industrial era and mass production and then from the era of mass production to 'flexible specialisation' (Piore and Sabel 1984). Each industrial period was associated with a distinct geography of local and regional development. Flexible specialisation heralded a return to the 'industrial districts' characterised by the regional industrial specialisation typical before the regional functional specialisation – captured in the spatial division of labour approach – of the mass production era. Concrete examples focused upon industrial districts in Emilia Romagna, Italy. Contrary to the rigid, inflexible and vertically integrated social organisation under mass production, densely localised networks of small firms could respond to differentiated and fast changing markets in flexible, specialised and adaptable ways (Hirst and Zeitlin 1991).

Vertical disintegration and agglomeration could reduce transaction costs, provide flexibility and reduce uncertainty between buyers and suppliers. In policy terms, the potential of flexible specialisation for local industrial renewal for labour and against monopolistic corporate power appealed to politically Left political administrations during the 1980s, for example the Greater London Council (Best 1991; Geddes and Newman 1999).

Transaction costs and 'new industrial spaces'

Building upon the transaction costs and external economies traditions of eminent economists Coase, Marshall and Williamson, transaction costs and neo-Marshallian theories of regional agglomeration and growth developed to explain the formation and success of the regional resurgence of 'new industrial spaces' (Scott 1988). In the context of the break-up of the mass production and consumption model of Fordism, increased market uncertainty and fragmentation coupled with technological change were interpreted as undermining internal economies of scale and scope (Storper and Walker 1989). Horizontal and vertical disintegration and the contracting-out or externalisation of production fostered the flexibility and adaptability necessary to cope with fast changing, differentiated demands and to avoid the rigidities of previous forms of social organisation.

Marshallian externalities – labour market pooling, specialist supplier availability and technological knowledge spillovers – provided local economic benefits for firms in similar industries and underpinned geographical agglomeration. This geographical concentration was particularly useful and efficient for transactions that were irregular, unpredictable and relied on face-to-face interaction. Such externalisation and agglomeration laid the foundations for the formation and development of 'territorial production complexes' or geographically concentrated production systems (Storper and Scott 1988). Cities and regions were interpreted as active and causal elements rather than passive backdrops in the economic growth process (Scott and Storper 2003). Local and regional development focused upon the extent to which localities and regions were exhibiting the characteristics of these growing and economically successful places. The ideas of increasing returns and positive externalities central to this approach are echoed in the extended neo-classical growth theories discussed below. Echoing the broader conception of 'development' and its increasingly international reach across developed, developing and transitional economies discussed in Chapter 2, agglomeration has since been promoted as 'a fundamental and ubiquitous constituent of successful development in economic systems at varying levels of GNP per capita' (Scott and Storper 2003: 581).

Regulation theory and the transition from Fordism

The macro-scale regulation approach interprets a transition from Fordism towards a more flexible era called variously neo-, post- or after-Fordism or even flexible accumulation and its co-stabilisation of a 'regime of accumulation' and 'mode of social regulation' (Scott 1988; Dunford 1990; Peck and Tickell 1995; Macleod 1997). As Table 3.7 details, the different eras have markedly different economic, social, political and institutional organisation. For regulation theory, it is the regulatory coupling between economic

Table 3.7 ***Fordism and flexible accumulation***

Fordist production (based on economies of scale)	*Just-in-time production (based on economies of scope)*
A The production process	
Mass production of homogeneous goods	Small batch production
Uniformity and standardisation	Flexibility and small batch production of a variety of product types
Large buffer stocks and inventory	No stocks
Testing quality ex-post (rejects and errors detected late)	Quality control part of process (immediate detection of errors)
Rejects are concealed in buffer stocks	Immediate reject of defective parts
Loss of production time because of long set-up times, defective parts, inventory bottlenecks etc.	Reduction of lost time, diminishing 'the porosity of the working day'
Resource driven	Demand driven (quasi-) vertical integration subcontracting
Vertical and (in some cases) horizontal integration	Learning-by-doing integrated into long-term planning
Cost reductions through wage control	
B Labour	
Single-task performance by worker	Multiple tasks
Payment per rate (based on job design criteria)	Personal payment (detailed bonus system)
High degree of job specialisation	Elimination of job demarcation
No or little on-the-job training	Long on-the-job training
Vertical labour organisation	More horizontal labour organisation
No learning experience	On-the-job learning
Emphasis on diminishing workers' responsibility (disciplining of labour force)	Emphasis on workers' co-responsibility
No job security	High employment security for core workers (lifetime employment)
	No job security and poor labour conditions for temporary workers
C Space	
Functional spatial specialisation (centralisation/decentralisation)	Spatial clustering and agglomeration
Spatial division of labour	Spatial integration
Homogenisation of regional markets (spatially segmented labour markets)	Labour market diversification (in-place labour market segmentation)
World-wide sourcing of components and subcontractors	Spatial proximity of vertically quasi-integrated firms
D State	
Regulation	Deregulation/re-regulation
Rigidity	Flexibility
Collective bargaining	Division/individualisation, local or firm-based negotiations
Socialisation of welfare (the welfare state)	Privatisation of collective needs and social security
International stability through multilateral agreements	International destabilisation; increased geopolitical tensions
Fordist production (based on economies of scale)	Just-in-time production (based on economies of scope)
Centralisation	Decentralisation and sharpened interregional/ intercity competition
Intercity intervention in markets through income and price policies	The 'entrepreneurial' state/city
Firm-financed research and development	Direct state intervention in markets through procurement
Industry-led innovation	'Territorial' regional policies (third party form) state-financed research and development
	State-led innovation
E Ideology	
Mass consumption of consumer durables: the consumption society	Individualised consumption: 'yuppie'-culture
Modernism	Postmodernism
Totality/structural reform	Specificity/adaptation
Socialisation	Individualisation: the 'spectacle' society

Source: Adapted from Harvey (1989b)

and extra-economic factors that institutionally embed and socially regularise capitalist development despite its inherent contradictions (Peck 2000). In the regulationist framework:

> national and regional growth rates are interpreted as depending fundamentally on the degree of correspondence between the organization of production and the regulatory institutional and social structures which support and regulate the economy.
>
> (Martin and Sunley 1998: 214)

In this regulationist theory, structural economic and social change has undermined the Fordist coupling of mass production and consumption regulated by national Keynesian demand management and welfarism (Martin and Sunley 1997). Strong national macroeconomic management and a supportive welfare state have waned across nation states. This break-up has underpinned regional growth rate divergence. Fordist industrial regions have declined. Post-Fordist 'flexible production complexes' have emerged, socially and geographically distinct from the Fordist growth centres (Storper and Scott 1988). Institutional and regulatory structures have shifted towards a more localised era of Schumpeterian Workfarism focused upon competitiveness and innovation (Jessop 2002). The state's role now encourages innovation and international competitiveness and subordinates social to economic policy aims. As we discuss below, national variants of regulatory regimes exist as different national varieties of capitalism with distinct institutional structures and histories mediate more generalised processes of economic, social, political and cultural change.

Transition theories and local and regional development policy

In policy terms, the shared emphasis in transition theories upon the resurgence of local and regional economies has stimulated interest in 'endogenous' – internal or within – or 'indigenous' – naturally occurring – 'development from below'. Informed by transition theories, a local and regional development policy repertoire has emerged. As detailed in Chapter 5, policy has focused upon locally decentralised production networks, local agglomeration economies and local networks of trust, cooperation and competition as well as the local capacity to promote social learning and adaptation, innovation, entrepreneurship (Stöhr 1990; Pyke and Sengenberger 1992; Cooke and Morgan 1998). The industrial districts model has been promoted in local and regional development policy because such districts are said to be flexible enough to adapt to the shifting context of heightened economic and technological change through their disintegrated production networks. As we discuss, these arguments are distinct from the new approaches to endogenous growth theory. However, although the district model has attracted considerable attention, it has been criticised for ignoring the need for policy learning sensitive to particular local and regional contexts rather than universalistic, 'off-the-shelf' and 'one-size-fits-all' policy transfer (Hudson *et al.* 1997; Storper 1997). Longstanding and thorny issues about the relative merits of specialisation or diversification in local and regional development continue to bedevil policy deliberation about the district model.

The critique of transition theories

Transition theories stimulated significant critique and debate during the 1980s and 1990s (Gertler 1992; Amin 1994). The more flexible nature of capitalist accumulation was largely agreed but not its conceptual and theoretical interpretation (Harvey 1989b). Given their particular forms of explanation and relatively limited repertoire of geographical expressions, transition theories have struggled to capture and explain the complexity and diversity of local and regional development. The value of the transition model focuses upon shifts in the macro-structural nature of capitalism and its geographies of local and regional growth have been steadily undermined (Sunley 2000). Transition models have been revealed as overly reliant upon the determining role of broader structures and unable adequately to explain continuity and change in local and regional development (Hudson 2001). For Sayer (1989), this is because of the simplistic and dualistic – 'before and after' – analysis deployed in transition models. The idea of clear breaks with their corresponding geographies has become much less convincing. The real world of local and regional development is much messier and more geographically uneven (Peck 2000).

In addition to criticising the tendency for transition models to generalise from a limited set of successful examples (Macleod 1997), further empirical evidence has qualified and/or challenged the nature and dynamics of such industrial agglomerations. In particular, criticism has questioned their reliance upon small firm dynamism, their relative ignorance of the role of larger firms, the geographical stretching of their social and productive relationships, the role of external and internal forces in shaping their evolution and the reality of their adaptive capabilities (Harrison 1994; Amin and Thrift 1995; Cooke and Morgan 1998). A clearer conception suggests there is a diversity of change and experience across different types of local and regional economy (Martin and Sunley 1998). This differentiation has been reflected in the recent development of district theory with a wider variety of typologies. The emergent concepts are more open to the contingency of particular circumstances and to the role of large firms, state actors, local fixed capital and skilled labour (Markusen 1996).

Institutionalism and socio-economics

Dissatisfaction with the macro-structural transition theories and their deterministic spatial implications has forged a recent change in focus towards the specific and particular attributes of localities and regions and how these relate to their development over time (Sunley 2000). For local and regional development, the emphasis has shifted to 'the contingent conditions of growth in particular regions, rather than on the long-term evolution of the entire regional system' (Martin and Sunley 1998: 202). Rather than focusing upon the aggregated and descriptive statistical summaries of the outcomes of growth, these kinds of theory seek to explain the underlying characteristics and form of growth. In particular, distinctive local assets and economic capabilities – indigenous and endogenous – are interpreted as the basis for development and the foundations for constructing and establishing local and regional competitiveness (Maskell *et al.* 1998).

Influenced by the broader 'old' institutionalism of Polanyi and Veblen (R. Martin 1999) and the 'socio-economics' of the 'new economic sociology' (Granovetter and Swedberg 1992; Grabher 1993), the emergent work has focused upon the embeddedness of social action in ongoing systems of social relations and the social and institutional context of local and regional growth (Grabher 1993; Pike *et al.* 2000; Macleod 2001; Wood and Valler 2001; Hess 2004). Formal (e.g. organisations, administrative systems) and informal (e.g. traditions, customs) institutions are interpreted as integral to reducing uncertainty and risk as well as promoting trust in economic relations. A distinction can be drawn between institutional environments and arrangements to explain the differing abilities of localities and regions to absorb or create technological progress which can, in turn, underpin disparities in economic performance (R. Martin 1999) (Table 3.8). Institutional context varies geographically with direct consequences for local and regional growth performance and development. As discussed in Chapter 4, institutional structures flow from and are influenced by multilevel systems of government and governance shaped by distinctive national 'varieties of capitalism' (Zysman 1996; Hall and Soskice 2001).

The institutionalist approach interprets particular forms of institutional organisation as the root causes and explanations of the conditions that promote or inhibit the growth and development of localities and regions. More abstractly, socio-economics argues that any conceptualisation or understanding of the 'economic' is explicitly 'social' and cannot be understood or explained except within its social context (Grabher 1993). Taking an institutionalist and socio-economic view, markets are not the free floating phenomena described in neo-classical growth theory. Instead, markets are interpreted as social constructs made and reproduced through frameworks of socially constructed institutions and conventions (Sunley 2000). Markets are therefore highly differentiated in their nature, form and local and regional development implications. While markets theoretically provide efficient allocation mechanisms for scarce resources as depicted

Table 3.8 ***Institutional environment and arrangements***

Theme	*Institutional regime*	*Nature of the systems*	*Examples of institutional expressions*
	Institutional environment	Informal conventions	Customs, norms and social routines
		Formal conventions	Structures of rules and regulations (usually legally enforced)
	Institutional arrangements	Organisational forms	Markets, firms, labour unions, welfare state, city councils

Sources: Dawley (2003: 104), adapted from R. Martin (1999)

by neo-classical theories, market failure is common as individual decisions by atomised agents may be individually rational and efficient but collectively irrational and inefficient. For example, from an individual perspective a firm may pay a premium to poach a skilled worker in a tight labour market but for their locality or industry such individually rational action is collectively irrational since it erodes the local skills base and creates inflationary wage pressure. Correcting market failure often requires collective institutions, for instance to underpin investment in public goods (e.g. skilled labour and vocational training), new generic technologies or patient capital markets for smaller firms.

Networks, trust and social capital

Stimulated by the interest in institutionalism and socio-economics, networks have received attention as intermediate and institutionalised forms of social organisation that are neither markets nor organisational hierarchies but are cooperative and potentially mutually beneficial (Cooke and Morgan 1998). Cooperative and reciprocal networks are founded upon trust-based relationships between participants. This enables information sharing and mutually beneficial action without the need for tightly prescribed contracts. 'High trust' localities and regions are interpreted as more capable of the rapid innovation and adaptation amenable to development due to collaboration to share costs and risks, exchange information and solve problems (Saxenian 1994). Trustful relations reduce monitoring and contracting costs for participants, for instance fostering the supply of cheap local credit and cooperative labour relations (Sunley 2000). In contrast, 'low trust' environments are characterised by distrustful relations and necessitate highly formalised contracts to govern market-based exchanges. The adaptive capabilities and local and regional development prospects of such places are consequently weaker.

The 'social capital' engendered by trust may underpin a local and regional collective 'intelligence' and capability to learn and adapt successfully to change (Cooke and Morgan 1998). There has been an explosion of recent writing on the topic of social capital and its role in development. While exhibiting a degree of common conceptual concern (albeit frequently confused), social capital has been operationalised in highly diverse ways, often reflecting differences between (and within) disciplines (including economics, sociology, anthropology, political science, education) and diverging normative concerns. Farr has attempted the following conceptual summary:

> In a way both compact and capacious, the concept of social capital boils down to networks, norms and trust. Upon inspection, networks prove dense and valuable, norms pervade individual actions and social relations, and trust appears psychologically complex . . . [Thus] social capital is complexly conceptualized as the network of associations, activities, or relations that bind people together as a community via certain norms and psychological capacities, notably trust, which are essential for civil society and productive of future collective action or goods, in the manner of other forms of capital.
>
> (Farr 2004: 8–9)

The use of the term 'capital' suggests the existence of an asset. Much of the literature on social capital suggests that this asset has substantial implications for economic development, notably by helping the innovation process through lowering of transaction costs in inter-firm networks, which can be a highly localised process involving the development of trust-based relationships (Maskell 2002). Such processes can have a down-side contributing to a lock-in of widely supported, but economically inefficient practices (Szreter 2002). For this reason the accumulation of local social capital may be insufficient to aid development. For development to proceed in poor communities, the initial benefits of intensive intra-community integration must give way over time to extensive extra-community linkages: too much or too little of either dimension at any given moment undermines economic advancement (Woolcock 1998).

This problem has been conceptualised as the relationship between bonding, bridging and linking capital. Bonding capital refers to networks formed from perceived shared identity relations. Bridging capital refers to networks of associations where the differentiating principle of shared social identity or status plays no necessary role in determining membership. Despite the analytical clarity of these concepts they have proved difficult to use in empirical work. Linking capital refers to relationships of exchange, like in the case of bridging capital, between differentiated parties, but in this case parties are also characterised by power asymmetries. The significance of this analytical distinction for policy is that development becomes not simply a question of empowering the poor, but also of managing the interaction resources held by 'external' agencies which are present in poor communities (Mohan and Mohan 2002). Accordingly, positive development outcomes occur:

> when people are willing and able to draw on nurturing social ties (i) within their local communities; (ii) between local communities and groups with external and more extensive social connections to civil groups with external and more extensive social connections to civil society; (iii) between civil society and macro-level institutions; and (iv) within corporate sector institutions. All four dimensions must be present for optimal developmental outcomes.
>
> (Woolcock 1998: 186–187)

In the context of dissatisfaction with the limitations of the conventional neo-classical theory discussed above, institutionalism and socio-economics seek to provide a means of integrating analysis of the intangible or 'softer' factors in explanations of local and regional growth and development. While they form key elements of recent economic geography (Barnes and Gertler 1999; Barnes and Sheppard 2000; Clark *et al.* 2000), such dimensions have remained outside the traditional focus of neo-classical approaches to local and regional development. 'Softer', less tangible, factors are difficult to measure, price and quantify and are often invisible to official data sources (Sunley 2000), militating against their aggregate quantitative analysis.

The economic and extra-economic approach of regulation theory discussed above has influenced the focus upon the institutional and regulatory supports and infrastructure of local and regional economies. In particular, Amin and Thrift's (1995) notion of 'institutional thickness' has been influential in explaining the shaping of local and regional development trajectories. Institutional thickness refers to a strong institutional presence locally, high levels of inter-institutional interaction, strong social structures and collective awareness of a common local and regional enterprise. Such institutional context may provide externalities that, depending upon their nature, can be central to the initial emergence, trajectory and adaptability of local and regional economies (Martin and Sunley 1998).

Historical trajectories and path dependency

Evolutionary theory is central to institutional and socio-economic approaches to local and regional development (Nelson and Winter 1985). In this evolutionary approach, the ways in which places change over time are understood in terms of historical trajectories. This conceptualisation is more able to address the indeterminate, complex and sometimes unpredictable nature of local and regional development: 'such gradualism allows a better understanding of the types of path and place dependency through which the historical geography of regions and cities shapes their future development' (Sunley 2000: 192). Path dependency, in particular, has proven an influential idea. It is a biological metaphor that refers to the ways in which the evolution of a system is conditioned by its past history (Arthur 1996). Trajectories are not predetermined, however, in the manner of some of the stage, cycle and wave theories discussed above. Local and regional development trajectories can be non-linear. Places can move forwards or backwards as well as remain static in economic and social terms. Places can change paths too, for example the rapid transition and fast growth of the late industrialising 'Asian Tigers' – including South Korea, discussed in Chapter 7 – from the 1970s (Storper *et al.* 1998). As we discussed in Chapter 2, the legacies of place can be decisive for local and regional development prospects and trajectories.

Institutionalism, socio-economics and local and regional development policy

Institutionalism and socio-economics have influenced local and regional development policy and, as we detail in Chapter 4, government and governance. In particular, the approaches emphasise the importance of local and regional institutions and their ability to develop especially indigenous assets and resources and their capability to foster adjustment to changing circumstances (Bennett *et al.* 1990; Campbell 1990; Storper and Scott 1992; Amin and Thrift 1995; Scott 2004). The emphasis upon recognising the distinctive structural problems and assets of localities, regions and nations and constructing context-sensitive development policy has historical roots (Hirschman 1958; Seers 1967). The promotion of networks has been identified as a route to growth in both prosperous and old industrial localities and regions (Cooke 1995; Cooke and Morgan 1998). Overlaps with the industrial district model promoted by transition theory are evident too given the key role that intermediate institutions can play in their economic performance. Institutions are thought to shape supply-side characteristics to allow localities and regions to create their own demand by gaining market share and investment from rival places, using agglomeration to create and sustain local indigenous potential (Sunley 2000). Such ideas borrow from Chinitz (1961) and, latterly, Porter's (2000) ideas of 'competitive advantage' and clustering discussed below. As Chapter 5 discusses, typical interventions are microeconomic and focus upon the supply-side including enterprise policy, small firm growth, innovation and skills development. Such forms of policy may be necessary and helpful but not sufficient for local and regional development.

The critique of institutionalism and socio-economics

Institutionalism and socio-economics are relatively new approaches to local and regional development. Much work remains to be undertaken in conceptual, theoretical and empirical terms to explore their ability to understand and explain local and regional development and policy (Wood and Valler 2001). These so-called 'heterodox' approaches are seldom as conceptually and theoretically coherent as they claim and have been subject to critique (Lovering 2001; Pike 2004). The virtues of networks for local and regional development have been questioned due to their generalisation from limited case study evidence (Sunley 2000), the limited attention given to the relative balances between cooperation and competition and rivalry between institutions within and between networks, and the uncertain adaptability of decentralised institutionalised structures to develop coordinated responses to economic change (Harrison 1994; Glasmeier 2000). The embeddedness of social relations in local and regional institutions can 'lock-in' localities and regions to trajectories of decline if the close and high-trust relationships that once fostered their earlier growth and innovation now inhibit their future adaptation, for example the old industrial region of the Ruhr in western Germany (Grabher 1993). Institutional context may be a necessary but not sufficient condition for local and regional development. The relationships and interactions between economic conditions and institutional effects remain poorly understood. The impact of institution-building upon economic performance is ambiguous and little is known about the feasibility of the geographical transfer of institutional frameworks between successful and lagging localities and regions (Hudson *et al.* 1997; Sunley 2000).

Innovation, knowledge and learning

Innovation, knowledge and learning have recently become central ideas in explaining and understanding contemporary local and regional development. Moving beyond the focus in neo-classical approaches upon static cost advantages and the 'black box' of technological progress discussed above, the new approaches have forged a connection to theories of innovation and, more latterly, knowledge and learning (Morgan 1997; Lundvall and Maskell 2000; MacKinnon *et al.* 2002). 'Development' is interpreted as the enhancement of the locality or region's ability to produce, absorb and utilise innovations and knowledge through learning processes. Parallel interest in the causal role of local and regional differentiation in technological transfer and spillovers is evident too in the extended neo-classical growth theories discussed below (Feldman 2000).

Innovation: from the linear to the interactive model

Approaches to innovation in local and regional development have sought to build upon the transition in understanding innovation from the linear to the interactive model (Lundvall 1992). The linear model emphasised the one-way flow of ideas and knowledge within public and private organisations from initial idea through design and development to production and sale. In contrast, as Figure 3.9 illustrates, the interactive model highlights the interactive and iterative nature of innovation among institutions between more closely interrelated stages of development. Influenced by the institutionalist and socio-economic theory discussed in the previous section, this approach sees innovation as a social process that occurs in a variety of geographically differentiated institutional settings (Wolfe and Gertler 2002).

The linear model was often mapped onto the kinds of functionally specialised hierarchies described in the spatial division of labour approach. Certain types of regions specialised in R&D activity with its positive knock-on implications for regional growth, occupational structure, wage levels and local and regional prosperity (Massey 1995). In contrast, the emergent interactive model emphasises the much closer interaction between the users and producers of knowledge, through spatially proximate co-location and/or facilitation by information and communication technologies (Howells and Wood 1993). Local and regional institutional contexts are therefore integral to explaining innovation potential and performance. Some places are evidently more innovative and capable of producing and adapting innovations than others, reflected in their differential levels of local and regional economic dynamism (Malecki 1997; Armstrong and Taylor 2000).

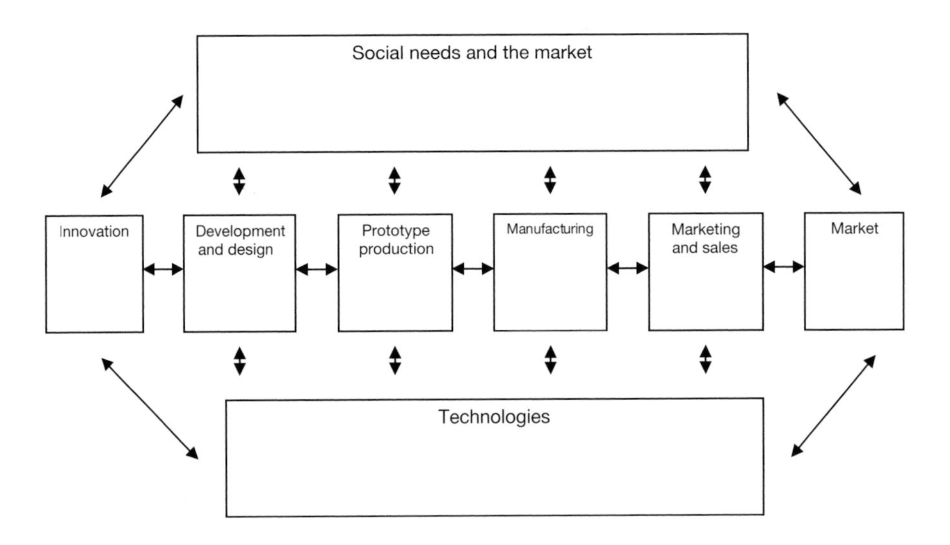

Figure 3.9 ***Interactive model of innovation***

Source: Adapted from Clark and Guy (1997: 8)

Regional innovation systems

A substantial literature has developed to understand and explain the geographical unevenness of innovation and its local and regional development implications. National innovation systems approaches have extended to the subnational level to examine the potential of regional innovation systems emerging from sustained institutional networks capable of regional learning that cohere, endure and adapt over time (Cooke and Morgan 1998; Lundvall and Maskell 2000). Table 3.9 outlines the constituent elements of 'strong' and 'weak' regional innovation systems. Here, regions are seen as 'externalized learning institutions' (Cooke and Morgan 1998: 66). In seeking to conceptualise this phenomena of locally and regionally rooted innovation potential and performance, other approaches have identified 'innovative milieux' (Camagni 1996), 'technopolis' (Castells and Hall 1994) or 'worlds of production' (Storper 1997). In explaining local and regional innovation, each approach shares a focus upon physical and technological infrastructures, such as industrial and university R&D and related industries and services, highly skilled local labour markets, risk capital availability as well as the supporting social context of ostensibly non-material factors such as regional technical culture and know-how and common representational systems (Storper 1997; Gertler 2004). As suggested by the institutionalist and socio-economic theories above, intermediate institutions play an integral role in reducing uncertainty and guiding the coordination of collective action in an explicitly social and geographical process of innovation.

The knowledge economy

The recent emphasis upon knowledge in the economy and its implications for local and regional development connects with the work on innovation. For some, echoing the stages theory discussed above, 'economic development is a process of moving from a set of assets based on primary products, exploited by unskilled labour, to a set of assets based on knowledge, exploited by skilled labour' (Amsden 2001: 2, cited in Cypher and

Table 3.9 ***Superstructural elements for strong and weak regional systems of innovation (RSI) potential***

	Institutions	*Firms*	*Policy*
Strong RSI potential	Cooperative culture Associative learning disposition Change orientation Public–private consensus	Trustful labour relations Workplace cooperation Worker-welfare orientation Mentoring Externalisation Innovation	Inclusive Monitoring Delegation Consultative Networking
Weak RSI potential	Competitive culture Individualistic 'Not invented here' Conservative Public–private dissension	Antagonistic labour relations Workplace division 'Sweating' 'Sink or swim' Internalisation Adaptation	Exclusive Reacting Centralisation Authoritarian 'Stand-alone'

Source: Adapted from Cooke *et al.* (1998: 1580)

Dietz 2004: 18). Indeed, within the pyramid structure outlined in Figure 3.10, information is seen as a critical commodity and knowledge is interpreted as the most scarce resource (Lundvall and Maskell 2000). In this approach, the production, utilisation and transmission of knowledge are considered integral in a more uncertain economic context introduced in Chapter 1 marked by rapid and often radical economic and technological change. Institutions – firms, public agencies and so on – can play a central role in fostering knowledge-rich local and regional environments. For localities and regions, the role they play within the knowledge economy and the management of their knowledge assets embodied in individuals and institutions has implications for their development trajectory and relative prosperity.

Learning and local and regional development

Central to the adaptive ability of localities and regions for development is the capability to learn (Lundvall 1992). Learning is understood as a collective, social and geographical process that effects a change in an individual or organisation's capability or understanding (Cooke and Morgan 1998). Learning is considered central to the continued innovation necessary in the changing context – detailed in Chapter 1 – of the pervasive technological change, particularly in communication and information systems, heightened uncertainty and volatility characteristic of contemporary forms of 'reflexive' capitalism:

> the centrality of learning for the innovation process stems from the recognition that the knowledge frontier is moving so rapidly in the current economy that

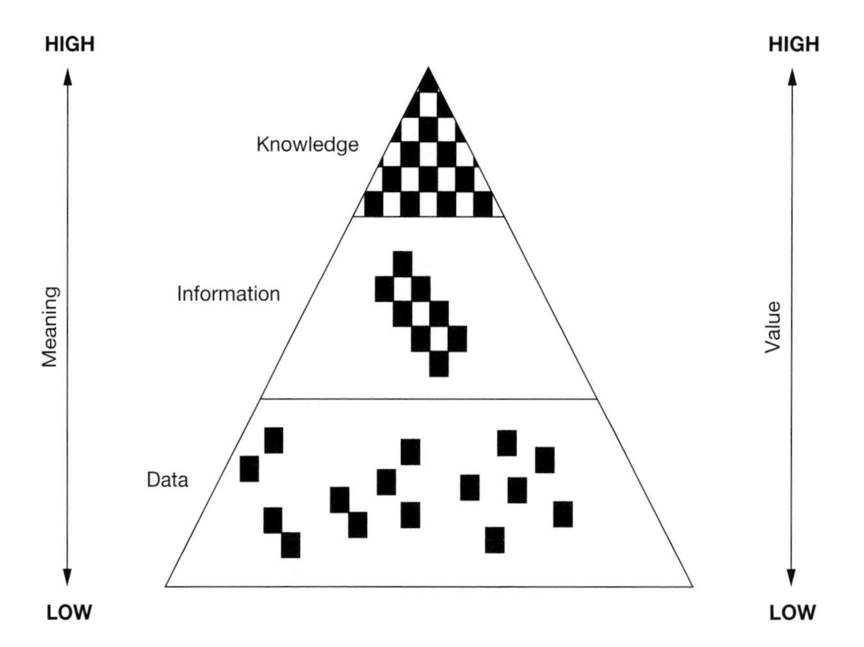

Figure 3.10 ***Data, information and knowledge***

Source: Burton-Jones (1999: 6)

> simply access to, or control over, knowledge assets affords merely a fleeting competitive advantage. It is the capacity to learn which is critical to the innovation process and essential for developing and maintaining a sustainable competitive advantage.
>
> (Wolfe and Gertler 2002: 2)

More self-aware or reflexive localities and regions are thought more likely to be able to adapt to economic change, often through an ability to recognise and discard outmoded and uncompetitive routines and practices (Cooke and Morgan 1998). Interactive relationships are evident between knowledge production, learning and forgetting for localities and regions. Crucially for local and regional development, learning is considered to be enhanced through local proximity as rapid knowledge transfer and application generate positive local externalities for firms and other institutions (Sunley 2000). Critical discussions about learning have animated debates about 'new regionalism' – discussed in Chapter 4 – and its emphasis upon the 'region' as the focus and causal factor in economic, social and political change (Amin 1999; Lovering 1999). Overlapping with the concepts of networks and embeddedness discussed above, Storper (1997) has focused attention upon the non-market interrelations or 'untraded interdependencies' between institutions as central to local systems of innovation, productivity growth and local and regional development. High levels of trust, tacit or uncodified knowledge and routine behaviours underpin sets of conventions and coordinating relations that are specific to the context of particular localities and regions. These untraded forms of interdependency constitute 'relational assets' that provide localities and regions with the capability to learn and to develop the unique and not easily reproducible competitive edges necessary to stay ahead of the forces of imitation in an increasingly globalised economy. As outlined in Table 3.10, learning and knowledge-creating regions are distinguished from the mass-production regions characteristic of Fordism in the transition models discussed above.

Alongside regions, cities and city-regions too are seen as key bases of growth in the knowledge and learning economy rather than just examples of its manifestation (Scott 2003). Scott and Storper (2003: 581) argue that 'urbanisation is less to be regarded as a problem to be reversed than as an essential condition of durable development'. Cities act as the foci for agglomeration, positive externalities and, for Florida (2002b), an emergent 'creative class' that some consider increasingly important for local, regional and national growth (for a critique see Chatterton 2000).

Innovation, knowledge and learning in local and regional development policy

As emergent ideas in explaining contemporary local and regional development, innovation, knowledge and learning currently occupy a central role in local and regional development policy. New policy forms – often described as heterodox rather than orthodox or conventional (Pike 2004) – represent a marked break from previous approaches. As we discussed in Chapter 1, local and regional policy increasingly involves the combination of 'hard' infrastructures, such as broadband telecommunication links

Table 3.10 ***From mass production to learning regions***

	Basis of competitiveness	*Production system*	*Manufacturing infrastructure*	*Human infrastructure*	*Physical and communication infrastructure*	*Industrial governance system*	*Policy system*
Mass-production region	Comparative advantage based upon natural resources and physical labour	Mass production: physical labour as source of value, separation of innovation and production	Arm's length supplier relations	Low-skill, low-cost labour, Taylorist workforce, Taylorist education and training	Domestically oriented	Adversarial relationships, top-down control	Specific retail policies
Learning/ knowledge-creating region	Sustainable advantage based upon knowledge creation and continuous improvement	Knowledge-based production: continuous creation, knowledge as a source of value, synthesis of innovation and production	Suppliers' systems as a source of innovation	Knowledge workers, continuous improvement of human resources, continuous education and training	Globally oriented	Mutually dependent relationships and network organisation	Systems/ infrastructure orientation

Source: Adapted from Florida (2000a)

Plate 3.1 ***High-technology growth poles: a micro- and nanotechnology centre under construction in Grenoble, France***

Source: Photograph by David Charles

or transport facilities, with 'soft' support for networking and knowledge transfer, to build innovation capacity and foster collective knowledge creation, application and learning (Morgan 1997). Self-sustaining growth and development are sought through building and developing indigenous and endogenous assets – linking to the 'development from below' policy approach discussed above. Experimentalism is encouraged too, marked by more interactive and consultative policy-making as well as new institutional forms such as task forces, to develop joint working and partnership to address shared problems (Morgan and Henderson 2002; Pike 2002b). Such policy design is often more context-sensitive and less universal (Storper 1997). Examples include the European Union's regional innovation strategies (Morgan and Nauwelaers 1999). In this context, the ability of development institutions to acquire, absorb and diffuse relevant information and knowledge is critical to local and regional prospects (Wolfe and Gertler 2002). As the institutionalist and socio-economic theories suggest, institutions are often integral parts of any explanation of how localities and regions have failed to adapt and surmount historically entrenched barriers to adjustment (Grabher 1993; Cooke 1997).

The critique of innovation, knowledge and learning in local and regional development

Criticism has accompanied the emergent approaches to innovation, knowledge and learning in local and regional development. For some, weak and 'fuzzy' conceptualisation has compromised clear thinking, theory building, standards of evidence and relevance to policy (see the debate between Hudson 2003; Lagendijk 2003; Markusen 2003; Peck 2003). The causal power of learning and knowledge accumulation is still the subject of theoretical development and debate. For Sunley (2000), there has been a tendency to overlook the continued importance of conventional price and cost conditions and exchange and market relations as well as a failure to establish the relative position of knowledge and learning as primary rather than contributory causes of economic growth.

The suspicion is that many of the claimed benefits of agglomeration and localised learning for local and regional development have been exaggerated and have yet to be questioned beyond the narrow evidence base of their supporting empirical examples (Amin 2000). The role of national central government regulation as well as policy, for example defence expenditure, has not always been given a sufficiently central role (Lovering 2001). The supporting role of culture has often been dealt with uncritically as a pre-given rather than something that is socially and geographically constructed and contested (Scott 2004). Academic co-option into the politics of policy-making around innovation, knowledge and learning and the 'globalisation-competitiveness' rhetoric of 'new regionalism' has attracted critical comment too (Lovering 2001). Since the production and application of knowledge has always been historically central to capital accumulation whether anything fundamentally new is happening has also been questioned (Hudson 1999).

Extended neo-classical theories: endogenous growth theory, geographical economics, competitive advantage and clusters

Endogenous growth theory

Dissatisfaction with the external or exogenous treatment of factors of production – population growth, savings rates, human capital and technological change – in traditional neo-classical growth theory discussed above has been addressed by a growing body of work on endogenous growth models (Martin and Sunley 1998; Stough 1998). These theories seek to incorporate formerly external and independent elements into their conceptualisation and explanation of economic growth. Connecting with the Keynesian theories of cumulative causation discussed above, the dynamics of regional convergence and divergence are the focus of endogenous approaches to local and regional development. 'Development' is conceived as the reduction in regional disparities. The theories attempt to introduce increasing returns into the neo-classical production function to determine long-run growth rates within – endogenously – the model (Martin and Sunley 1998). The subnational entity of the 'region' is the geographical focus. The theories retain core elements of the neo-classical approach and language.

Table 3.11 ***A typology of 'new' growth theories***

	Type of growth theory			
	Augmented neo-classical	*Endogenous broad capital*	*Intentional human capital*	*Schumpeterian endogenous innovation*
'Engine of growth' convergence?	Physical and human capital, exogenous technical progress universally available. Slow and conditional convergence within clubs of countries with similar socio-economic structures.	Capital investment, constant returns through knowledge spillovers. Cumulative divergence, but shaped by government spending and taxation.	Spillovers from education and training investments by individual agents. Convergence dependent on returns to investment, public policy, and patterns of industrial and trade specialisation.	Technological innovation by oligopolistic producers, with technological diffusion, transfer and imitation. Multiple steady states and persistent divergence likely. Possible club convergence and catch-up.

Source: Adapted from Martin and Sunley (1998: 209)

As Table 3.11 shows, different sorts of increasing returns and externalities are envisaged in the different models. Endogenous 'broad capital' models either emphasise the externalities generated by capital investment or human capital and the 'learning by doing' and knowledge spillover effects of technological change (Crafts 1996; Martin and Sunley 1998). In common with the neo-classical model, capital stock investment is interpreted as a driving force of growth. Endogenous innovation models draw upon Schumpeterian notions of innovation in their emphasis upon the potentially monopolistic returns generated by innovations and technological developments by producers (Armstrong and Taylor 2000). For these models, human capital investment produces positive spillover effects that boost both capital and labour productivity. Both sorts of models have been criticised. Evidence suggests periods of rapid growth in localities and regions may be preceded rather than followed by high rates of fixed capital investment. Technological progress is considered the result of deliberate choices and actions rather than a coincidental effect of other activities (Romer 1994; Blomstrom *et al.* 1996). Both models struggle to convince on their central argument that returns may be constant and increasing rather than diminishing (Martin and Sunley 1998).

Innovation, technological change and geographical spillovers

Most attention has focused upon making technological change and innovation endogenous to the economic growth model. As we discussed above, neo-classical theory interprets technological progress as necessarily driving output growth per capita but fails to identify the causes of technological progress. The underlying explanation of growth is not spelt out (Armstrong and Taylor 2000). Endogenous growth theory explicitly seeks to explain the causes of technological progress. It specifies the relationships of technological change and innovation to the growth process. Technological progress is seen as both cause and effect of economic growth. It is endogenous rather than exogenous to the growth process. Put simply, individuals and institutions see the incentive to produce new ideas for sale and profit, technological progress is therefore internalised within the growth process: 'The economy's technological frontier is automatically pushed outwards because of the profits to be earned in the knowledge-producing industry' (Armstrong and Taylor 2000: 76).

Explanations of technological progress focus upon the number of workers in knowledge-producing industries, the existing stock of knowledge and technological transfer and diffusion (Romer 1990). Exogenously or externally produced technology embodied in capital goods can be bought-in and determines a region's technology by its capital stock vintage. In contrast, disembodied technological progress contributes to regional growth disparities independently of capital stock. As Figure 3.11 illustrates, such disembodied knowledge is more likely to be produced in knowledge-rich and creative environments, vary between regions and influence regional growth rates (Armstrong and Taylor 2000). In common with the institutionalist and socio-economic theories discussed above, divergence between 'leading' and 'following' localities and regions with differing social capabilities for connecting innovation and growth is a potential outcome.

Endogenous growth theory and local and regional development

Endogenous growth theory has directly influenced local and regional development theory. The geographically uneven rates of regional convergence and the spatial clustering of high- and slow-growth regions are explained by the new economic growth

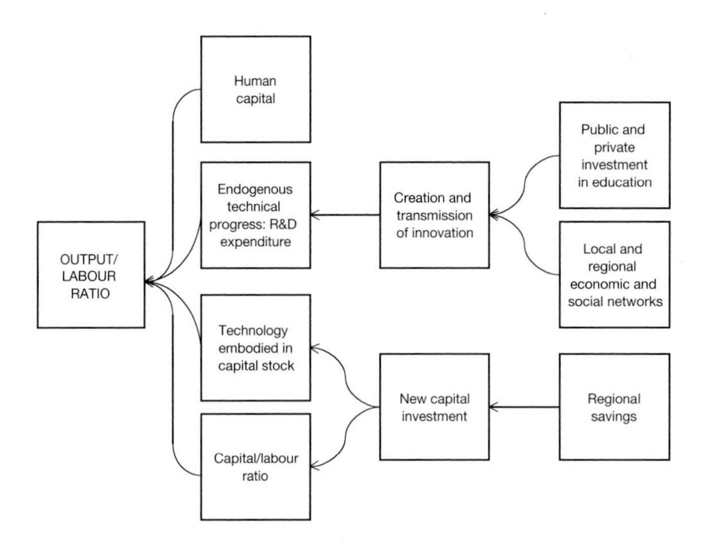

Figure 3.11 ***Endogenous growth theory: the determination of labour productivity***

Source: Adpated from Armstrong and Taylor (2000: 88)

theories. Neo-Schumpeterian approaches emphasise the role of technological spillovers in increasing technological mobility at the inter-regional and international scales (Rodríguez-Pose 2001). Such mobility is not costless, however (Audretsch and Feldman 1996). Despite the partly non-rival (non-competing) and non-excludable (non-exclusive) nature of technology and innovation (Storper 1997), the returns from the transition of knowledge are geographically bounded and the costs of transmission increase with distance (Jaffe *et al.* 1993). In combination with the traditional agglomeration economies and externalities discussed above, the emphasis upon human capital and technological leadership in endogenous growth theory suggests:

> Together, these types of increasing returns imply that regional development is highly path dependent; temporary conditions and shocks, as well as historical 'accidents', may have permanent effects as patterns of specialisation, of economic success or economic backwardness, become 'locked-in' through external and self-reinforcing effects.
>
> (Martin and Sunley 1998: 211)

The extent to which increasing returns and spillovers are geographically based at particular geographical scales implies a role for institutions and policy to capture and shape the kinds of investment that might influence these elements of local and regional growth. Fiscal policies and public infrastructure as well as the resources and incentives for technologically innovative sectors have consequently received attention (Martin and Sunley 1998). In common with the institutionalist and socio-economic theories discussed above and the content of Chapter 5, emphasis upon the endogenous dimensions of economic growth has shifted the focus back towards the mobilisation of indigenous potential at the local and regional levels (Goddard *et al.* 1979). Geographical differentiation marks the uneven appropriability of techology and innovation. Lagging regions can suffer from a 'growth limbo' between insufficient size and capability to generate returns and spillovers from investment and a limited capacity to appropriate spillovers from more advanced localities and regions (Rodríguez-Pose 2001). The context and extra-local connections and flows are critical, however:

> Endogenous growth theory makes the key factors to growth, including human capital, technology, and externalities, internal to the production function, not to local or even national economies. On the contrary, the theory underlines the importance of national and international (global) flows of goods and knowledge.
>
> (Martin and Sunley 1998: 219)

Endogenous growth theory and local and regional development policy

Theoretical emphasis upon releasing the development potential within localities and regions for their own and the national good has refocused the aims of local and regional policy:

> many countries are also concentrating their public expenditures on their most dynamic agglomerations at the expense of basic equity issues both within these agglomerations and between them and other areas of the national territory.
>
> (Scott and Storper 2003: 588)

Local and regional development policy is often no longer considered a response to the needs of 'problem' localities and regions that may require redistribution from a national centre. This can be described as the donor-recipient model, redirecting growth from growing to lagging regions (Figure 3.12). This model has been criticised for its failure to address structural problems in lagging regions, its high cost and its inability to redistribute growth within the national economy in the context of an internationalising or global economy. The new growth-oriented regional policy seeks to raise the economic performance of growing *and* under-performing regions to contribute to the growth of each region and their national economy (Figure 3.12). New economic policies for regions

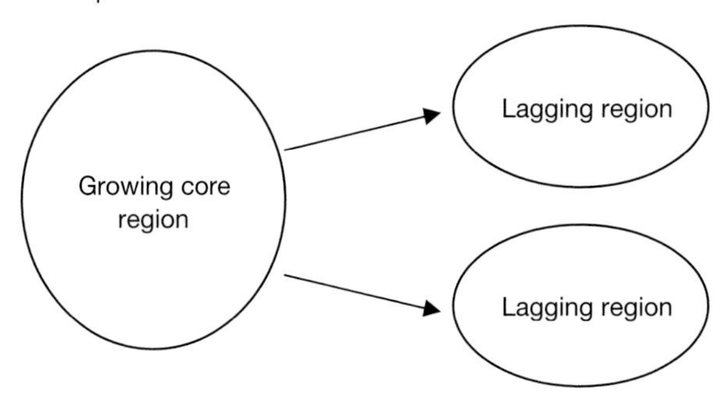

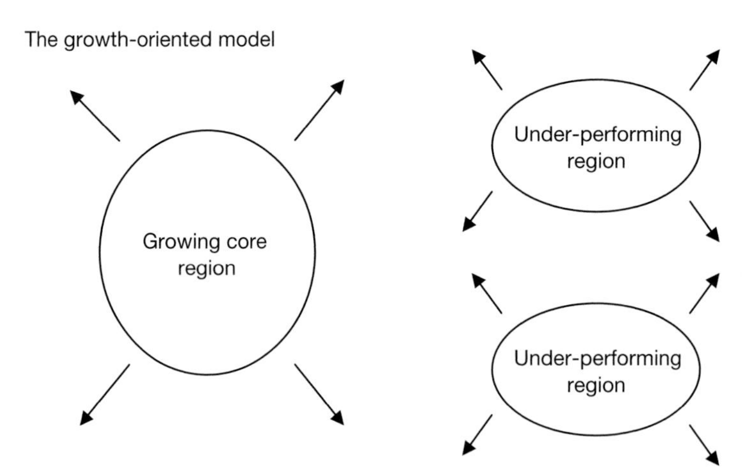

Figure 3.12 ***Donor-recipient and growth-oriented models of regional policy***

Source: Authors' own research

are focused on all regions within national economies (Aufhauser *et al.* 2003; Scott and Storper 2003; Fothergill 2005). 'Levelling-up' the economic performance of each territory is considered the key to enhanced economic outcomes at the local, regional and national levels. This approach is often contrasted with the traditional regional policy of redistributing growth from prosperous to lagging regions – latterly characterised as 'levelling down'.

The critique of endogenous growth theory

Endogenous economic growth theory is not without its critics, particularly due to its adherence to the neo-classical equilibrium framework discussed above (Martin and Sunley 1998). Endogenous theory remains wedded to the standard neo-classical assumptions about economically rational agents fully knowledgeable of alternative choices and the consequences of their decisions. The Keynesian and Kaldorian critique discussed above is relevant too. Endogenous theory focuses on the supply-side and gives relatively little attention to the demand-side issues of exports and balance of payments constraints on employment and productivity (McCombie and Thirlwall 1997). The Verdoorn effect is largely ignored as a source of increasing returns as rising output generates scale economies and raises productivity (Kaldor 1981). Other problems in relating endogenous growth theory to local and regional development concern the limited empirical evidence of how increasing returns operate in specific industries and places, the inability to address historical change and to account for shifts and reversals in rates of regional convergence (Martin and Sunley 1998). Endogenous theories can also be weak in addressing the (historical) social and institutional contexts – conditioned by geography and place – that shape the operation of economic growth processes. As we discussed above, such concerns are central to the institutionalist and socio-economic theories of local and regional development.

Geographical economics

Drawing upon a new Keynesian critique of the neo-classical approach, geographical economics focuses upon the role of localities and regions in shaping the trading performance of industries within particular nations. Geographical economics is concerned with national economic prosperity and trade and their implications for uneven local and regional development (Meardon 2000; Brakman and Garretsen 2003). 'Development' is interpreted as increased income and prosperity through enhanced regional and national competitiveness (Kitson *et al.* 2004). The model critiques existing neo-classical of the variety discussed above approaches but relies upon its core assumptions of methodological individualism, perfect information, economically rational individuals, profit-maximising firms and exchange (Dymski 1996).

New trade theory

For geographical economics, the Keynesian and Kaldorian notions of imperfect competition, increasing returns and external economies we detailed above – combined with the growth in intra-industry and intra-corporate trade – are interpreted as undermining the neo-classical model of comparative advantage and trade specialisation. Ricardo's traditional model assumed perfect competition and the relative immobility of significant factors of production (Armstrong and Taylor 2000). Nations specialised in those industries in which they held comparative factor advantages, for instance quality raw materials or cheaper labour. International trade mutually benefited nations holding dissimilar advantages. Indigenous factor endowments determined international trade and specialisation.

In contrast, the new trade theory central to geographical economics emphasises how regional industrial specialisation and concentration can influence and, in turn, be shaped by trade (Martin and Sunley 1996). In the context of imperfect competition, increased specialisation has resulted from increasing returns to scale rather than the exploitation of differential national factor endowments (Krugman 1990). External economies driving increasing returns – the Marshallian externalities of labour market pooling, specialist supplier availability and technological knowledge spillovers – are likely to be realised at the local and regional scales rather than the national and international levels. Indeed, urbanisation economies from the general infrastructure and common externalities arise from different industries locating in urban areas. These growth spillovers underpin the localisation of industry and shape the relative competitiveness of the constituent firms within regional agglomerations (Krugman 1993). Pecuniary economies that materially affect prices in market exchanges are produced by the agglomeration of firms from different industries. Geographical concentration makes a difference to the economics of firms and industries. These economies underpin the growth of urban locations, with large and diversified markets supporting output growth (Krugman 1991). As Martin and Sunley (1998: 207) argue, the spatial clustering of regions with similar growth rates suggests the spillover effects of labour, capital, technology and other influences on growth are geographically localised rather than perfectly mobile as suggested by neo-classical growth theory.

Echoing the institutionalist and socio-economic approaches, trade specialisation is seen as history dependent. Established patterns of specialisation get 'locked-in' by the cumulative gains from trade. These effects impart strong path dependence upon local and regional development trajectories (Krugman 1990). Patterns of uneven local and regional development, once established, can exhibit strong degrees of persistence over time that may support or inhibit growth. Local and regional development is likely to be characterised by geographical unevenness (Krugman 1995). The divergence of output and income between centres and peripheries and multiple possible equilibrium positions are likely rather than the long-run convergence proposed by orthodox neo-classical economics (Krugman 1991).

Strategic trade policy and local and regional development

New trade theory emphasises how the geography of trade is shaped by states, trade regimes and increasing intra-industry trade between similarly endowed countries (Krugman 1986; Drache and Gertler 1991; Noponen *et al.* 1993). It provides a theoretical argument for strategic trade policy. Comparative advantage may be shaped by supporting specialised export sectors and, given their necessary localisation, localities and regions where external scale economies and technological spillovers may provide sources of monopolistic rents (Martin and Sunley 1996). Strategic high value or sunrise sectors and the localities and regions in which they are concentrated can be identified, targeted and prioritised given their potential to raise national and regional incomes (Reich 1991). Support may include trade protection often through non-tariff barriers such as technical standards, export subsidies and tax incentives for R&D investment.

For local and regional development policy, the strategic choices made about which geographical concentrations of regional industrial specialisation to support are critical. Indeed, such clusters not only provide empirical evidence of external economies, but also help to define which industries should be supported (Martin and Sunley 1996). Some industries and places have the potential for greater growth and productivity increases. Conversely, other lower-value and sunset sectors and places may receive less or no priority. New trade theory policy debates focus upon the relative merits of specialisation or diversification (Geroski 1989). Specialisation may provide the externalities and potential for productivity and output growth but risks regional instability and structural changes through over-concentration in a narrow range of industrial sectors. Diversification may not provide the dynamic externalities and boost to export growth and productivity but may insulate the local and regional economy from adverse demand shocks and structural crisis by widening the sectoral mix of its industrial base.

The critique of geographical economics

Alongside its emergent contributions to explaining local and regional development, critiques of geographical economics focus upon the ways in which its particular approach tends to neglect real people and places in their real historical, social and cultural settings (R. Martin 1999). An inadequate sense of geographical and historical context is provided. The approach tends to reduce the region to a receptacle rather than a potential motor of economic activity (Scott 2004). Despite its stated importance, the historical grounding of the model remains unclear and clouded in ambiguity (Martin and Sunley 1996). The emphasis upon simplifying assumptions and formal mathematical modelling produces a partial analysis of the potential diversity of the externalities central to local and regional growth. Geographical economics fails to consider the influence of local institutional, social and cultural structures in facilitating or constraining local and regional development, for example the innovation and learning and the role of local and regional institutional agency discussed above (Martin and Sunley 1996; Scott 2004).

Competitive advantage and clusters

Business economist Michael Porter has developed an influential new economics of competitive advantage to explain the role and dynamics of the geographical clustering of industries within national economies and their potential contribution to productivity growth and trading competitiveness (Porter 1990, 1998). 'Development' is understood as the enhanced competitive advantage of firms, clusters and national economies within international markets. In common with geographical economics and distinct from traditional neo-classical conceptions of Ricardian comparative advantage based upon initial factor endowments (Kitson *et al.* 2004), Porter's initial microeconomic analysis argued that competitive advantage could be actively created through the strategic management and upgrading of corporate activities or 'value chains' (Porter 1985). This initial work concluded, however, that 'competitive success cannot solely depend on managerial and company attributes when many successful firms in a given field are concentrated in just a few locations' (Porter 2000: 254). Such geographical concentrations or clusters were interpreted as containing a nation's most competitive industries. Clusters therefore became central to the theory:

> clusters are geographic concentrations of interconnected companies, specialised suppliers and service providers, firms in related industries, and associated institutions (e.g. universities, standards agencies, and trade associations) in particular fields that compete but also cooperate.
>
> (Porter 2000: 253)

The commonalities and complementarities between cluster institutions are interpreted as providing localised externalities and spillovers that could make positive contributions to the competitive advantage and trading performance of cluster participants. The competitive advantage of leading firms and industries could be reinforced and intensified by their geographical concentration. For Porter, the effect of location upon competition has four interrelated analytical elements, captured metaphorically in the notion of the 'diamond' depicted in Figure 3.13.

The main benefits of clusters for competitiveness comprise, first, boosts to static productivity growth through access to specialised inputs and labour, information and knowledge, institutions and public goods as well as localised complementarities and incentives to performance enhancement. Second, clusters can foster innovation through clear and rapid perception of buyer needs as well as promoting early and consistent learning about evolving industry trends, technologies and other knowledge vital to ongoing competitiveness. Third, clusters can promote new business formation and innovative inter-organisational forms through inducements and relatively lower barriers to entry as well as new experiments in collaboration and partnering. Increasing returns and the spillover effects of externalities characteristic of the new endogenous growth theories discussed above are integral to the dynamism and growth potential of clusters. Successful clusters can forge 'first mover' advantages and benefit from externalities and increasing returns to establish their competitive advantage at the expense of other localities and regions.

Extending from the firm level, Porter's research initially focused upon the national level and then, in search of a fuller explanation, on the local and regional levels (Porter 1990). Some work even explored the competitive advantage of the 'inner city' (Porter 1995). Clusters can range from a city or state to a country or even a group of neighbouring countries in their geographical scope (Enright 1993). Echoing elements of institutionalism and socio-economic theories, clusters are seen as capable of providing an

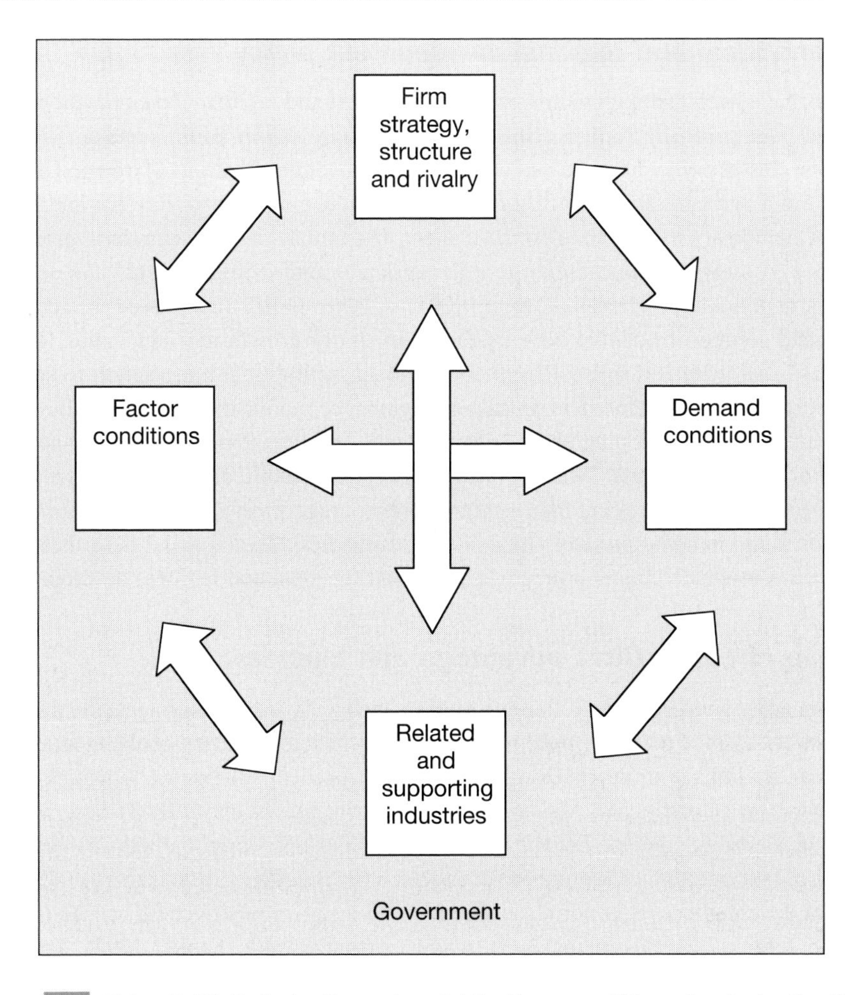

Figure 3.13 ***Porter's diamond model for the competitive advantage of nations***

Source: Adapted from Porter (1990: 258)

intermediate organisational form and means of coordination in the continuum between markets and hierarchies:

> Repeated interactions and informal contracts within a cluster structure result from living and working in a circumscribed geographic area and foster trust, open communication, and lower the costs of severing and recombining market relationships.
>
> (Porter 2000: 264)

Indeed, Porter's work forms part of an intellectual lineage common to some of the approaches to industrial agglomeration within economic geography and regional science discussed above, including agglomeration theory using Marshallian and transaction costs approaches, transition theories and industrial districts, innovative milieux and socio-economics (Gordon and McCann 2000; Martin and Sunley 2003).

Clusters and local and regional development policy

Clusters have been taken up with some gusto for local and regional development policy. Inspired – and sold – by Porter's accessible and narrative academic work and consultancy business, cluster policy has received considerable academic attention (Martin and Sunley 2003). For Lagendijk and Cornford (2000), the practice of clusters has become something of an industry itself among policy-makers. The OECD, in particular, have promoted clusters as a means of contributing to internationally competitive regional and national systems of innovation (Bergman *et al.* 2001). Cluster policy has proved attractive as a potential source of positive benefits for productivity growth and innovation. It also provides a rationale and role for local and regional institutional intervention to support cluster creation and development. Local and regional economic development policy finds a clear purpose in developing cluster potential and, by extension, in supporting national industrial competitiveness. Much policy activity has focused upon identifying and mapping clusters and seeking interventions to encourage their growth and contribution to regional and national productivity and competitiveness. Example 3.3 describes some of the ways in which cluster policy has been used for local and regional development.

The critique of competitive advantage and clusters

The popularity and influence of Porter's clusters theory have prompted substantial reflection and criticism. First, the conceptual clarity of clusters has been challenged and, in particular, its linkage to the diversity of existing theoretical approaches to geographical agglomeration (Gordon and McCann 2000). Martin and Sunley (2003) describe the cluster as a chaotic concept. Second, Porter's emphasis upon firm and industry-oriented notions of competition and competitiveness has been questioned in relation to local and regional development. It is not clear whether and how competitiveness can be territorial and defined in terms of localities, regions or nations (O'Donnell 1997). To what extent can places conceive of themselves as being in competition with each other? More recent analytical review has sought to identify the interrelated factors that drive local and regional competitiveness (Gardiner *et al.* 2004) (Figure 3.14). Indeed, given its focus upon the existing workforce, higher levels of competitiveness can be compatible with job loss and greater economic inequality and contrary to local and regional development (Sunley 2000). Although, its proponents argue, enhanced competitiveness and productivity may increase economic growth, prosperity and income.

Third, the scale and levels at which clusters form, operate and extend have not been clearly specified. The key geographical concepts of space, scale, place and territory introduced in Chapter 2 remain underdeveloped, specifically in Porter's version of the theory. Fourth, Porter's theory gives limited attention to the social dimensions of cluster formation and dynamics (Martin and Sunley 2003). Last, the Porter brand of clusters has become tainted to a degree by commercial promotion and consultancy coupled with fashionable policy transfer and faddish adoption by international, national, regional and local development institutions (Martin and Sunley 2003). Critical evaluation of the actual impacts of cluster policy upon local and regional development has been limited. Universal models, such as clusters, may only work when adapted to particular local and regional contexts (Hudson *et al.* 1997).

Example 3.3 Cluster policy for local and regional development

Local and regional development policy has been attracted to cluster policy as a means of promoting national, regional and local competitiveness, growth and innovation. Many policy frameworks have utilised Michael Porter's ideas as the standard concept in cluster policy. As we discuss in Chapter 5, decentralised approaches to local and regional policy emphasising indigenous strengths and endogenous growth have reinforced this trend. Cluster policy focuses upon the supply-side and often aims to provide public goods formerly absent due to market failures. These public goods often include cooperative networks between cluster participants, collective marketing of specialised skills and knowledge, local business services (e.g. finance, legal, marketing, design) and diagnosis and responses to remedying cluster weaknesses. For local and regional development institutions, cluster policy development typically comprises several activities. First, the process begins with mapping and categorising clusters within local and regional economies. Mapping identifies what the clusters are and their geographies. Typologies to group similar clusters together may link to their stage of development, for instance embryonic, growing or declining. Second, analysis is undertaken of the regional and/or national scale and significance of the clusters, for example assessing their relative shares of exports, employment or R&D investment. Here, cluster depth – the mix and range of industries present in the cluster, dynamism and contribution to regional and national competitiveness – may be assessed. Third, the strengths, weaknesses and needs of clusters are examined to identify the priorities for cluster development policy by local and regional development institutions. Despite the popularity of cluster policy in local and regional development circles, critics have noted the tension between wanting to include as many firms as possible in clusters rather than being selective and prioritising and the need for targeting for cost effective public policy. A further problem is that similar types of clusters have often been identified in different regions as each seeks to capture the growth potential of knowledge-based, high-tech and/or creative activities. A UK-based commentator bemoaned that rather than seeking to identify distinctive regional assets every region now appears to be seeking to develop the same clusters of 'ballet and biotech'. This approach to clusters is problematic since it undermines the central notion of the need to build upon distinctive, indigenous regional strengths.

Source: Martin and Sunley (2003); Porter (2003); Trends Business Research (2003)

Sustainable development

Sustainable development has arguably become the central influence in local and regional development in recent years (Angel 2000; Gibbs 2002; Haughton and Counsell 2004; Morgan 2004; Roberts 2004). Traditional forms of local and regional development have been challenged as overly economistic and too focused upon economic growth (Morgan 2004). Amid enduring social inequalities and the increasing impact and awareness of the ecological and environmental problems of existing patterns of resource use, forms

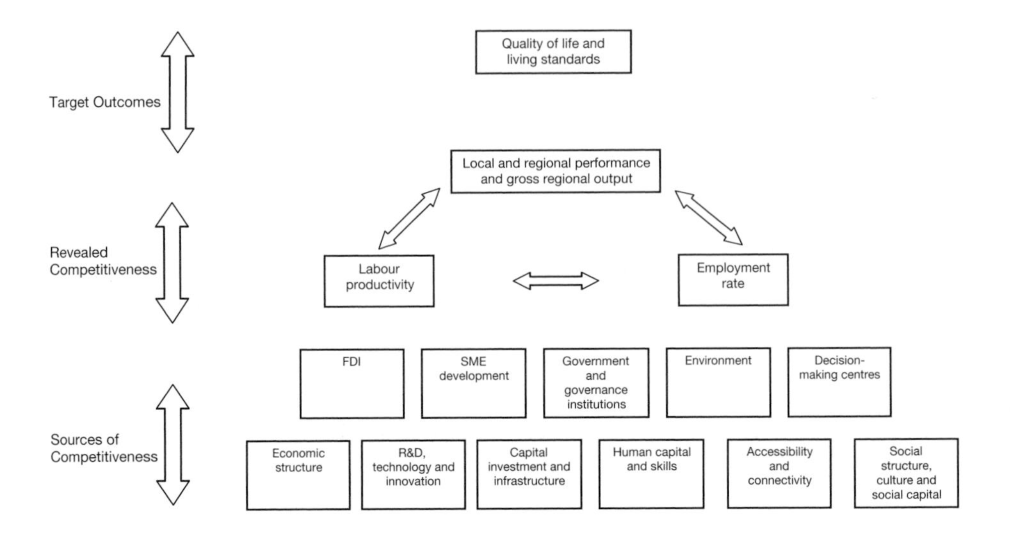

Figure 3.14 ***The 'pyramid model' of local and regional competitiveness***

Source: Adapted from Gardiner *et al.* (2004: 1045–1067)

of local and regional development have been sought that might prove more sustainable – in some sense longer term, more durable and/or less damaging – in economic, social and environmental terms.

As we discussed in Chapter 2, new metrics for local and regional development have been sought that reflect a broader notion of 'development' encompassing health, well-being and quality of life in localities and regions (Morgan 2004). For example, local and regional quality of life can vary substantially even when places appear to have similar levels of GDP per capita and income:

> the regions of the Mezzogiorno [the south of Italy] are as poor as Wales in terms of income, but they do not suffer from such debilitating rates of long-term limiting illness, partly because they have access to a much healthier diet. Poor health is both a cause and a consequence of a weak labour market in Wales because high rates of limiting long-term illness are part of the explanation for high levels of economic inactivity.
>
> (Morgan 2004: 884)

Moving on from the initially environment-dominated and sometimes anti-growth concerns in the 1970s, debate now focuses upon the fundamental questioning of economic growth as an end in itself or as an inevitable means to achieve higher standards of living (Sunley 2000).

Recent approaches to sustainable local and regional development seek to integrate economic, environmental and social outcomes together rather than compromise through trade-offs and balances (Haughton and Counsell 2004). Distinctions are drawn between the appropriate priority given to intrinsically significant things – such as health, well-being and education – and instrumentally significant things – such as jobs and income (Morgan 2004). Definitions of sustainable development often build upon the World Commission on Environment and Development's (1987: 8, 43) version of 'development that meets the needs of the present without compromising the ability of future generations to meet their own needs'. In conceptual and theoretical terms, however, sustainable development remains difficult, slippery and elusive (Williams and Millington 2004). As Table 3.12 illustrates, a ladder of sustainable development has been developed to identify the different elements of the specific approaches to sustainable development, ranging from the status quo of the 'Treadmill' through weak and strong versions to the 'Ideal Model' (Chatterton 2002; Williams and Millington 2004).

'Weak' and 'strong' sustainable local and regional development

'Weaker' forms of sustainable development – often derided as 'shallow environmentalism' – interpret nature in human or anthropocentric terms as a resource and economic growth as progress. Expanding the stock of resources through technological solutions without challenging existing capitalist structures is central. Principles comprise the use of renewable energy, substitutes for non-renewables and more efficient resource utilisation. Ecological modernisation has gained ground as a means of promoting more enlightened and sensitive approaches to sustainable economic growth and development. Environmental justice is a weaker form of sustainable development that seeks further economic growth but with a more equitable redistribution of costs and benefits. In particular, the approach seeks intra- and intergenerational equity (Hudson and Weaver 1997).

Stronger forms of sustainable development connect with notions of deep and political ecology that challenge prevailing capitalist social organisation (Harvey 1996). In this approach, the people–nature relationship is reversed and interpreted as human adaptation to finite nature (Williams and Millington 2004). Indeed, 'biocentric egalitarianism' seeks to endow nature with biotic rights to prevent its exploitation. Notions of wealth are understood in a non-material way as well-being and harmonious co-habitation within the biosphere. Stronger sustainable development seeks to reduce the demand and consumption of resources.

Sustainable local and regional development policy

Of the weak sustainable development approaches, 'ecological modernisation' has most influenced local and regional development policy. Examples include the promotion of more efficient economic growth that uses fewer natural resources, regulated markets and using environmental practices as an economic driver (Gibbs 2002; Roberts 2004). Concrete policy initiatives include environmental clusters (e.g. air-pollution control) and industrial ecology that connects and utilises waste resource flows from locally proximate industries to yield 'wealth from waste'. Environmental justice approaches have influenced local and regional development policy in seeking positive distributional outcomes, including the remediation of degraded environments and the recycling

Table 3.12 *Principles for sustainable regeneration*

Approach to sustainable development	*Treadmill*	*Weak sustainable development*	*Strong sustainable development*	*'Ideal model' of sustainable development*
Philosophy	Anthropocentric	⟶	Ecocentric and biocentric	
Role of economy and nature of growth	Exponential growth	Market-reliant environmental policy, changes in patterns of consumption	Environmentally regulated market, changes in patterns of production and consumption	Right livelihood, meeting needs not wants, changes in patterns and levels of consumption
Geographical focus	Global markets and global economy	Initial moves to local economic self-sufficiency, minor initiatives to alleviate global market power	Heightened local economic self-sufficiency in the context of global markets	Bioregionalism, extensive local self-sufficiency
Nature	Resource exploitation	Replacing finite resources with capital; exploitation of renewable resources	Environmental management and protection	Promoting and protecting biodiversity
Policies and sectoral integration	No change	Sector-driven approach	Environmental policy integration across sectors	Holistic intersectoral integration
Technology	Capital-intensive, progressive automation	End-of-pipe technical solutions, mixed labour- and capital-intensive technology	Clean technology, product life cycle management, mixed labour- and capital-intensive technology	Labour-intensive appropriate technology
Institutions	No change	Minimal amendments	Some restructuring	Decentralisation of political, legal, social and economic institutions
Policy instruments and tools	Conventional accounting	Token use of environmental indicators, limited range of market-led policy tools	Advanced use of sustainability indicators, wide range of policy tools	Full range of policy tools, sophisticated use of indicators extending to social dimensions
Redistribution	Equity not an issue	Equity a marginal issue	Strengthened	Inter- and intra-generational equity
Civil society	Very limited dialogue between state and environmental movements	Top-down initiatives, limited state-environmental movements dialogue	Open-ended dialogue and envisioning	Bottom-up community structures and control, new approach to valuing work

Source: Adapted from Baker *et al.* (1997)

potential of 'demanufacturing'. Sustainable approaches to regeneration have overlapped and influenced discussion in local and regional development around the contrasts between 'top-down' and 'grass-roots' approaches. Strong sustainable development has promoted small-scale, decentralised and localised forms of social organisation that promote self-reliance and mutual aid (Chatterton 2002). Local and regional development examples include local trading networks and ecological taxes on energy, resource use and pollution (Hines 2000).

The critique of sustainable local and regional development

Sustainable approaches to local and regional development have been subject to criticism. For 'weaker' sustainable development, criticism focuses upon the reformism and limited contribution of such ideas to sustainability and the possibility of actually achieving economic, environmental and socially integrated approaches to local and regional development (Harvey 1996; Haughton and Counsell 2004). Stronger sustainability in local and regional development has attracted criticism for its potentially unrealistic search for ideological purity, practical lack of feasibility and limited, small-scale examples. Given the magnitude of changes required to put local and regional development onto a more sustainable footing, the relatively small-scale initiatives introduced to date often seem limited relative to the scale of the problem. In terms of government and governance, local and regional institutions may lack the power and resources within a multilevel polity to deliver sustainable development (Morgan 2004). Notwithstanding such issues, sustainable development is a key concern for local and regional development explored throughout the book.

Post-developmentalism

As we saw in Chapter 2, the notion of 'development' has been questioned in the light of post-structuralist debates in social theory (Peet 2002). Post-structuralism is 'a theoretical approach to knowledge and society that embraces the ultimate undecidability of meaning, the constitutive power of discourse, and the political effectivity of theory and research' (Gibson-Graham 2000: 95). At its heart is a critique of modernism and its epistemology or theory of knowledge. For post-structuralists, modernist thinking sees knowledge as singular, cumulative and neutral. Post-structuralism interprets knowledge as multiple, contradictory and powerful. 'Development' is understood as a specific discourse – a socially constructed narrative assembled and promoted by certain interests – that organises knowledge of economic change in a particular way. For Gibson-Graham (2000: 103), 'development' is 'the story of growth along a universal social trajectory in which regions or nations characterised by "backwardness" are seen to progress towards modernity, maturity, and the full realization of their potential'.

The post-structuralist critique focuses upon the modernist model of change in post-war developmentalism discussed in Chapter 2 (see Figure 2.1). It criticises its 'Eurocentrism' and representation as the 'one-best-way' route to 'development' tried and tested by the industrialised and developed North. *Post*-development theorists influenced by post-structuralist thinking argue that this inappropriate and externally determined model has been foisted upon the 'developing' South, often by international lending agencies such as the IMF and the World Bank. The 'development' model is seen as intimately connected to the global extension of neo-liberalism through its adherence to *laissez-faire* multilateral free trade, ensuring that 'developing' country markets remain open to the exports and investments of 'developed' world producers. The 'development' discourse is interpreted as having had devaluing and disabling effects upon the 'less developed' (Escobar 1995). The post-structuralist approach interprets the prevailing discourse of 'development' as further reinforcing the colonial legacy of unequal relations between the 'developed' and 'developing'.

Post-development in localities and regions

The critique of modernist 'development' has stimulated thinking about 'post-developmentalism' and is having some influence upon questions of local and regional development (Edwards 1989; Rahnema and Bawtree 1997; Gibson-Graham 2000). Post-structuralist analysis interprets 'economic rationalities as socially constructed' and takes 'diverse historical forms, have distinct geographies and produce specific regional forms of development' (Peet 1998: 2). Central to this post-structuralist approach are the strategies of deconstruction, genealogy and discourse analysis to trace the historical construction of what appear to be mainstream narratives of 'development'. Questioning prevailing wisdom – articulated in often competing discourses of modernity based in different social and political imaginations – then provides the ability to challenge such conceptions and develop alternative constructions. The objects of 'development' might then be repositioned 'outside a discourse that produces subservience, victimhood, and economic impotence' (Gibson-Graham and Ruccio 2001; see also Gibson-Graham 2000: 104).

Post-development in local and regional policy

Post-developmentalism seeks a theory of development determined by those to be 'developed' or, crucially, those who choose not to be 'developed' in a particular way. Empowered, grass-roots leadership and nationally, regionally and locally appropriate forms of development are the aspiration of this approach. As discussed in Chapter 2, rather than have a model of 'development' socially constructed and imposed by other interests, post-developmentalism encourages localities and regions to seek their own answers to the question of what kind of local and regional development and for whom. In local and regional development policy, post-development ideas have gained ground since the 1990s. Example 3.4 describes Gibson-Graham's approach to capitalism, non-capitalism and community economies. Attention has focused upon the potential economic and social development benefits of 'alternative' and more 'diverse' economies better connected to the social needs and aspirations of localities and regions, including initiatives such as Local Exchange Trading Schemes, social enterprise and intermediate markets for labour, goods and services (Leyshon *et al.* 2003).

Example 3.4 Capitalism, non-capitalism and community economies

Gibson-Graham's (2000) 'post-structuralist' approach to local and regional economies seeks to challenge the ways in which the capitalism–non-capitalism relationship has provided an economic discourse that constitutes capitalism as a necessary and dominant form of the economy. They argue that non-capitalist forms are typically understood relative to capitalism only as equivalent, opposite, complementary or subsumed. In particular, non-capitalism is often seen as subservient, weaker and less reproducible. Non-capitalist economic practices may include the household, the informal economy, alternative economic experiments or cooperatives. Such activities may constitute a substantial part of the 'economy' but are currently treated as invisible – the hidden part of the 'iceberg' (Figure 3.15).

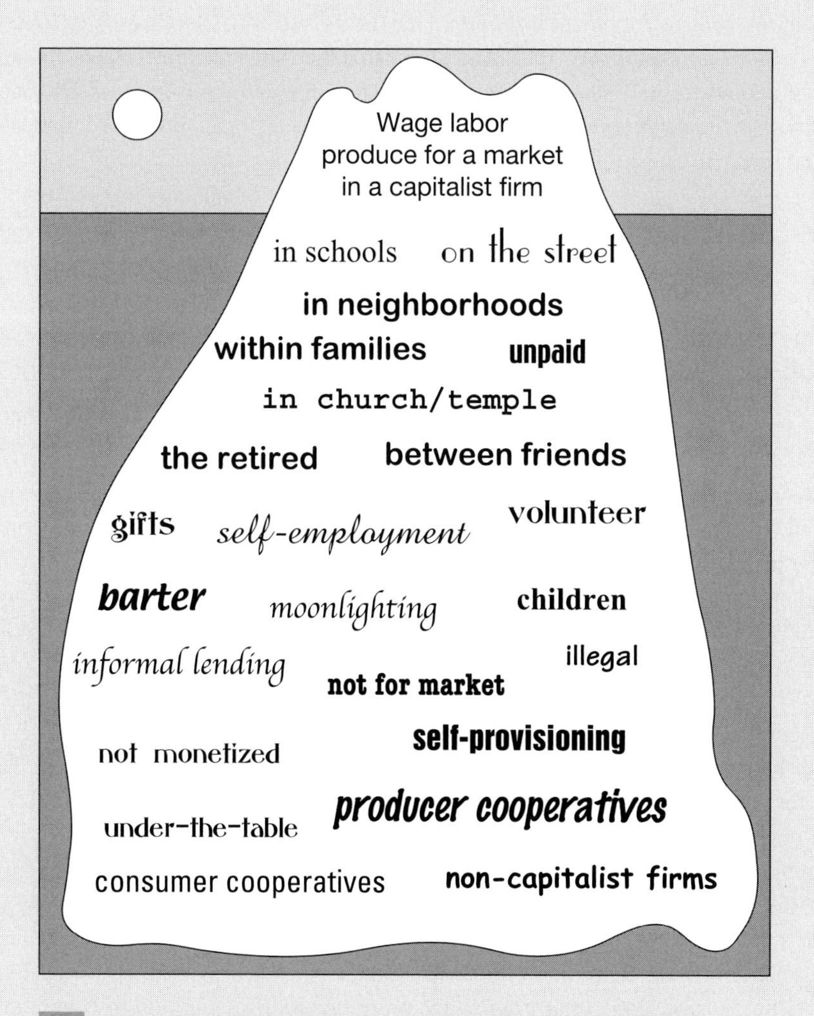

Figure 3.15 ***The 'economic' in capitalism and non-capitalism***

Source: Community Economies (http://www.communityeconomies.org/)

Gibson-Graham seek to challenge or 'destabilise' the 'mainstream' development discourse that focuses upon capitalism as dominant. They seek to represent the realm of non-capitalism positively as a potential array of diverse economic practices that could be resilient and capable of generative local and regional growth. In policy and practice, this approach has informed community-based action research at the regional level that seeks to promote a discourse of economic diversity and economic possibility, often in localities facing the deindustrialisation and decline of formerly dominant capitalist firms and industries. Since discursive constructions are interpreted as having material and symbolic effects in post-structuralist thinking, innovative forms of language and representation are deployed to challenge the 'mainstream discourse of "development" that "had positioned the region as entirely dependent on investment by capitalist firms, which might or might not be attracted by various blandishments"' (Gibson-Graham 2003: 108). 'Destabilising' the existing identities of localities and regions in this way is seen as a means 'to produce new models of regional development that exceed the theory and practice of capitalist industrialisation' (Gibson-Graham 2000: 108). The extent to which such alternative conceptions can challenge and/or accommodate the broader capitalist economy and grow small-scale experiments into sustainable and context-sensitive local and regional 'post-development' remains to be seen.

Source: Gibson-Graham (2000)

The critique of post-developmentalism

Perhaps unsurprisingly given their radical intent and early stage of evolution, post-structuralist approaches to local and regional development have attracted criticism. Critiques of post-structuralism emphasise its philosophical relativism. It deliberately lacks and rejects meta-theories, such as Marxism, and foundational or universal principles (Harvey 1996). Rather than any notion of trans-historical relations and processes that endure over space in place and over time, the guiding ideas and values of post-structuralism are the products of particular interests, places and times. We return to this very important issue of relativism in the conclusions in Chapter 8. For Scott (2004), post-structuralist analysis displays a naive relativism, philosophical idealism and political voluntarism that fails adequately to recognise the often determining power of external forces and underlying structures shaping local and regional development. Peet (1998) forwards a counter-critique of post-structuralism's negative attitude and rejection of development, modernity and economic progress and interprets the critique of 'development' as an attack upon the modern and progressive idea of rational social intervention for the improvement of human existence and emancipation. In his view, the need for a modernist theory and practice of development alternatives to the current neoliberal order is critical. In addition, post-structuralist approaches often ignore or deliberately fail to acknowledge their links to the more 'modernist' tradition of Community Economic Development and its core themes of local control and empowerment we discussed in Chapter 2 (see Example 2.6).

Conclusion

This chapter has reviewed the main concepts and theories that seek to help us to understand, interpret and explain local and regional development. Neo-classical theories focus upon explaining disparities in regional growth and their long-run reduction and convergence. Despite critique based upon its simplifying assumptions and contrary empirical evidence, this conventional theory remains influential for 'free-market' versions of local and regional development policy. In contrast, Keynesian theories emphasise regional divergence and the ways in which unfettered markets tend to reinforce rather than reduce regional disparities. The extent of polarisation between core and peripheral regions explained by the theory remains the subject of debate and an important influence for more interventionist forms of local and regional development policy.

Theories of structural and temporal change interpret local and regional development as historical and evolutionary processes that may incorporate periods of structural or systemic change. Stage, cycle and wave theories use temporal frameworks to explain regional development and its particular historical evolution in specific types of places. Marxism and radical political economy use the 'spatial division of labour' to reveal the hierarchical relations between places and to explain regional growth as episodic and capable of periods of convergence and divergence. Transition theories – institutionalist, transaction costs and regulationist – interpret local and regional development in the context of substantive shifts in the nature of capitalism. Social, technological and institutional characteristics are central to explaining the resurgence of specific types of local and regional economies. Despite critique of their reliance upon macro-structural change and their failure to explain the diversity of local and regional development experiences, transition theories have stimulated a resilient policy focus upon indigenous assets and 'development from below'.

Institutionalism and socio-economics emphasise social and institutional context to explain uneven local and regional development. Specific and particular attributes of localities and regions are central to explaining development trajectories over time, especially the role of intermediate institutions between markets and hierarchies. Theories of innovation, knowledge and learning seek to open the 'black box' of technological progress integral to neo-classical explanations of local and regional growth. Geographical unevenness in innovation, knowledge and learning is explained by differentiated social and institutional structures and contexts with marked implications for local and regional development. In response, local and regional development policy informed by these ideas has sought to foster the capability of institutions to build innovation capacity and foster collective knowledge creation, application and social learning.

Extended neo-classical theories of local and regional development seek to address the problems of the conventional neo-classical approach. Endogenous theories incorporate formerly external or exogenous factors – population growth, saving rates, human capital and technological progress – within their models to explain regional convergence and divergence and the spatial concentration of high and low growth regions. Despite criticisms, endogenous theory has reshaped the focus of local and regional development policy towards levelling-up the economic performance of all localities and regions to enhance local, regional and ultimately national development. Geographical economics emphasises imperfect competition, increasing returns and external economies combined with intra-industry and intra-corporate trade in its conceptual and theoretical framework. Rather than the long-run convergence focus of conventional neo-classical theory, it explains local and regional divergence in terms of the development of multiple cores and peripheries. Strategic trade policy seeks to support the specialised, internationally competitive and geographically concentrated export sectors with local and regional development implications. Competitive advantage theory explains the role, dynamics and competitiveness enhancing potential of the geographical 'clustering' of industries at the local and regional levels within national economies. Cluster policy is highly influential for local and regional development in fostering the benefits of 'clustering' to competitiveness and providing a role for local and regional institutions.

Sustainable development seeks to understand and explain longer term, more durable and less damaging forms of local and regional development that integrate economic, social and environmental concerns. Sustainable forms of local and regional development have become a central challenge for national, regional and local institutions and policy. 'Weaker' forms of sustainability have most influenced local and regional development thinking and policy experimentation with concrete initiatives. Post-developmentalism draws upon post-structuralist theory to critique the prevailing mainstream discourses of 'development' and to promote alternative, locally determined, social constructions of 'development' based upon both capitalist and non-capitalist economic activities. To complete this section on frameworks of understanding, the next chapter engages with the institutions of government and governance of local and regional development.

Further reading

For a complementary and general overview, see Armstrong, H. and Taylor, J. (2000) *Regional Economics and Policy* (3rd edn). Oxford: Blackwell.

For wide-ranging collections, see Barnes, T. and Gertler, M. (eds) (1999) *The New Industrial Geography: Regions, Regulation and Institutions*. London and New York: Routledge; Barnes, T.J. and Sheppard, E. (eds) (2000) *A Companion to Economic Geography*. Oxford: Blackwell; Clark, G.L., Feldman, M. and Gertler, M. (2000) *The Oxford Handbook of Economic Geography*. Oxford: Oxford University Press.

For critical reviews of recent conceptual and theoretical approaches, see Scott, A.J. (2004) 'A perspective of economic geography', *Journal of Economic Geography* 4: 479–499; Scott, A.J. and Storper, M. (2003) 'Regions, globalization, development', *Regional Studies* 37(6–7): 579–593; Sunley, P. (2000) 'Urban and regional growth', in T.J. Barnes and E. Sheppard (eds) *A Companion to Economic Geography*. Oxford: Blackwell.

For synoptic reviews of key conceptual developments in local and regional development, see Martin, R. (1999) 'Institutional approaches in economic geography', in T.J. Barnes and E. Sheppard (eds) *A Companion to Economic Geography*. Oxford: Blackwell; Martin, R. and Sunley, P. (1996) 'Paul Krugman's geographical economics and its implications for regional development theory: a critical assessment', *Economic Geography* 72: 259–292; Martin, R. and Sunley, P. (1998) 'Slow convergence? Post neo-classical endogenous growth theory and regional development', *Economic Geography* 74(3): 201–227; Martin, R. and Sunley, P. (2003) 'Deconstructing clusters: chaotic concept or policy panacea?', *Journal of Economic Geography* 3(1): 5–35.

On the conceptual linkages to local and regional development policy and practice, see Glasmeier, A. (2000) 'Economic geography in practice: local economic development policy', in G.L. Clark, M. Feldman and M. Gertler (eds) *The Oxford Handbook of Economic Geography*. Oxford: Oxford University Press; Cooke, P. and Morgan, K. (1998) *The Associational Economy: Firms, Regions and Innovation*. Oxford: Oxford University Press.

INSTITUTIONS: GOVERNMENT AND GOVERNANCE

4

Introduction: States and local and regional development

The role of the state in sponsoring industrialisation can be traced back to the nineteenth century and beyond. Such intervention in the economy was frequently a means of nation-building. Since the Second World War governments have intervened in their economies to ensure local and regional development, often reflecting a commitment to limit the growth of inter-regional disparities and promote the development of rural areas. Although having some success in shaping patterns of local and regional development, as we noted in Chapter 1, centralised or top-down forms of intervention, pursued by national planning and development authorities, were criticised for their heavy concentration on the provision of physical infrastructure and, as we shall consider in Chapter 6, an overemphasis on the attraction of mobile investment. Such an approach often failed to close the development gaps between prosperous and lagging regions. The perceived failure of such approaches, together with the challenges of globalisation, has led to a growing emphasis upon bottom-up approaches to the promotion of local and regional development. Such interventions, in theory at least, tend to require strong institutions of local and regional governance and to be based on local and regional participation and dialogue. As we discussed in Chapter 1, bottom-up approaches are concerned with integrated territorial development, focusing upon the mobilisation of local resources and competitive advantages that are locally owned and managed.

The concern of this chapter is the government and governance of local and regional development. That is, the development of institutions which are responsible for the design, implementation and monitoring of strategies of development. This involves the vertical and horizontal coordination of different levels of government and local public and private actors and raises important issues of governance that need to be addressed by new institutional forms. Local and regional development is typically subject to increasingly complex governance systems often involving new forms of cooperation and coordination. Such developments can serve to empower populations and assist individuals and communities to develop their own syntheses of what kind of local and regional development and for whom that we discussed in Chapter 2. Potentially, new forms of governance can also foster the mobilisation of civil society and promote the formation of networks and partnerships that can provide a basis for economic and social progress.

However, such developments are not guaranteed and the 'new regional governance' also brings with it new problems and challenges that we shall consider below.

This chapter investigates changes in the form of the state and their implications for local and regional development. It examines the arguments that, first, we are moving from an era of government to one of governance, and second, to a decentralised era of devolution and 'new regionalism', or, even, 'new localism'. The emergence of multilevel governance provides a framework encompassing multilayered institutional contexts for local and regional development at a range of scales from the supranational to the neighbourhood. Finally, the chapter addresses the relationship between democracy and local and regional development that we return to in the conclusions in Chapter 8.

The role of government in the management of local and economic development was noted in Chapter 2. Major changes in the role of government marked the shift from the era of 'developmentalism' to that of 'globalism' (see Table 2.1). In federal states – such as Brazil, Germany and the United States – subnational or state governments have always played a role in relation to local and regional development. As a general rule though, during this period national governments tended to take the leading role in planning economic development, reflecting the central role of the nation state in politics and government in the modern era (Dunford 1988; Le Galès and Lequesne 1998). Powerful arguments were advanced that the state should act to regulate markets and limit their excesses (Polanyi 1944; Moggridge 1973), while others saw the growing role of the state in economic management as a threat to liberty (Hayek 1944).

Modern state politics were often a reflection of class interests as expressed through national political parties, and through corporatist structures of interest representation by organisations such trade unions and business associations, which directly or indirectly were able to influence the direction of government policy. Such class interests tended to be underpinned by supporting ideologies such as, in Europe for instance, social democracy and Christian democracy (Therborn 1995). Politics during this era was contained largely within a national territory and focused on the control of state power within clear boundaries and rules of sovereignty (Taylor and Flint 2000).

Typically during this period, national governments of various political hues sought actively to intervene in the management of the economy and the promotion of industry. Informed by the Keynesian macroeconomic theories outlined in Chapter 3, interventions included the use of tax and public expenditure decisions on the part of governments to maintain high levels of effective demand in the economy and, therefore, full employment. In Europe, such interventions usually took place alongside the expansion of the welfare state. In Latin America and Asia government interventions were guided by strategies of import substitution. Governments also supported the creation of new industrial capacity by subsidising the development of key sectors of the economy. In the United Kingdom, for instance, this bundle of state economic measures was usually known as 'Keynesianism' and marked the approach to national economic management after the Second World War (Chisholm 1990). The Keynesian revolution placed the emphasis on full employment as the key means of reconciling capitalism and democracy. This form of economic regulation accompanied the emergence of national varieties of Fordist mass production and consumption. Indeed, this form of economic management underpinned the sustained period of economic growth from the 1950s to the 1970s (Jessop 1997). Crouch summarises the situation thus:

> In those industrial societies which did not become communist, a certain social compromise was reached between capitalist business interests and working people. In exchange for the survival of the capitalist system and the general quietening of protest against the inequalities it produced, business interests learned to accept limitations on their capacity to use their power. And democratic political capacity concentrated at the level of the nation state was able to guarantee those limitations, as firms were largely subordinate to the authority of nation states.
>
> (Crouch 2004: 7–8)

As part of these measures, governments across the world frequently sought to shape the geography of economic activity at the subnational scale. In pursuit of geographical equity states tended to share the goal of redistributing resources to lagging or peripheral regions and sought to promote spatially balanced forms of development, a goal sometimes described as 'spatial Keynesianism' (Martin and Sunley 1997). In this perspective, achieving regional development contributed to national efficiency by ensuring that all the economy's resources were utilised. Governments in newly industrialising countries sought to direct the development of particular regions often as a component of national modernisation strategies (Dicken 2003). By providing infrastructure and financial incentives, or, by implanting state controlled or owned businesses into particular regions, governments sought to promote the development of lagging regions and deal with the decline of traditional industries. The policy mechanisms used by national governments during this period are sketched out in Table 2.1. For practical purposes, governments often created specialised agencies to promote local and regional development such as DATAR (*Délégation à l'aménagement du territoire et à l'action régionale*) in France, *la Cassa per il Mezzogiorno* in Italy, the Tennessee Valley Authority in the United States and the Japan Regional Development Corporation.

The broader activities of government can have implications for local and regional development in mixed economies. In some countries, notably the United Kingdom in recent years, arguments about the geographical distribution of public expenditure and its local and regional development impacts have proved politically contentious (see MacKay 2001; McLean 2005).

Fiscal policy has local and regional impacts. Generally, the interaction of progressive taxation and public spending has a redistributive effect, acting as an automatic stabiliser of the economy. Thus, for instance, governments increase expenditure on unemployment benefits when economies are in recession and, to the extent that unemployment is distributed unevenly between localities and regions, resources are channelled into those regions with higher unemployment. Government transfers, then, can act as regional stabilisers (Armstrong and Taylor 2000). Figure 4.1 shows the case of Italy where regions with the highest levels of public expenditure in relation to GDP have the lowest levels of GDP per head, indicating that a degree of redistribution is occurring. But the provision of public goods and the formula used in their allocation can produce regional tensions. These tensions are visible in the United Kingdom where there are differences in the per capita provision of public goods, such as education and health services, which primarily are driven by population size and are only loosely related to needs (McLean 2005). Such tensions may be accentuated where inter-regional transfers lack transparency and where devolved forms of government provide political voice to challenge the distribution of public expenditure. By contrast, federal states such as Australia and Germany generally have more explicit mechanisms for distributing resources between subnational authorities and ensuring equity and territorial solidarity (Smith 1994; McLean 2004).

Macroeconomic policies aimed at effecting overall levels of economic activity may also have local and regional impacts. For instance, a central bank with a remit to maintain stable prices may raise interest rates in order to dampen inflationary pressures in a fast growing region, but the effects of this might also be felt in lagging regions, where economic conditions may in fact call for a demand stimulus and the competitiveness of exporting business may be adversely affected by exchange rate appreciation. Such a scenario characterised the United Kingdom during the 1990s. A single national currency may operate across a territory which contains marked regional differences in economic conditions, but the introduction of monetary union within the European Union, involving twelve Member States, extends a single currency over a range of economies with quite significant differences in terms of industrial structure. The critics of monetary union argue that a single interest rate poses particular challenges for lagging regions, which lack the capacity to adjust to the new economic conditions (see Amin and Tomaney 1995a).

'Non-spatial' government policies can have local and regional effects. In the United States and the United Kingdom, for instance, government expenditure on defence equipment has had pronounced regional effects generally favouring prosperous high-tech regions as the R&D intensity and sophistication of defence products and services

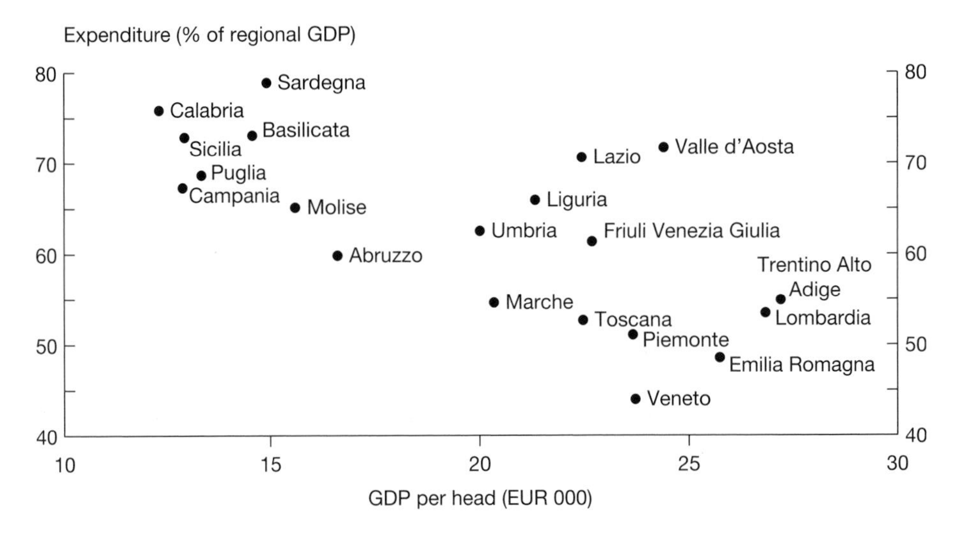

Figure 4.1 ***Public expenditure and GDP per head by region in Italy, 2000***

Source: European Commission (2004: 90)

has increased. In such cases, regional policy can act as a kind of 'counter-regional policy', accentuating rather than ameliorating regional inequalities (Lovering 1991; Markusen 1991).

In sum, there is more to the government action in localities and regions than simply local and regional development policy. Virtually all the actions of governments – even when governments themselves are unaware of it – have local and regional impacts.

Traditional forms of government intervention presupposed a sequestered national economy and sovereign state – a delineated political-economic space that contained the effects of policy interventions and nationally focused demand management. The internationalisation of both economies and governance marked the period from the end of the 1970s (Held 1995; Michie and Grieve Smith 1995; Dicken 2003). Thus, contemporary governments face new challenges. Mulgan has argued:

> Now the pertinence of the national levels of political economy has been reduced by a globalizing economy, by localism and by the failure of national governments to solve the problems they set themselves. Indeed, although the nation clearly remains the pre-eminent political entity, it is in secular retreat.
>
> (Mulgan 1994: 12)

Such tendencies have unleashed profound changes in the government and governance of local and regional development.

From government to governance

There are at least three important transformations in the state which have a bearing on the governance of local and regional development. First, there is a shift from government to governance on various territorial scales and across various functional domains (Jessop 1997). The central role of official state apparatus in securing state-sponsored economic and social projects and political hegemony has diminished in relative terms. Instead, there is an emphasis on partnerships between governmental, para-governmental and non-governmental organisations in which the state apparatus is often only 'first among equals'. This involves the complex art of steering multiple agencies – what Jessop (1997) calls 'meta-governance' – which are operationally autonomous from one another but loosely linked together and reliant on each other. Many activities previously undertaken directly by the state have been devolved to varying extents to arm's-length agencies. The role of political authorities is to steer the self organisation of partnerships and networks – a kind of 'governance without government', according to Rhodes (1996: see Example 4.1).

Second, there is a general trend towards the 'hollowing out' of the nation state with old and new state capacities being reorganised territorially and functionally on subnational, supranational and trans-territorial lines. State power disperses upwards to supranational institutions like the European Union or IMF, downwards to regional and local bodies and sideways to trans-territorial networks as attempts are made by state managers at different territorial scales to enhance their respective operational

Example 4.1 Characteristics of 'governance'

Rhodes (1996) identifies a number of characteristics of 'governance' including interdependence between organisations. Governance is broader than government, covering non-state actors. Changing the boundaries of the state has meant the boundaries between public, private and voluntary sectors have become shifting and opaque. Continuing interactions between network members are caused by the need to exchange resources and negotiate shared purposes. Game-like interactions, rooted in trust and regulated by the rules of the game, are negotiated and agreed by network participants. Such networks have a significant degree of autonomy from the state. Networks are not accountable to the state; they are self-organising. Although the state does not occupy a privileged, sovereign position, it can indirectly and imperfectly steer networks.

Source: Rhodes (1996)

autonomies and strategic capacities. Finally, according to Jessop, the international context of domestic state action has grown in importance leading to a situation where economic and social policies become more concerned with 'international competitiveness'. At the same time, the roles of national government are now increasingly affected by the regulatory functions of international bodies. Such bodies include the IMF, World Bank and WTO, and organisations such as the European Union, NAFTA, ASEAN and Mercosur. Thus, both the globalisation of the economy – especially the rise of global production networks and global financial markets – and the growth of international institutions seeking to address issues beyond the scope of individual nation states (e.g. trade, environment, terrorism) have occurred simultaneously and have placed constraints on the activities of national governments in the area of economic regulation. Jessop (1997) codifies these changes under the rubric of the shift from a Keynesian welfare state to a Schumpeterian workfare state.

This transformation of the state, incorporating the shift from government to governance, can be identified in many different states, including 'transition' countries and those formerly regarded as 'newly industrialising' (Dicken 2003). Above all the notion of governance refers to governing styles in which the boundaries between and within public and private sectors have become blurred. The focus is on governing mechanisms which do not rest on recourse to classical ideas of state authority and sanctions, but on the interaction of multiple actors. Such forms of working are generally seen as requiring high levels of trust between the actors involved to be effective, and the task of governance becomes that of 'steering networks' (Stoker 1995). Such forms of governance, which are concerned with managing an increasingly fragmented field of public policy, have been linked by some to similar processes of fragmentation in post-modern society (Bogason 2004a).

One danger in the discussion of the shift from government to governance is that this development is seen as natural and inevitable, rather than the product of decisions taken often, above all, by national state actors. But even those who implicitly welcome this shift recognise that it is associated with 'dilemmas' including a divorce between the complex reality of decision-making under the new governance and the language used to justify government, the blurring of responsibilities that can lead to avoidance or 'scapegoating' and problems of mutual power dependence that can exacerbate the problem of unintended consequences. 'Self-governing networks' raise difficulties over accountability, which means that in practice steering governance networks is likely to prove difficult (see Stoker 1995). Above all, any uneven shifts towards governance must continue to recognise the integral role of government, especially at the national level.

The practical difficulties of the emergent networks of governance are revealed when we think about partnerships. Partnerships are a tool by which the new regional governance is managed. On the face of it, the term 'partnership' holds out the prospect of more inclusive decision-making. However, reviewing evidence from across Europe, Geddes (2000) has identified some dangers in the use of 'partnerships' as the vehicle for local and regional development. He suggests that partnerships often exclude the very groups at which they are targeted. While the aim of partnerships is to engage non-state actors in governance, frequently partnerships are dominated by the public sector which has the capacity and resources to devote to the task. Partnerships, moreover, are more often a vehicle for managing distrust rather than fomenting trust. Marginalised groups find it as difficult to influence partnerships as they do state actors. Stoker raises a set of questions about the efficacy of partnerships:

> Is it possible for elected officials to exercise some control over the partnership networks that constitute the emerging system of local governance? Can the achievements of partnership be evaluated or does it run the risk of becoming an end in itself? Can the dynamics of governance be reconciled with the traditional concerns about accountability and propriety in public affairs? Do the partnerships developed through governance undermine democracy by restricting access to 'insider' groups, leaving other interests underrepresented and excluded?
>
> (Stoker 1997: 48–49)

The answers to these questions cannot be given *a priori*. They are a matter for empirical verification, but there is enough evidence to suggest that the move to partnership forms of governance contains new difficulties rather than simple solutions to problems of representation and governance.

There is nothing inherently democratic about the emergence of network forms of governance which occur 'in the shadow of hierarchy' (Jessop 1997: 575), that is, under the domination of traditional forms of authority. Moreover, much discourse on the subject of governance can be reduced merely to a justification of public sector reform along the lines of the 'new public management' which is principally concerned with the introduction of market principles for the provision of public services, along with the privatisation and 'agencification' of public administration (Rhodes 1996; see also Kjær 2004). Such developments were a response initially, mainly by conservative governments to the apparent decline in the effectiveness of democratic institutions and the state and sought a solution to this problem in deregulating the economy and rolling back the frontiers of the state (see Gamble 1994). Keating (2005), indeed, argues that '"Governance" is a notoriously loose concept, and is often used to hide . . . critical questions about the balance of power, the representation of interests and the direction of policy. However conceptualised, governance is not a substitute for government' (Keating 2005: 208).

The role envisaged for the state by writers such as Bogason and Stoker is much more restrictive than the more directive role governments in many states throughout the period after the Second World War. Institutions of local and regional government are seen as playing important roles in the new governance system. But according to Stoker (1995), their role is 'system management' by providing leadership, forming partnerships and regulating the overall environment by defining situations, identifying stakeholders and managing relationships between parties. The so-called 'new regionalism' and 'new localism' are frequently seen as corollaries of these emergent forms of governance. Indeed, the experiments with new forms of governance have often emerged alongside the devolution of political or administrative power in many states (Bogason 2004b).

'New regionalism'?

> Is there something perhaps about population units of around three to eight million (the size of many US states, regions of larger European states, or the small European nation states themselves) that make policy-making between public authorities and business organisations particularly useful and flexible?
>
> (Crouch and Marquand 1989: x)

Trends towards the decentralisation of government are discernible in different political contexts including Asia, Europe, Australia, Latin America and North America (Keating 1998; Javed Burki *et al.* 1999; Brenner 2002; Rozman 2002; Rainnie and Grobelaar 2004) and in developed and developing countries (Bardhan 2002). Despite important differences between different national circumstances some writers have identified the emergence of a 'new regionalism' (Keating *et al.* 2003) and/or 'new localism' (Goetz and Clarke 1993) linked to a newly attached importance for subnational government by international organisations such as the OECD and World Bank (e.g. OECD 2001, 2004a). As we discussed in Chapter 3, 'new regionalism' connects with the renewed emphasis upon the region as the locus for economic, social and political action and the roles of institutions in local and regional development.

In the global context of the 'new regionalism', the trend towards decentralisation of government is driven to a large extent by political factors at the national and subnational level (Rodríguez-Pose and Gill 2004). In countries such as Spain and Brazil, the creation or strengthening of structures of subnational governance occurred alongside the restoration of democracy following long periods of authoritarian government (Rodríguez-Pose 1999; Rodríguez-Pose and Tomaney 1999). A similar story can be found in South Korea (see Chapter 7). In the United Kingdom, devolution to Scotland and Wales was a response to the rise of nationalist sentiments in a multinational state (Tomaney 2000). Elsewhere, in Europe, 'historical regions' have asserted their cultural identities, while emergent or 'new', 'economic regions' inspired by the 'new regionalism' have also claimed the right to act directly in their own political interest (Harvie 1994; Keating

1998; Loughlin 2001). In the United States, the rise of 'metropolitan regionalism' is a response to 'the rapid growth of suburbs, the creation of satellite cities, new modes of transportation, increased mobility of citizens, and easier forms of communication which require new structures between the city, county and state' (Miller 2002: 1; see also Benjamin and Nathan 2001). More generally, the trend to devolution may reflect a rejection of centralised forms of politics and government at the national level and declining faith in the nation state as an instrument for solving social and economic problems – issues to which we will return. As Table 4.1 illustrates, regional structures of governance, then, are a feature of an increasing number of states.

Although the nature and sources of the new regionalism vary significantly between societies, its growth reflects a widely held belief that the regional scale (or the small nation state) represents a particularly salient scale at which to organise policy interventions in the economy (e.g. Crouch and Marquand 1989). Such a view rests on the idea that local and regional supply-side conditions help to determine the competitiveness of firms. Thus, Scharpf (1991) argues smaller nation states, in Europe at least, appear to enjoy higher levels of political approval and economic 'success' than larger states, perhaps because of:

> their ability to conduct policy discourses that are based on those policy alternatives that are based on a realistic understanding of their own capabilities and constraints and to focus debates on those policy alternatives that might be

Plate 4.1 ***Devolved government: the Scottish Parliament in Edinburgh***

Source: © Scottish Parliamentary Corporate Body (2006)

Table 4.1 ***Regional and subnational government in OECD countries and in South Africa***

	Population (millions) 1997	*Number of regions*	*Population per region (millions)*	*Notes*
States with a full system of elected regional government				
United States	266.8	50	5.3	Fifty federated states
Japan	126.2	8/59	15.8 or 2.1	Eight regions, subdivided into 59 prefectures (47) and cities (12) with the same status as prefectures
Mexico	94.2	32	2.9	Thirty-two federated states
Germany	82.1	16	5.1	Sixteen Bundesländer
Turkey	63.7	80	0.8	Eighty provinces
France	58.6	22	2.7	Twenty-two regions in metropolitan France (excludes overseas possessions)
Italy	56.9	20	2.8	Twenty regions
Korea	46.0	8	5.8	Eight provinces
South Africa	43.4	9	4.8	Nine federated states
Spain	39.3	17	2.3	Seventeen autonomous communities
Poland	38.7	8	4.8	Eight provinces
Canada	30.3	13	2.3	Thirteen federated states (10) and territories (3)
Australia	18.5	7	2.6	Seven federated states
Netherlands	15.6	12	1.3	Twelve provinces
Greece	10.5	51	0.2	Fifty-one prefectures
Czech Republic	10.3	8	1.3	Eight regions
Belgium	10.2	3	3.4	Flanders, Wallonia and Brussels
Hungary	10.2	40	0.3	Forty counties and municipal counties, including Budapest
Sweden	8.8	8	1.1	Eight regions
Austria	8.1	9	0.9	Nine Bundesländer
Switzerland	7.1	26	0.3	Twenty-six cantons
Denmark	5.3	14	0.4	Fourteen counties
Finland	5.1	19	0.3	Nineteen regions
States with no, or limited, elected regional government				
Portugal	10.0			Principle of regional government rejected in a referendum
United Kingdom	59.0			Elected regional governments in London, Scotland, Wales and Northern Ireland

Source: Authors' own research

Note: OECD countries with populations exceeding 5 millions.

> feasible and effective in an international policy environment that is characterised by high degrees of institutional integration, economic interdependence and regulatory competition.
>
> (Scharpf 1991: 120)

Indeed, as we discussed in Chapter 3, effective governance has been seen as a factor in underpinning the performance of successful regions by providing support for the formation of skills, technological change and the nurturing of the natural and built environment and through the promotion of 'untraded interdependencies' that include labour markets, public institutions, rules of action, understandings and values (Keating 1998: 137). It is important, however, not to assume an easy causal relationship between 'good' governance and economic success. Structures of governance undoubtedly help to shape patterns of local and economic development but, conversely, effective forms of governance are often underpinned by strong economic performance.

As we explained in Chapter 3, regional level institutions, therefore, are increasingly seen as a necessary ingredient of bottom-up forms of regional policy in an era in which localities and regions are more directly exposed to the international economy. This changing context seeks to mobilise localities and regions as agents of their own development. Harvey (1989a: 6) argues that the role of the local state has shifted from one of local management of the welfare state aimed at ensuring redistribution to one based on 'entrepreneurialism', whereby cities and regions directly compete with each other. Such entrepreneurialism needs to be examined 'at a variety of spatial scales – local, neighbourhood and community, central city and suburb, metropolitan region, region, nation state, and the like' (Harvey 1989a: 6).

According to Keating, the new regional governance tends to be underpinned by development coalitions which comprise a cross-class, place-based alliance of social and political actors dedicated to economic growth in a specific location, its composition varying from one place to another (Keating 1998; Keating *et al.* 2003). Such coalitions tend to be embodied in regional institutions which provide public goods and foster dialogue and communication among economic actors in the region. Local and regional political leaders become important figures in shaping development coalitions, albeit compelled to act in concert with other actors within multilevel and multi-agent governance structures (e.g. Harvie 1994).

Political structures provide democratic legitimacy for regional governance arrangements, but even within Europe, for instance, there is great diversity in the forms of political structures and the powers and responsibilities of local and regional governments (Keating 1998). There are some common trends in the process of decentralisation that can be observed across the world. Many local and regional governments make their economic interventions through local and regional development agencies. The rise of the locally and regionally governed development agency is a notable feature of the period since the 1970s. According to Danson *et al.* (1998), the growth of regional development agencies (RDAs) draws upon a bottom-up approach to regional development (Table 4.2). RDAs can be seen as a practical manifestation of the shift from managerialism to entrepreneurialism in the transformation of regional governance. The development of regional

Table 4.2 ***Traditional top-down and new model bottom-up policies and institutions***

Characteristics	*Traditional top-down*	*New model bottom-up*
Organisation	National	Regional
	Government department	Semi-autonomous body (agency, partnership)
	Bureaucracy	Business-led
	Generalist qualifications	Specific expertise
	Administrative hierarchy and infrastructure	Task-led projects and teams
Political control	Directly through government department and ministerial responsibility	Indirectly through sponsor government departments and weak accountability structures
Operational freedom	Limited	Arm's length earned autonomy and target-based flexibility
Economic objectives	Inter-regional equality	Inter-regional competitiveness and raising economic performance
	Growth of national economy	Growth of regional economy
	Redistributed growth	Indigenous/imported growth
Mode of operation	Non-selective	Selective
	Automatic/discretionary	Discretionary
	Reactive	Proactive
Policy instruments	Bureaucratic regulation	Financial inducements
	Financial inducements	Advisory services
	Advisory services	Public provision
	Public provision	

Source: Adapted from Halkier *et al.* (1998)

level jurisdictions in policy-making affects the way policy objectives are determined and the methods by which they are implemented (e.g. Tomaney 1996; Armstrong 1997). This focus has perhaps accelerated the shift away from redistributive policies which are aimed at tackling inter-regional inequality towards growth-oriented policies focused on 'regional competitiveness' discussed in Chapter 3.

The activities and accountability of RDAs are shaped by the types of political structures – or their absence – that operate at the regional level. One example of these relationships is provided by ERVET (*Ente Regionale per la Valorizzazione Economica de Territorio*), a development agency that was established by the regional government of Emilia Romagna in 1973 following the first devolution of political power to the Italian regions in 1970 (Garmise 1994; Bellini and Pasquini 1998) (Figure 4.2). The weakness and ineffectiveness of the nation state in Italy, in part, was a stimulus to action at the regional level. The centralised nature of the Italian polity was transformed during the final decades of the twentieth century as the powers and profiles of regional governments grew (Keating 1998). ERVET has attracted much international attention because its creation was associated with the rapid economic growth in Emilia Romagna, part of the wider phenomenon of the growth of the Third Italy based on geographically concentrated and highly productive clusters of small, family-owned firms, known as industrial districts that we discussed in Chapter 3. Emilia Romagna, a region of 4 million people,

Figure 4.2 ***Emilia Romagna, Italy***

has a distinctive economic structure with 98 per cent of firms employing fewer than fifty people, sometimes referred to as the 'Emilian model' (Brusco 1982). Bellini and Pasquini (1998) define the Emilian model as a combination of progressive government, social integration and entrepreneurial success. Thus, while the region is noted as a location of successful entrepreneurship traditionally it has been dominated politically by the Italian Communist Party and its successors. Indeed, the creation of ERVET was part of an effort by the Emilian Communists to cement their political relationship with firms in the industrial districts. Garmise (1994) sees the Communist Party and its successors in Emilia Romagna as forming the heart of a 'regional productivity coalition', through the role of 'mediator and coordinator' of groups including unions, the cooperative movement, artisan associations and employers.

The main task of ERVET is to nurture the development of the specialised networks of small firms that underpin the region's prosperity. Figure 4.3 shows the relationship between the regional government and ERVET is an arm's-length one with a division of labour between 'political' strategy and 'technical' execution. In practice, according to Bellini and Pasquini (1998), ERVET itself has strongly influenced the direction of economic development policy because of its accumulated expertise and its arm's-length relationship to the political system makes it a forum for the resolution of contentious issues. ERVET conforms to the model of a public–private partnership in which the regional government takes a majority stake, but in which private sector organisations

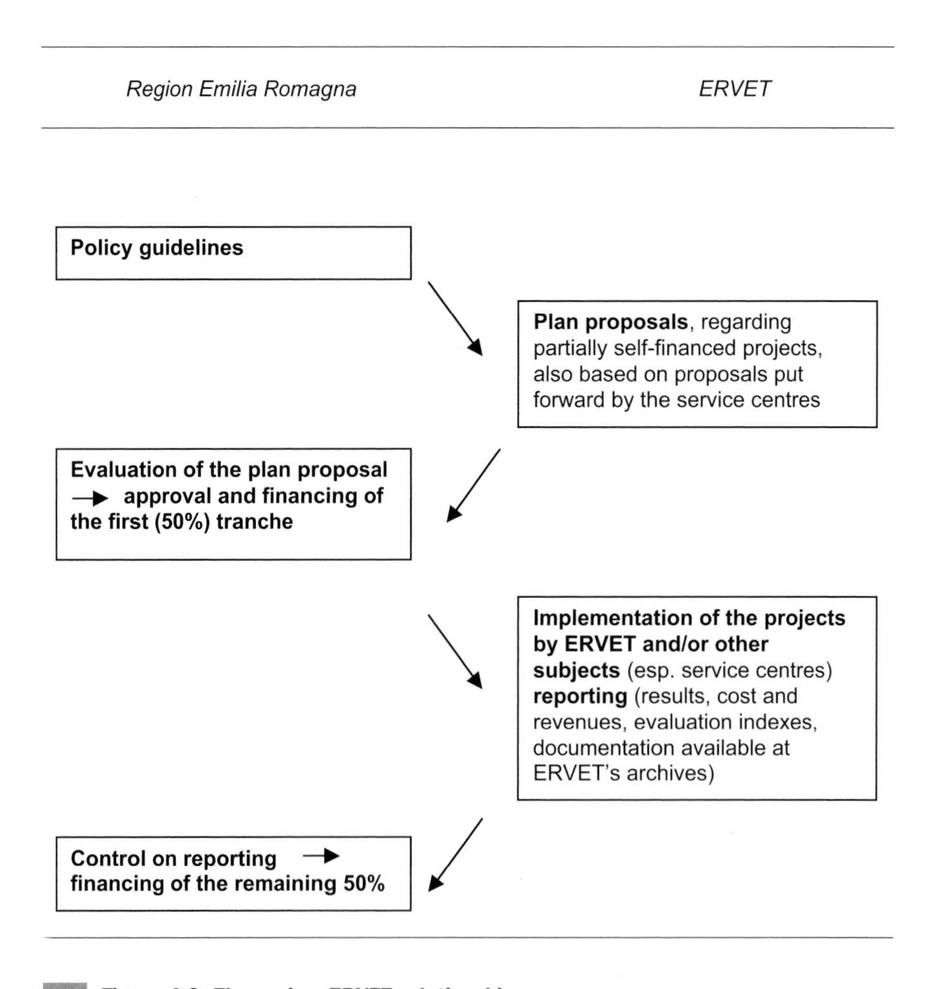

Figure 4.3 ***The region–ERVET relationship***

Source: Adapted from ERVET

also participate. Moreover, ERVET is the hub of a network of specialised agencies that support the development of key sectors of the regional economy by giving the access to 'real services' such as technology and marketing support.

Globalisation has posed new challenges for the Emilian model of small family firms. ERVET itself was criticised within the region for failing to respond quickly enough to the new competitive conditions, even as it was being lauded outside the region for its innovative approach. The regional government instituted a reform of the ERVET network in the 1990s which gave a greater role to the private sector in shaping its strategies and which gave its activities a stronger commercial orientation. The governance of local and regional development in Emilia Romagna can be seen as a classic example of the European variant of 'new regionalism', in which elected regional governments play an important role, albeit in the context of encroaching commercial pressures. While much writing about Emilia Romagna has focused on the distinctive character of its industrial structure, Garmise concludes that:

> The real lesson of the Emilian model for other regions in Europe can be summed up in two words: progressive government. Implementing informed social and economic policy and working through a productivity coalition with social and economic players, Emilia-Romagna has been able to effectively juggle the competing demands for a prosperous economy, social justice and legitimate political interventionism.
>
> (Garmise 1994: 158)

The role of regional governance in the ABC region – comprising the municipalities of Santo André, São Bernardo and São Caetano – in Brazil provides an example of the development of new structures of regional governances in a very different socio-economic and political context (Rodríguez-Pose and Tomaney 1999; Rodríguez-Pose *et al.* 2001). The greater ABC region, comprising seven municipalities, contains 2 million people in the south-eastern suburbs of São Paulo (Figure 4.4). The context for local and regional development is quite different from that in Emilia Romagna. Brazil is a federal state, albeit one that has experienced a series of authoritarian governments throughout the twentieth century, finally ridding itself of military dictatorship in 1985. The move to democracy saw a resurgence of the power of the Brazilian states. Brazil is a vast and populous country of 170 million people and federalism represents a practical solution to the problem of its government. The state of São Paulo itself has a population of 37 million and the city as a whole a population of 18 million. The ABC region was the economic powerhouse of Brazil during its industrialisation from the 1960s through import-substitution, notable for its concentration of very large automotive plants and their suppliers. The region was also the heartland of the Brazilian metalworkers' union, noted for its militancy, and the Brazilian Workers' Party, which led the opposition to military dictatorship in the 1970s and 1980s.

The economic fortunes of the region were transformed during the 1990s. The decision of the Cardoso government to open up the Brazilian economy to foreign investment and competition in the 1990s, together with growing 'negative externalities' such as shortages and rising costs in land and labour markets and worsening pollution, led

Figure 4.4 ***The ABC region, Brazil***

to an economic crisis and rising unemployment in the region as large firms began to investigate the possibilities of relocation to other parts of Brazil. This development was accelerated by the aggressive fiscal competition strategies pursued by some states in Brazil in order to attract manufacturing investment described in Chapter 2.

In this context, political, social and economic actors responded by attempting to develop a new strategy for the region, drawing on the support of a wide range of stakeholders, including unions, employers, NGOs and political parties. In part, these efforts were inspired and informed by knowledge of the 'new regionalism' and the role of new regional governance arrangements and regional development agencies in Europe and North America. The outcome of these deliberations was the decision by the mayors of seven municipalities to promote a new regional institution, the Chamber of the Greater

ABC region, which in turn sponsored the creation of a regional development agency (*Agência de Desenvolvimento Econômico do Grande ABC*) in 1998. The creation of the regional development agency signalled a decisive alteration of regional economic policy in the region, involving a shift in concern from the needs of large firms to the provision of 'real services' to networks of small and medium-sized firms. The effects of these developments will be felt only over the long term, but new governance arrangements themselves are novel in a region traditionally characterised by intense class conflicts (Rodríguez-Pose and Tomaney 1999; Rodríguez-Pose *et al.* 2001).

The experiences of Emilia Romagna and the Greater ABC region demonstrate how the pursuit of the 'new regional governance' is affecting quite different parts of the world. The unifying feature of the two examples is a belief that regional institutions can play an important role in organising the supply-side of the economy despite differences in national context. Both stories provide examples of the shift from government to governance by drawing new actors into the process of public policy formation, albeit with differences in degree and nature. Both are linked to changes in the external environment, the growing economic integration of the European Union and the emergence of Mercosur, as well as the growth of global firms. They are by no means the only examples of such trends.

The geographical reorganisation or 'rescaling' of the state in pursuit of effective local and regional governance can in some instances involve the transgression of national frontiers. A good example of this type of development is the creation of the cross-border Øresund region which links the region of Skåne in southern Sweden and Zealand in Denmark (see Maskell and Törnqvist 1999; Berg *et al.* 2000; OECD 2003b). Of symbolic and practical significance in the construction of this cross-border region was the opening of a bridge, which crosses the Øresund strait, funded by the Danish and Swedish governments, in order to link Malmö and Copenhagen (Figure 4.5).

The region is governed by the Øresund Committee – a gathering of political representatives from the local and regional authorities on both sides of the Øresund. The Committee is a classic example of 'governance without government':

> The committee is composed of local and regional political bodies from both sides of the sound and – which is quite exceptional for transnational regionalism – by the two national ministries . . . The process of integration in Øresund is therefore achieved not through the set up of an additional government layer but through the voluntary coordination of policies of its members.
>
> (OECD 2003b: 160)

The key institutions are the Copenhagen regional authority and the region of Skåne, which was created in 1999 as one of a number of 'experimental regions' within Sweden, which took over responsibilities from the central government in the field of economic development.

The political impetus to create the cross-border region came from the strong economic complementarities between Zealand and Skåne in sectors such as pharmaceuticals, biotechnology and medical equipment and IT and telecommunications industries. In practical terms, the Øresund region is characterised by cross-border collaboration in

Figure 4.5 ***The Øresund region, Denmark and Sweden***

research and education and the development of networking associations in the region's key industries. Thus, in order to support the growth of these sectors universities in the region cooperate through the Øresund University – a network involving all the existing institutions of higher education in the region. The economic effects of the creation of the Øresund region are likely to be long-term ones, but the new region quickly established an identity for itself: the strait separating the two regions is no longer a barrier but a means of communication and integration. Moreover, there is evidence that it has

Plate 4.2 ***Transnational infrastructure connections: the Øresund Bridge between Denmark and Sweden***

Source: Øresundsbro Konsortiet

stimulated new investment in the key hi-tech sectors in the region and that there has been a modest growth in cross-border commuting (OECD 2003b).

The examples of the development of new forms of governance in diverse political conditions demonstrate the apparently wide appeal of decentralised solutions to the challenge of local and regional development. The different examples of Emilia Romagna, the ABC region and Øresund each demonstrate, to varying degrees, evidence of the shift from government to governance. In the latter case, the emergent network forms of governance extend across national borders. Determining the actual impact of the new regional governance is fraught with difficulty because it requires us to isolate the effects of one actor or set of actors among others. Moreover, as we noted earlier, while there is some evidence that devolution has been accompanied by a growth of regional inequality, there can be little doubt about the general trend towards decentralisation. The regional scale has emerged as an important one in the governance of local and regional development.

The growth of the region as a sphere of governance, however, has not been universally applauded. Some writers see the trend towards devolution as little more than an instrument of neo-liberalism. Lovering argues that 'new regional structures' must be seen as part of a project 'to dismantle national redistributive structures and hollow out the democratic content of economic governance' (Lovering 1999: 392). Morgan, however, cautions against a 'functionalist and reductionist view of regionalism' (Morgan

2004: 874). The local and regional scales of activity, then, are arenas of conflict and contest in much the way that the national and supranational scale remains characterised by politics and negotiation. Moreover, the local and regional levels remain an important terrain for action to create progressive, inclusive and sustainable forms of policy – themes which we take up in the conclusions in Chapter 8. Local and regional governments have frequently proved to be pioneers in the development of new fields of public policy. For instance, local and regional governments have been at the fore-front of the search for more sustainable forms of local and regional development (Example 4.2).

Example 4.2 The constitutional duty to promote sustainable development in Wales

The National Assembly for Wales was established as part of the devolution and constitutional reform in the United Kingdom during the late 1990s. Amidst hopes of political renewal, an opportunity has been taken to embed sustainable development in the very workings of this new political institution. Legally, this initiative became manifest in the obligation to promote sustainable development imposed on the Assembly through Section 121 of the Government of Wales Act in 1998. This constitutional duty was a first in the European context. An aspirational strategy – *Learning to Live Differently* – was published in 2000. The aim has been to approach the economic, social and environmental dimensions of sustainable local and regional development in a more integrated way. Early initiatives focused upon health, public procurement and sustainable agriculture. For example, the Welsh Assembly Government (WAG) – the Assembly's executive – has been attempting to shift the basis of the National Health Service in Wales towards health and well-being rather than illness and treatment, and to encourage a shared responsibility between the institutions that make up the public realm and individuals:

> because it recognises that the solution to problems like childhood obesity, for example, has less to do with the health service per se and more to do with providing nutritious school meals and safe routes to school, thereby helping children to acquire healthy eating habits *and* encouraging them to walk in car-free environments.
>
> (Morgan 2004: 884–885)

While laudable, this constitutional innovation for a subnational political institution has raised questions over whether the regional scale of government has the competence, especially in legal terms, to promote and deliver on the potentially radical changes suggested by more sustainable forms of local and regional development. European and national level policies and regulations provide both the barriers and opportunities to frame the delivery of more sustainable forms of development at the local and regional scales. For the WAG, its relatively modest powers may need to be deployed in tandem with institutions at the European and national scale within a multilayered system of government and governance.

Sources: National Assembly for Wales (2000); Morgan (2004)

One danger of the debate about novel forms of local and regional governance, though, is that they detract attention from the enduring role of the nation state and national governments in the framing of public policy. It has been an assumption of some writing on the 'new regionalism' that the combined consequences of globalisation and devolution have rendered the nation state redundant (Ohmae 1990). Yet such an assumption is not borne out in reality. Despite processes of globalisation, privatisation and declining trust, there are strong reasons to believe that governments retain an important role in shaping social and economic structures.

Globalisation creates new challenges which cannot be dealt with by single nation states acting alone. According to Held *et al.* (1999), 'national communities of fate' now form an element of 'overlapping communities of fate', where the prospects of nations are increasingly intertwined with one another. This does not mean that states and national governments are irrelevant but, instead, remain powerful stratifying forces in the world, especially in terms of managing the distributive consequences of globalisation.

The activities of governments, manifest in different welfare state regimes, continue to have a central impact on economic growth, on inequality within nation states and between nation states. Thus, national politics still count and hold the potential to make a difference, as does effective state management and intelligent leadership. National, regional and local social partnerships still play an important role in limiting inequality, empowering citizens through education and learning and improving competitiveness (Ó Riain 2004). We return to these political questions in the conclusions in Chapter 8.

At the same time, the role of international regulatory bodies has grown in importance along with questions about the accountability, transparency and efficacy of their activities (Stiglitz 2002). Such bodies have been central to constructing the liberalised global economy resting on the assumption that this brings general benefits to all economies. But, as we discussed in Chapter 1, the impact of globalisation has been highly uneven for different social groups and countries. There is evidence that liberalisation can be generally good for the world's poorest countries; but that rapidly liberalising capital flows can damage economic prosperity, increase inequality and limit the life chances of the poor. In this context, new regulatory policies at the international level are important, but so also are the activities of states to intervene to promote local and regional development and build their capacity to manage their national economies, invest in infrastructure, improve human capital and practise selectivity in their approach to foreign direct investment. Interventions, instruments and policies are the subject of Part III of this book.

Multilevel governance and local and regional development

Local and regional development now occurs within the context of a multilevel polity in which local, regional, national and supranational authorities and institutions all play a role (see Hooghe and Marks 2001; Bache and Flinders 2004). According to one analysis:

> the dispersion of governance across jurisdictions is both more efficient than, and normatively superior to, central state monopoly. They claim that

> governance must operate at multiple scales in order to capture variations in the territorial reach of policy externalities. Because externalities arising from the provision of public goods vary immensely – from planet-wide in the case of global warming to local in the case of many city services – so should the scale of governance. To internalize externalities, governance must be multilevel.
>
> (Marks and Hooghe 2004: 16)

There is a large academic literature addressing the emergence of multilevel governance, much of it concerned with its evolution in context of the twin political processes of European integration and devolution to subnational authorities, notably in fields such as regional policy (Benz and Burkhard 1999). The emergence of multilevel governance is linked in some analyses to the declining authority of the state and is 'manifested in a growing number of exchanges between subnational and transnational institutions, seemingly bypassing the state' and which have a 'non-hierarchical' character and which reputedly offer new opportunities for cities and regions (Peters and Pierre 2001: 131–132). It is a development which has been linked to the emergence of 'new public management', which allows 'each level of government to separate the political-democratic element from the managerial-service-producing-sector of government, and partly because these reforms have tended to relax the "command and control nature" of previous intergovernmental relationships' (Peters and Pierre 2001: 132). The complex nature of the emergent hierarchies of relations between institutions working at different scales is illustrated in Figure 4.6. Such developments raise new challenges and questions

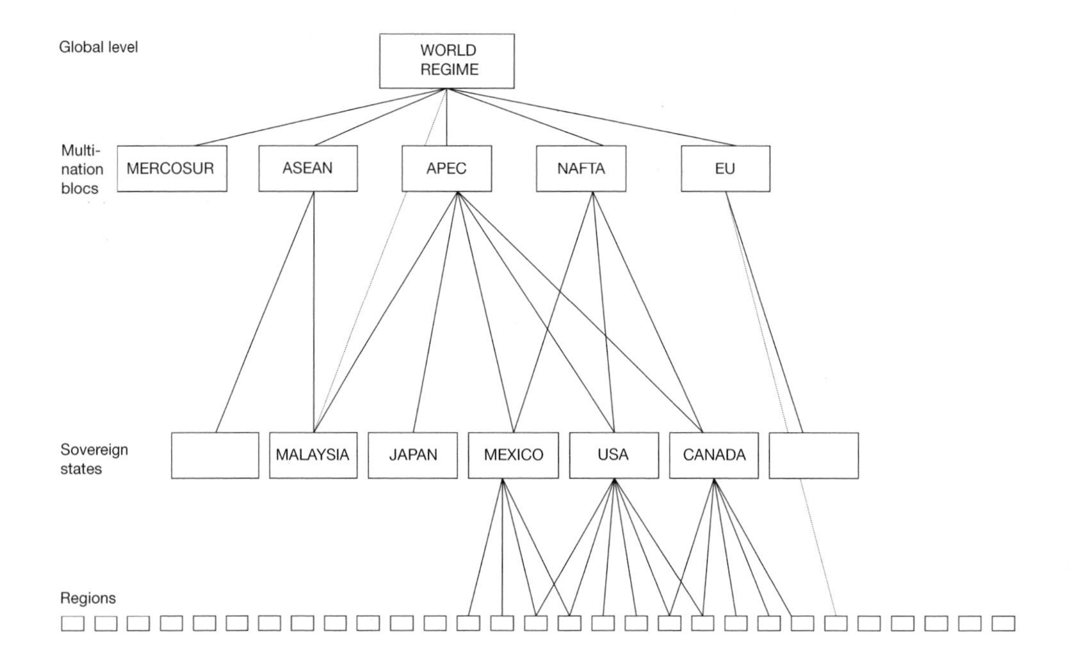

Figure 4.6 ***Fragment of an emerging global hierarchy of economic and political relations***
Source: Scott (1998: 138)

for governmental actors, although these vary between countries and sectors even within the European Union (Scharpf 1991).

At the supranational level, the regulatory framework of international trade is shaped by organisations such as the WTO or G8 (The 'Group of Eight' wealthy nations including Canada, France, Germany, Italy, Japan, Russia, the United Kingdom and the United States). At the same time, the European Union directly operates its own regional policy and plays a role in regulating the regional and other policies of its Member States (Amin and Tomaney 1995a; Rodríguez-Pose 2002b). However, national governments continue to play an important role notably in areas such as taxation, public expenditure and labour market regulation. In these areas the actions of central government can shape the prospects for local and regional development. As discussed above, the national tax and public expenditure system has the potential still to act as an important means of financial redistribution between richer and poorer regions. Some federal states such as Germany, Canada and Australia operate with explicit systems of fiscal equalisation, which transfer resources between prosperous and lagging regions (Owen Smith 1994; McLean 2004). Indeed, in states which lack transparent systems of regional expenditure allocation, such as in the United Kingdom, public expenditure patterns continue to have important regional effects and generate disputes about geographical equity (McLean 2005).

In this context, the 'new regional governance' plays a particular economic role in orchestrating the supply-side of the economy:

> an industrial policy oriented towards structural changes in the economy and the promotion of producers' adaptability to the conditions of domestic and international competition must focus on meso-level structures in the given economy and its social environment. Therefore, the creation of intermediate level structures that would facilitate economic restructuring is the top priority of industrial policy and the goal of the economic strategy.
>
> (Hausner 1995: 265; see also Tomaney 1996)

These interventions become especially important in places where, or time periods when, competitive success is not a simple function of low-cost and flexible labour, but instead requires constant upgrading of the production system. Thus:

> Where institutional guidance of market processes affects the level and character of the employment that is generated, full employment requires more than just aggregate demand management plus flexible markets; indeed it may critically depend on the state being able to conduct, for instance a fine-targeted active labour market policy of industrial policy. For this, the structure of public administration must be such that it can reach down into the networks that mediate exchanges in civil society, putting these to effective use. Where the state is not well-equipped for this purpose, the conscious design or reform of appropriate state, or para state, institutions may assume critical significance for employment.
>
> (Matzner and Streeck 1991: 8)

Example 4.3 The limits of local and regional entrepreneurialism

In a classic analysis, David Harvey sketched out the dangers of local entrepreneurialism:

> Many of the innovations and investments designed to make particular cities more attractive as cultural consumer centres have quickly been imitated elsewhere, thus rendering any competitive advantage within a system of cities ephemeral. How many successful convention centres, sport stadia, disney-worlds, harbour places and spectacular shopping malls can there be? Success is often short-lived or rendered moot by parallel or alternative innovations arising elsewhere. Local coalitions have no option, given the coercive laws of competition, except to keep ahead of the game thus engendering leap-frogging innovations in life styles, cultural forms, products and service mixes, even institutional and political forms if they are to survive. The result is a stimulating if often destructive maelstrom of urban-based cultural, political, production and consumption innovations. It is at this point we can identify albeit subterranean but nonetheless vital connection between the rise of urban entrepreneurialism and the post-modern penchant for design of urban fragments rather than comprehensive urban planning, for ephemerality and eclecticism of fashion and style rather than the search for enduring values, for quotation and fiction rather than invention and function, and, finally, for medium over message and image over substance.

Source: Harvey (1989a: 12–13)

The strengthening of local and regional governance can thus add to the capacity of public authorities to make effective interventions to promote economic development. Local and regional governance then plays an important role especially in ensuring improvements in the quality of the supply-side of the economy. But without national or increasingly supranational forms of regulation there is a danger that the 'new regionalism' may accelerate wasteful forms of territorial competition between places seeking to promote their own development at the expense of other places. Indeed, one danger of the general trend towards devolution is that it can create weak subnational governments that can do little more than engage in local or regional 'entrepreneurialism' that fuels territorial competition in a context where TNCs, ostensibly at least, can move investments easily between regions in search of the best returns (Example 4.3).

Democracy and local and regional development

The emergent forms of government and governance of local and regional development are also in part a response to more broadly perceived deficiencies in centralised forms of government and governance based on the nation state. Part of the attraction of devolution is that it appears to bring government closer to the people and open spaces for new actors to influence and shape the priorities of local and regional development policy (Humphrey and Shaw 2004; O'Brien *et al.* 2004; O'Brien 2004; Pike and Tomaney 2004; Rogers 2004). This seems important at a time when public faith in the capacities of national governments seems to be declining. The limitations of traditional models of representative democracy are alleged to be reflected in falling voter turnout in elections and declining levels of trust in politicians and political institutions (Pharr and Putnam 2000; Cohen and Rogers 2003; Fung and Wright 2003). In this context, devolution is linked to a wider rethinking of democracy, which seeks to augment or move beyond the limits of representative democracy and in the direction of more participatory forms of government (Held 1993). In this respect, devolved government can be seen as a space in which new forms of accountability and even a 'new politics' might be tested (e.g. Morgan and Mungham 2000; Humphrey and Shaw 2004). In an influential peroration on contemporary politics, Giddens has argued:

> The downward pressure of globalization introduces not only the possibility but also the necessity of forms of democracy other than the orthodox voting process. Government can re-establish more direct contact with citizens and citizens with government through 'experiments with democracy' – local direct democracy, electronic referenda, citizens' juries and other possibilities. These won't substitute for the normal voting mechanism in local and central government, but could be an enduring complement to them.
>
> (Giddens 1998: 75)

The shift from government to governance implies the influence of new actors over policy decisions as networks are extended to include new representations. Governance requires inputs from a wide range of actors, including non-governmental organisations, in an attempt to achieve policy goals. Social partnerships, involving statutory organisations and voluntary and community sectors, representatives and active citizens, are central to this concept of governance and have helped to set social and economic priorities (McCall and Williamson 2001).

The nature of contemporary politics, however, presents structural difficulties in the achievement of sustainable local and regional development. Most efforts to promote local and regional governance are likely to be effective over the long term, maybe taking a generation to have an effect. Yet political cycles have a shorter-term character, typically four or five years. One means of tackling this problem is to create planning mechanisms which stand outside – at least directly – the normal political cycle. The creation of social partnership arrangements charged with developing long-term economic strategies is a feature of some nations and regions. As we discuss in the case studies in Chapter 7, the role of 'national partnership' arrangements is credited in Ireland by some commentators with an important role in the emergence of the 'Celtic Tiger' phenomenon of sustained high growth in the 1990s (O'Donnell 2004). However, such governance arrangements can also be criticised for removing key decisions even further from direct democratic control. While the Irish authorities have responded to this problem by extending the national partnership to a wider range of civic actors, the danger remains that the governance of economic development retreats into technocratic networks which are even more impenetrable to the citizen. This raises questions about the value and function of the democratic leadership traditionally provided by politicians and political parties (Olsson 1998).

As Robert Dahl (2000: 113) notes, representative democracy has always had a 'dark side', which involves citizens delegating discretionary authority over decisions of extraordinary importance to political and bureaucratic elites. The institutions of representative democracy, such as parliaments and regional and local assemblies, have traditionally placed limits on elite bargaining as the main motor of politics. However, developments such as the 'new public management' and 'multilevel governance' may mean that more political power is removed from the purview of democratically accountable representatives. Crouch (2004) goes so far as to identify the emergence of an era of 'post-democracy' in the industrial countries, in which structures of formal democracy remain in place but the growing role of the private sector in the provision of government services means that important political decisions move beyond the direct influence of the citizen (see also Ringen 2004). We return to these important themes in the conclusions in Chapter 8.

Without strong democratic leadership, an additional danger is that debate about social, economic and environmental priorities are increasingly shaped by the growth of powerful news media. Indeed, according to Mayer (2002), we can observe the emergence of a 'media democracy' in which the media 'colonise politics' and politicians collude in the process. Between them politicians and the media are then reduced to adopting a narrow range of well-tested formulas in constructing their responses to policy challenges. Political debate becomes fixed in a closed world of media-savvy political elites, pollsters and media executives driven by the exigencies of twenty-four-hour news gathering. The traditional forms of democratic deliberation such as political parties, business associations and trade unions and, indeed, parliaments find themselves increasingly marginalised in the political process (see also Crouch 2004). Moreover, the media have been accused of not reporting debate and dissent, but 'manufacturing' it in cahoots with short-lived or single-issue campaigns (Milne 2005). Democratic deliberation, then, is a casualty of media democracy.

Mayer is one of many who argue that democratic societies need urgently to find new means of deliberation that generate long-term and sustainable solutions to problems of development in ways which enhance democracy (see also Crouch 2004). Fung and Wright maintain that the traditional form of representative democracy:

> seems ineffective in accomplishing the central ideals of democratic politics: facilitating active political involvement of the citizenry, forging political consensus through dialogue, devising and implementing public policies that ground a productive economy and a healthy society, and, in a more radical egalitarian version of the democratic ideal, assuring that all citizens benefit from the nation's wealth.
>
> (Fung and Wright 2003: 3)

This type of thinking has generated some experiments in what Fung and Wright (2003) describe as 'empowered participatory governance'. These experiments 'are participatory because they rely upon the commitment and capacities of ordinary people to make sensible decisions through reasoned deliberation and empowered because they attempt to tie action to discussion' (Fung and Wright 2003: 5).

Among the well-known examples of these new forms of governance is the participatory budgeting system that was pioneered in Porto Alegre, a city of 1.3 million and the capital of the state of Rio Grande do Sul in Brazil. The system was introduced by the Workers' Party after it gained political control of the city in the 1990s. Some sixteen local assemblies meet twice a year to settle budgetary issues. The assemblies involve municipal officials, community and youth groups and any interested citizen. The municipal government facilitates the process, but it involves numerous neighbourhood meetings which discuss the priorities for the city budget. A Participatory Budgeting Council processes this information and formulates a city-wide budget from the local agendas. The role of municipal officials is to inform and assist in the process but citizens and local groups determine the budget which is then submitted to the mayor for their approval (Fung and Wright 2003). The success of participatory budgeting has seen its introduction in other parts of Brazil such as Santo André in the ABC region (Acioly 2002).

In India, similar decentralisation initiatives – this time to villages in rural areas (or panchayats) – are the hallmarks of states such as West Bengal and Kerala, with populations of 80 million and 32 million respectively. Here, responsibility for many development programmes and their budgets have been decentralised to village assemblies. In Kerala, this process is assisted by a 'Voluntary Technical Corps' comprising mainly retired professionals who assist in the preparation of development plans. In both Brazil and India, such participatory forms of local governance have not been problem free. Some communities prove more adept at adapting to participatory governance than others and there are unresolved tensions between the power of technocracies and citizens. Nevertheless, they represent radical innovations that have assisted the development of at least some localities, by ensuring that local policies are more genuinely encompassing of local and regional needs and aspirations.

Efforts to develop deliberative local institutions are not restricted to developing or newly industrialising countries. In the United States, the Center on Wisconsin Strategy (COWS) and the linked Wisconsin Regional Training Partnership (WRTP) provide an example of deliberative policy-making (Vigor 2002). COWS is a 'think-and-do-tank' based at the University of Wisconsin-Madison, which advocates the 'high road' economic strategy we discussed in Chapter 2. It seeks to achieve productivity improvements in industry through workplace democratisation and diversified, quality production rather than competition based on low skills and price. The state's development priorities are based on a strategy developed by researchers, unions, business and the state government. Such cooperative behaviour between state, unions and business is unusual in the US context and COWS as a result has attracted much attention. In part the initiative emerged as a response to the 'workfare' agenda of Wisconsin's Republican governor Tommy G. Thompson during the 1990s, which was viewed by the trade unions as eroding the position of working people.

The model of enhanced participatory democracy is underpinned by a set of principles that, among other things, place a value on communicative forms of action and deliberation, on enhancing civic engagement through the involvement of secondary associations in policy development and a 'rethinking of democracy' in ways which make it more participatory. The new forms of participatory governance exhibit a strong practical orientation and place an emphasis on genuinely bottom-up forms of participation and deliberative forms of decision-making in which participants listen to each other's positions and generate choices after due consideration (Fung and Wright 2003). Achieving this form of decision-making is challenging, particularly in long-established systems of government and governance. Moreover, much of the experimentation with these new forms of participatory government appears to be taking place at the local or regional level, where questions about the quality of development seem pressing.

Conclusion

Rumours of the death of government have been grossly exaggerated, notwithstanding claims about 'governing without government'. The role of governments, especially at the national level, in shaping patterns of local and regional development remains very important, particularly in relation to decisions about taxing, spending and redistribution. Example 4.4 provides some reflections on importance of government from the United Kingdom's recent experience. The internationalisation of the economy, hollowing out and new regionalism places constraints on governments and shapes what they can achieve and most national governments must work alongside supranational and subnational governments in systems of multilevel governance, but this does not mean that government no longer matters. Likewise, partnerships have become more important as a mode of governance, but the nature of these partnerships is shaped in large measure by choices made by governments. Partnerships can range from little more than political cover for strategies of privatisation to genuine efforts to empower communities and citizens.

Government is about political choices. Although there are limits to the choices governments can make, this is nothing new. Moreover, the range of choices available to governments is greater than that implied by some of the simpler versions of the globalisation thesis. The largest challenge to government comes from the disillusion that many citizens feel about the operation of traditional forms of representative democracy, fuelled by the idea that all governments are the same and are too closely linked to powerful vested interests to have the concerns of citizens at heart. The German novelist and Nobel Laureate Günter Grass, reviewing the state of German democracy, has argued:

> Now, I believe that our freely elected members of parliament are no longer free to decide. The customary party pressures, for which there may well be reasons, are not critical here; it is, rather, the ring of lobbyists with their multifarious interests that constricts and influences the Federal parliament and its democratically elected members, placing them under pressure and forcing them into disharmony, even when framing and deciding the content of laws. Favours

Example 4.4 The importance of government

Reflecting on his role as a key adviser to the New Labour government in the United Kingdom between 1997 and 2005, Geoff Mulgan has argued:

> It is widely assumed that governments have lost power–upwards to a globalised market or Brussels, downwards to the people, or outwards to the private sector and the media. This is one of the reasons why social democratic governments have reined in their ambitions, and I expected to leave government more conscious of its constraints than of its possibilities. But instead I came away convinced that the perception of powerlessness is an illusion. Strong forces do limit government's room for manoeuvre: global markets and treaties impose limits on economic policy, and the media and business constrain government as much as churches and trade unions did a few decades ago. Yet the basic powers of governments have not diminished. The capacity to tax, for example, remains in rude health. Across the OECD, governments' share in GDP has risen over the past few decades; even the tax take (as opposed to the rates) on profits has gone up. Many of the world's most competitive economies are overseen by relatively big governments. Moreover, governments' ability to deal with problems like pollution and organised crime has been enhanced, not diminished, by globalisation. And while governments have reduced their roles in running economies – the vast bureaucracies that a generation ago were running nationalised industries have melted away – this retreat has been matched by a growing role in health, old age, childhood and security. The idea that governments have become impotent is an illusion, albeit one that can provide a useful alibi.

Source: Mulgan (2005: 24)

> minor and major smooth the way. Reprehensible scams are dismissed as sorry misdemeanours. No one any longer takes serious exception to what is now a sophisticated system, operating on the basis of reciprocal backhanders. Consequently, parliament is no longer sovereign in its decisions. It depends on powerful pressure groups – the banks and multinationals – which are not subject to any democratic control.
>
> (Grass 2005: 3)

While this disillusion is most evident in Europe and North America, there are signs that it is a malaise that is affecting even newer democracies. Finding a means of deepening and extending democratic processes is important for the health of society and can play a role in shaping the kinds of principles and values that underpin local and regional development strategies that we introduced in Chapter 2, although this task is far from straightforward. In the end, however, nothing replaces the importance of narratives of the 'good society' with their competing visions of social justice and environmental sustainability, whether at the international, national or regional and local level. We return to such issues in the conclusions in Chapter 8. Part III of the book addresses the interventions – instruments and policies – of local and regional development. We begin by addressing the mobilisation of indigenous potential in the next chapter.

Further reading

For a review of the changing nature of the state, government and governance, see Martin, R. and Sunley, P. (1997) 'The Post-Keynesian state and the space-economy', in R. Lee and J. Wills (eds) *Geographies of Economies*. London: Edward Arnold; Jessop, B. (1997) 'Capitalism and its future: remarks on regulation, government and governance', *Review of International Political Economy* 4(3): 561–581.

On nation states in the context of globalisation, see Dicken, P. (2003) *Global Shift: Reshaping the Global Economic Map in the 21st Century* (4th edn). London: Sage.

For a review of new regionalism, see Keating, M. (1998) *The New Regionalism in Western Europe: Territorial Restructuring and Political Change*. Cheltenham: Edward Elgar.

On the relationship between devolved government and governance and regional inequality, see Rodríguez-Pose, A. and Gill, N. (2005) 'On the "economic dividend" of devolution', *Regional Studies* 39(4): 405–420; Pike, A. and Tomaney, J. (2004) 'Sub-national governance and economic and social development', *Environment and Planning A* 36(12): 2091–2096.

On multilevel governance, see Bache, I. and Flinders, M. (2004) *Multi-level Governance*. Oxford: Oxford University Press.

On democratic renewal, see Fung, A. and Wright, E.O. (2003) 'Thinking about empowered participatory governance', in A. Fung and E.O. Wright (eds) *Deepening Democracy: Institutional Innovations in Empowered Participatory Governance*. London: Verso.

PART III

Interventions: instruments and policies

MOBILISING INDIGENOUS POTENTIAL

5

Introduction

Indigenous development is based upon the naturally occurring sources of economic potential growing from within localities and regions. Indigenous approaches are a means of nurturing 'home-grown' assets and resources that may be more locally and regionally embedded, perhaps more committed and less willing to divest, and more capable of making enduring and sustainable contributions to local and regional development. Strategies of indigenous local and regional development may seek to make places less dependent upon exogenous or external economic interests. Indigenous interventions connect directly to the bottom-up approach detailed in the Introduction in Chapter 1. They seek to work with existing assets and resources from the ground up to explore and unleash their potential for local and regional development. Beyond the superficial attractions of lower cost factors of production such as land and labour, traditional top-down and centralised approaches often overlooked or ignored the assets and resources deeply embedded in localities and regions. In the context of the heightened globalisation discussed in Chapter 1, the enhanced mobility of factors of production has arguably increased the significance of indigenous strategies that recognise the importance of places and their embedded characteristics. If capital and labour can, in theory, locate anywhere across the globe each may become more sensitive to differences in assets and resources between places.

Central to the indigenous approach is the idea of latent or somehow underutilised assets and resources that require mobilisation or stimulation to make or increase their substantive contributions to local and regional development (Goddard *et al.* 1979). New business ideas that go undeveloped for the want of advice, encouragement and support, firms that fail to grow and develop due to insufficient managerial expertise, and aspirations for education and training frustrated by the lack of local resources are examples of unfulfilled economic potential. Marshalling such resources in places can be fulfilling for individuals, households and social groups as well as significant for local and regional development. Realising the often untapped or underdeveloped potential of indigenous resources in localities and regions can be a formidable task. Barriers can include insufficient access to capital, limited local and regional markets as well as cultural traditions weakly disposed to entrepreneurialism, business formation and further education and learning.

This chapter examines the rationale for indigenous approaches to local and regional development and the tools aimed at capitalising upon indigenous or naturally occurring economic potential and promoting endogenous growth from within localities and regions. Connecting to the different 'Frameworks of understanding' in Part II, instruments and policies are addressed for establishing new businesses, growing and sustaining existing businesses and developing and upgrading labour. Conclusions are offered that reflect upon the potential and limitations of the indigenous approach.

Indigenous approaches to local and regional development

The indigenous approach connects to the concepts and theories of local and regional development discussed in Chapter 3. In their emphasis upon the resurgence of local and regional economies, transition models characterise this perspective as 'development from below' (Stöhr 1990). This view interprets indigenous approaches as bottom-up ways of growing and nurturing economic activities that are embedded in localities and regions. It seeks to draw upon the distinctive local and regional strengths and characteristics of dynamic industrial districts, especially their flexibility and adjustment capability. Policy interventions can promote the development of locally decentralised production networks, local agglomeration economies and local networks of trust, cooperation and competition as well as the local capacity to promote social learning and adaptation, innovation, entrepreneurship (Stöhr 1990; Pyke and Sengenberger 1992; Amin and Thrift 1995; Cooke and Morgan 1998; Crouch *et al.* 2001). Problems have emerged in attempts to implant simplistic versions of the industrial district model rather than adapting it to particular local and regional circumstances (Hudson *et al.* 1997; Storper 1997).

Institutionalist and socio-economic approaches overlap with theories of innovation, knowledge and learning to emphasise the importance of local and regional institutions in developing indigenous assets and resources and promoting adjustment capabilities in localities and regions (Bennett *et al.* 1990; Campbell 1990; Storper and Scott 1992; Amin and Thrift 1995; Scott 2004). Institutions – both formal such as organisations and informal such as networks – can mobilise potential assets, promote innovation and shape local and regional supply-side characteristics (Cooke and Morgan 1998). Institutions are also central to explanations of why localities and regions have failed to adapt and overcome the 'lock-ins' that can inhibit growth trajectories (Grabher 1993; Cooke 1997; Wolfe and Gertler 2002). Recognising particular structural problems and distinctive assets in place, and developing context-sensitive policy instruments have long traditions (Hirschman 1958; Seers 1967). Experiments are encouraged, drawing upon more interactive and consultative forms of policy-making, and new institutions are built to develop joint working and partnership to address shared problems (Morgan and Henderson 2002). Interventions may be microeconomic and focus upon the supply-side in combining 'hard' infrastructures, such as broadband telecommunications links, with 'soft' support for networking and knowledge transfer to build innovation capacity, encourage new business establishment and existing business growth, and to foster collective knowledge creation, application and learning (Morgan 1997).

The new endogenous growth theories have contributed much to the renewal of thinking about the potential of indigenous local and regional development. The emphasis upon endogenous growth *from within* localities and regions opens up opportunities for policy intervention. Indigenous approaches have begun to explore the ways policy may shape the local and regional external economies central to increasing returns and endogenous growth. These include Alfred Marshall's labour market pooling, specialist supplier availability and technological knowledge spillovers that were discussed in Chapter 3. The possibility of creating and building competitive advantage and localised clusters of economic activity (Porter 2000), rather than simply relying upon the comparative advantage of inherited factor endowments, has furthered interest in indigenous approaches. Policy interventions guided by these concepts target market failures for land, capital and labour (Bennett and Robson 2000), seek to ensure the clear communication and response to market signals (Acs and Storey 2004), and emphasise human capital, innovation and technological development. Endogenous approaches highlight the need for appropriate balances between localisation to promote externalities and agglomeration economies *and* external connection to national and international flows of goods, services and knowledge (Martin and Sunley 1998). In addition, as we discussed in Chapters 3 and 4, the growth-oriented focus of current 'new regionalist' local and regional policy places a premium upon the contribution of indigenous development in increasing the economic performance of every region and raises the issue of territorial equity and balanced local and regional development (Scott and Storper 2003).

Sustainable development resonates strongly with indigenous and grass-roots local and regional development (Haughton and Counsell 2004; Morgan 2004). The development of policy interventions more sensitive to the relationships between economic, social and ecological issues and with a longer-term outlook is increasingly evident. Business can be interpreted as a contributor to economic growth and employment creation within mainstream markets and/or a 'social enterprise' with broader social, economic and environmental aims capable of tackling localised disadvantage (Amin *et al.* 2002; Beer *et al.* 2003). The promotion of more sustainable stewardship of indigenous assets and resources in ways that encourage locally and regionally appropriate and sustainable forms of local and regional development has been sought. 'Weak' sustainable development policy interventions include the use of environmental regulation and standards to develop new businesses, local trading networks and ecological taxes on energy, resource use and pollution (Hines 2000; Gibbs 2002; Roberts 2004). 'Strong' sustainable development has promoted policy interventions seeking small-scale, decentralised and localised forms of social organisation that promote self-reliance and mutual aid (Chatterton 2002).

The aspirations of post-development in localities and regions chimes with indigenous approaches in its emphasis upon empowered, grass-roots leadership and nationally, regionally and locally appropriate and determined forms of development (Gibson-Graham 2000). Post-development proposes radical bottom-up and grass-roots approaches to local and regional development that present alternatives to the imposition of top-down models developed and imposed by external interests. Locally and regionally derived and led policy interventions have sought to nurture more 'diverse' or varied economies better suited to the particular economic and social needs and aspirations of localities and regions (Gibson-Graham 2004). Initiatives include Local Exchange Trading Schemes, time-banks, social enterprises and intermediate and secondary markets for labour, goods and services (Leyshon *et al.* 2003).

In sum, indigenous approaches attempt to work with the grain of local and regional economies rather than attempting to implant unconnected assets and resources with little existing linkages within particular places. Sustainable development is central to indigenous interventions in recognising the relationships between economy, society and ecology, the benefits of smaller-scale and incremental development and embedding longer-term approaches more appropriate to the development aspirations and needs of localities and regions. Local and regional institutions play a central role in indigenous policy in gathering and interpreting local and regional knowledge of economic and social needs and contexts, building relationships, working closely and providing ongoing support as well as building networks and mutual learning and cooperation among peers. Formal institutions can include the public sector as well as the social partners of organised labour unions and business associations, and civil society.

To provide the foundations for more context-sensitive policy, indigenous interventions are founded upon detailed assessment and understanding of how particular local and regional economies actually work. They retain the potential for selectivity and targeting in developing initiatives to address particular needs, for example tapping into the entrepreneurialism of black and ethnic minority groups, addressing the vocational training needs of young people or meeting the needs of specific disadvantaged communities. Using the distinction from Chapter 2, the objects of indigenous interventions are individuals, entrepreneurs, micro-businesses and SMEs and the subjects are establishing new businesses, growing and sustaining existing businesses and developing and upgrading labour. The following sections examine the instruments and policies deployed in the indigenous approach to local and regional development.

Establishing new businesses

The creation of new businesses is a fundamental element of indigenous local and regional development. Establishing new businesses is an important way of fostering economic activity and tapping into underutilised resources in localities and regions. Enterprise – the readiness to embark upon new ventures with boldness and enthusiasm – and entrepreneurialism – the ability to seek profits through risk and initiative – are important assets and resources with the potential to contribute to economic growth, income generation and job creation (Armstrong and Taylor 2000). Entrepreneurs can identify opportunities and resources that are currently yielding low or non-existent returns and shift them into higher-return activities through establishing new businesses, increasing efficiency through more optimal resource allocation and intensifying competition (Acs and Storey 2004). Social enterprise with broader social, economic and/or environmental aims, for example bringing formerly marginalised groups into education, training and work or improving degraded landscapes, has become increasingly important as a route to new business creation (Beer *et al.* 2003). Self-employment can be an important source of 'self-help' in places where labour market opportunities among existing employers are limited or low quality.

The geography of business start-ups is highly uneven, hampering local and regional development prospects particularly in disadvantaged places. Table 5.1 shows the spatial disparities in 'start-up intensity' across industries by ranking the top and bottom ten localities and regions in Germany. The most prosperous places of Frankfurt, Munich, Hamburg and Düsseldorf and their surrounding regions have the most 'entrepreneurship capital' and perform markedly better than the less prosperous places with less entrepreneurship capital in the older industrial areas and former eastern Länder (Audretsch and Keilbach 2004). International comparisons reveal that new firm formation rates were broadly similar with rates in the highest performing regions at between two and four times that of the lowest performing regions (Reynolds *et al.* 1994). The highest rates of new firm formation are evident in urban regions with high proportions of employment in small firms and high rates of in-migration (Acs and Storey 2004). Prosperous and densely populated areas offer large and potentially diverse markets for goods and services, supporting a diversity of opportunities for new business establishment. Local traditions of business start-up and toleration of failure can encourage others and attract entrepreneurs from beyond the locality or region.

Table 5.1 ***Start-up intensity by locality and region in Germany, 1989–1992***

Rank	*Locality/region*	*Start-up intensity*
1	Munich, surrounding area	24.6
2	Düsseldorf, city	20.2
3	Hamburg, city	19.7
4	Offenbach, surrounding area	18.6
5	Wiesbaden, city	17.7
6	Starnberg	17.1
7	Munich, city	16.1
8	Frankfurt am Main, city	16.0
9	Hochtaunuskreis	15.9
10	Speyer, city	15.4
318	Lichtenfels	5.6
319	Trier-Saarburg	5.5
320	Herne, city	5.5
321	Graftschaft Bentheim	5.4
322	Höxter	5.3
323	Bremerhaven, city	5.3
324	Tirschenreuth	5.2
325	Coburg	5.2
326	Cuxhaven	5.2
327	Kusel	4.8

Source: Adapted from Audretsch and Keilbach (2004: 956)

Note: Start-ups per 1,000 population for all industries.

While analysis of the evidence of their impacts remains ambiguous and unclear (Acs and Storey 2004), policy interventions seeking to encourage the formation and establishment of new businesses extend across the range of factors of production central to getting new economic entities off the ground. For the international development policy organisation the OECD, entrepreneurship in localities and regions is shaped by the strength of the entrepreneurial culture, the set of framework conditions and the presence and quality of public support institutions and programmes (Table 5.2). Support services have moved towards more integrated programmes, rather than disconnected arrays of instruments, often using the 'one-stop-shop' approach where entrepreneurs and fledgling firms can access a single institution for advice, information and networks (Armstrong and Taylor 2000).

In seeking to influence the sources of new entrepreneurs and businesses, the local and regional context of enterprise and entrepreneurialism is often the most difficult to shape. Places often have deep-rooted legacies and traditions that shape people's attitudes and beliefs, influencing their disposition towards starting new firms and being an employer with its inherent risks and responsibilities. Entrepreneurial propensity varies geographically (Acs and Storey 2004). As we saw in Chapter 2, for example, more associative or cooperative rather than individualistic forms of entrepreneurialism are deemed more appropriate and likely to succeed in Wales due to its particular economic and social history (Cato 2004). Indeed, encouraging business start-ups is difficult in old industrial

Table 5.2 ***Foundations of entrepreneurial vitality***

Local entrepreneurial culture	*Attitudes to employment and enterprise*
Local framework conditions	Existence of role models of entrepreneurial behaviour
	Entrepreneurship skills
	Access to finance
	Education and training
	Exchange and cooperation networks
	Bureaucratic and administrative barriers
	Infrastructure, such as business sites and premises
Existing public policies	Influencing attitudes and motivation
	Advice, consultancy and information
	Education
	Training
	Access to finance
	Business ideas
	Facilitating acquisition of businesses from retiring entrepreneurs
	Premises provision
	Sales and export assistance
	Counselling, mentoring and peer support groups
	Networks, clusters and strategic alliance programmes
	Support for innovation, including university–industry linkages
	Regulatory and tax climate (including special zones)
	Community development

Source: Adapted from OECD (2005c: 4)

regions in which an 'employee' rather than 'employer' culture predominates due to the historical labour market dominance of large industrial employers. Drawing on the experience of Glasgow in Scotland, Checkland (1976) used the metaphor of the Upas Tree to describe this effect – an African tree whose wide branches prevent the sunlight reaching and nurturing growth beneath its canopy. Echoing these sentiments, research in the United Kingdom suggests policy measures encouraging individuals with limited human capital to start businesses may have negligible effects in low-enterprise areas (van Stel and Storey 2004). Such 'destructive' entrepreneurship can displace existing businesses through low-price competition but ultimately prove unviable (Acs and Storey 2004).

Identifying and stimulating potential openings for new businesses and encouraging individuals and groups to develop their business ideas are central to entrepreneurship policy (Armstrong and Taylor 2000). Unmet local and regional needs in new market segments or areas where current provision is poorly performing or weak may provide the opportunities for new businesses. Exploiting the benefits of new innovations and technologies may be another source of new economic activities. Linking to the supply chain needs of new inward investment projects may stimulate market demands for goods and services. Examples may include affordable childcare provision, component subcontracting, personal services, recycling and internet-based services. Recent policy attention has also focused on the potential to create local markets through the procurement of goods and services from expenditure by public bodies such as local governments, hospitals and schools (Morgan 2004). Social businesses in disadvantaged communities can be established with targeted support to tap into such public sector contracts.

Gaining access to capital is a critical starting point for new businesses. The fundamental elements of premises, equipment, materials, staff and working capital all require initial financing. Mirroring the uneven geographies of business start-ups, access to capital is marked by local and regional disparities and finance gaps for specific levels, sizes and sectors of firms and types of funding (Mason and Harrison 2002). Table 5.3 illustrates the situation in the United Kingdom and the manifestation of the high degree of spatial centralisation of financial institutions in and around London and its south-eastern bias in the regional distribution of investment funds (Klagge and Martin 2005). Across each of the early stages of business development, the share of venture capital investment in London and the neighbouring South East region is above the level that would be expected by the respective shares of the national stock of businesses in these regions. Scotland benefits from its relatively autonomous subnational financial system but the lagging northern and western regions experience a dearth of venture capital relative to their business base. Prosperous regions often benefit from higher value land and property prices that can act as collateral against which to borrow investment capital. In declining localities and regions, redundancy payments have supported the enterprise of 'reluctant entrepreneurs' often pushed into self-employment by the lack of alternative jobs (Turner and Gregory 1996).

Where capital is lacking and/or the aspirations and purpose of the business is both economic *and* social, more diverse ownership structures can provide routes to indigenous local and regional development. Cooperatives, mutuals, community and employee ownership are forms of collective ownership of assets that can be used to exercise local

Table 5.3 ***Regional structure of the venture capital market in the United Kingdom, 1998–2002***

	Location quotient			
Region	*Early stage*	*Expansions*	*MBO/MBI*	*Total*
London	2.07	1.56	2.03	2.02
South East	1.37	1.12	1.20	1.17
Eastern	1.18	0.62	0.63	0.70
South West	0.44	0.35	0.50	0.41
East Midlands	0.46	1.21	1.31	0.99
West Midlands	0.48	0.90	1.00	0.90
Yorkshire–Humberside	0.47	0.64	0.64	0.61
North West–Merseyside	0.75	1.80	0.75	0.83
North East	0.40	0.64	0.60	0.54
Wales	0.10	0.28	0.21	0.18
Scotland	1.24	1.25	0.55	1.02
Northern Ireland	0.39	0.02	0.09	0.15
Total	1.00	1.00	1.00	1.00

Source: Adapted from Klagge and Martin's (2005: 405) analysis of British Venture Capital Association data

Notes: Location quotient (LQ) defined as a region's share of national venture capital investment divided by region's share of the national stock of VAT-registered businesses (for 2001). Values greater than unity indicate a relative concentration of venture capital investment in the regions concerned. MBO – Management buy-out; MBI – management buy-in.

control, retain surpluses and orient businesses towards the workforce and needs of localities and regions (Wills 1998; Cato 2004). Without the need to distribute returns as dividends to external shareholders, financial surpluses can be recirculated among potentially more committed local owners and employees and reinvested in growing and sustaining local businesses.

Other alternative approaches to stimulating and establishing new economic activities outside the mainstream monetary or cash economy include Local Exchange Trading Schemes. LETS are labour exchange systems established by local communities to facilitate economic activity in the absence or shortage of the national currency (Williams 1996). Members of a LETS agree to provide their service to one another, such as accountancy, babysitting or painting and decorating. When they perform the service their labour is logged in a central accounting system as a credit under their name – they are then owed by the rest of the members of the system the number of hours they performed. Each member chooses the type of labour in which they wish to be remunerated. Skilled and unskilled labour may be weighted differently depending upon the preferences of the organisers and participants of the system. LETS have the potential to benefit time-rich but perhaps cash-poor local businesses by improving their access to human capital. LETS and other local currency systems have experienced some successes and contributed to local and regional development (North 2005). Problems may include initially low participation rates and longer-term sustainability.

The right type and skill level of labour are critical to new businesses. At the outset, entrepreneurs and owner-managers may be able to muddle through the early stages

Plate 5.1 ***Growing indigenous businesses: a small firm incubator in Eindhoven, the Netherlands***

Source: Photograph by David Charles

of development but as businesses expand functional divisions of labour and technical expertise become more important (Campbell *et al.* 1998). Local labour market information and matching as well as professional support and skills upgrading can be supported by targeted policy interventions. However, existing entrepreneurship strategies commonly favour those individuals who already possess superior financial, human and social assets. Indigenous labour market policy interventions are explored in more detail below.

To root and support new business formation in place, policy interventions have utilised property-based approaches. The provision of appropriate sites and premises linked to the vital infrastructures of telecommunications and transport can provide the local and regional spaces for entrepreneurship. As part of the capital stock in neo-classical theory, such elements are integral factors of production. Business incubators, for example, aim to improve survival rates for new business by providing common services, infrastructure and peer group support in affordable, often subsidised, and shared accommodation. Once established, businesses then leave for more appropriate premises and the incubation process begins again. Incubators connect with indigenous local and regional development by targeting particular activities, places and/or social groups, for example marginalised groups such as women, ethnic minorities or youth and places such as inner cities, peripheral housing estates or rural areas. Example 5.1 describes an incubator initiative for black and ethnic minority women in Cincinnati in the United States.

Example 5.1 The Cincinnati Minority and Female Business Incubator, United States

The Cincinnati Minority and Female Business Incubator was first organised in 1989 to address the small business development needs of minority and women small business owners. It was launched by the city's economic development corporation inside a pre-existing, specially demarcated 'empowerment zone' where resident businesses are eligible to access targeted tax incentives, loans and federally supported business programmes. The incubator provides young entrepreneurs with simple but essential business support services including developing company letterheads, sending email, and creating websites for inter-facing with clients. It also helps them find financial resources through public and private sector programmes. These services are delivered to tenants through a modular workshop series whose titles include 'Creating a Profit', 'Business Skills' and 'Building a Profit'. Classes are available in a traditional classroom setting with an instructor, or as 'tele-classes' over the phone and on-line at the entrepreneur's convenience through a business develop-ment website. The programme uses indicators to monitor service delivery to allow programme administrators constantly to improve the programme based on a steady stream of feedback. The economic development corporation's evaluation of the scheme reveals that within the last decade of operation the incubator has created 504 new jobs filled by low-income residents, 84 per cent of those businesses were minority owned and 43 per cent were female owned. This targeted initiative has achieved some success but business survival remains a concern and the scheme may reach the limits of the local pool of potential entrepreneurs over time.

Source: www.cbincubator.org

Overlapping with innovation and technology support, larger-scale property-based initiatives comprise research, science or technology parks for new businesses. Such policy initiatives are often themed in a bid to attract and develop the kinds of specialised 'clusters' of businesses discussed in Porter's theory of competitive advantage in Chapter 3, for example in biotechnology or aeronautics. For new businesses, these interventions are typically connected to the commercialisation and knowledge-transfer activities from universities and research centres (Kominos 2002). Common services may include high-quality infrastructures and shared facilities, consulting and commercial advice and tech-nology monitoring. Such initiatives have not always delivered, and may become diluted in their focus if a flow of tenant individuals and businesses in the targeted sectors fails to materialise (Massey *et al.* 1992).

The growing sophistication and knowledge-intensity of economic activities is reflected in the increasing emphasis upon innovation and technology support within indigenous approaches to new business formation. Connecting existing businesses to local and regional innovation systems of the kind discussed in Chapter 3 have provided a focus for policy interventions (Braczyk *et al.* 1998). Securing intellectual property rights, patents, licensing opportunities and other legal safeguards are key policy elements in supporting the exploitation of returns from new innovations. In the context of the priority given to innovation, knowledge and learning discussed in Chapter 3, spin-off businesses from universities and research centres, for example, are a key focus (Armstrong and Taylor 2000). Exploiting the potential of new technologies for local development is central. Policy initiatives have sought to capitalise on the success of emergent industrial districts such as 'Silicon Alley' in Manhattan, New York, based upon the new and fast growing service and media businesses developing from the digitisation and convergence of information and communication technologies through the internet (Indergaard 2004). The geographies of the new economy are not just spatially concentrated, however. More diverse and geographically dispersed geographies are evident, making such sectors less amenable to spatially targeted development policy (Cornford *et al.* 2000).

In parallel with the uneven geography of business start-ups, failure rates for new enterprises are locally and regionally uneven (Armstrong and Taylor 2000). High start-up and failure rates can suggest a buoyant local and regional economy with a stream of new business opportunities emerging and being explored. The net change and balance between start-up and failure rates reveals whether entrepreneurial activity creates net additions or subtractions to the stock of local and regional businesses. Once new firms are established they face the challenge of surviving through growing and sustaining the business. Indigenous policy has focused support on these stages of growth in a bid to maximise the contributions of fledgling and expanding businesses to local and regional development.

Growing and sustaining existing businesses

Approaches to indigenous development in localities and regions prioritise the development and expansion of existing businesses, especially micro and small and medium-sized enterprises. Table 5.4 illustrates the size ranges of such firms in terms of employment. Echoing our discussion in Chapter 3, small firms are central to the locally decentralised production networks in transition theory, the external economies in new endogenous growth theory, the small-scale and local forms of sustainable development and the diversity of firm and size and ownership structures envisaged in post-development.

Positive assessments of the role of micro-businesses and SMEs underline their dynamism and potential for employment growth, their ability rapidly to pick up on new demands and occupy market niches and their agility, flexibility and adaptability due to simpler organisational and decision-making structures (Birch 1981; Armstrong and Taylor 2000; Acs and Storey 2004). Micro-businesses and SMEs may be locally owned

Table 5.4 ***Firm sizes by employment***

Size	*Employment*
Micro	0–4
Small	5–49
Medium	50–249
Large	Over 250

Source: Adapted from Armstrong and Taylor (2000: 266)

and controlled, reducing dependence upon external economic interests, and they may have potentially higher degrees of local and regional embeddedness and commitment and loyalty to locality and region. At the local and regional level, a varied size structure of micro-businesses and SMEs can underpin diversified economic structures better able to absorb fluctuations in business and industrial cycles (Armstrong and Taylor 2000).

More critical views of micro-businesses and SMEs have emphasised their lack of scale and market power, their vulnerability to external shocks and market shifts, their subordinate positions within hierarchical supply chains dominated by larger firms, their limited financial resources and inability to invest in long-term projects such as R&D, and the often limited geographical scope of their markets (Harrison 1994; Rodríguez-Pose and Refolo 2003). Qualitatively, the nature and sometimes low quality of employment in small firms has been questioned (Acs and Storey 2004). In addition, some micro- and small firms may be 'lifestyle businesses' with limited growth ambitions beyond generating income streams for the business owners or making personal interests or hobbies financially sustainable (Dejonckheere *et al.* 2003). These businesses may deliver welcome but ultimately constrained contributions to indigenous local and regional development.

Local and regional policy interventions aimed at indigenous development have developed increasingly sophisticated approaches to supporting micro-business and SME growth and expansion. Policy is bedevilled by the unclear evidence of its impacts, however, and, without geographical targeting, its potentially spatially regressive effects in benefiting the already more prosperous places ahead of less prosperous localities and regions (Acs and Storey 2004). In the context of the heightened global competition outlined in Chapter 1, enhancing competitiveness has become dominant since the 1990s. This often combines initiatives to reduce costs – for example substituting technology for labour, rationalising organisational systems and processes – and expand markets – for example by supporting the development of more sophisticated and higher value-added goods and services. Cluster policy of the kind described in Chapter 3 has been used to support both types of approach. The ethos of intervention can encompass varying degrees of alignment with or shaping of market structures (Bennett and Robson 2000), for example supporting SMEs to make their own decisions about business strategy in their specific market contexts or encouraging radical innovation to influence market trends.

Research and the local and regional knowledge base on micro-, small and medium-sized business needs is critical in designing support programmes. Business support services can offer both generic advice, for example financial management and planning, as well as access to specialised expertise, for example on technological issues or bespoke business services. Policy can be selective and targeted towards specific areas and/or social groups. 'Development from below' approaches emphasise the importance of such 'real' services delivered by local and regional institutions. These services – such as those pioneered by the regional development agency ERVET in Emilia Romagna, Italy, discussed in Chapter 4 – are close to the ground and the user community and more capable of making tangible and meaningful contributions to local and regional business development and growth (Amin and Thrift 1995). Support for mainstream, market-oriented business is often distinguished from that for social enterprises with broader economic, social and environmental purposes.

The neo-classical vision of the capital market providing finance equally across space and different sizes and types of businesses we discussed in Chapter 3 is often markedly divergent from the experience of SMEs in localities and regions (Armstrong and Taylor 2000). Capital market failure and imperfection mean that growing and expanding micro-businesses and SMEs can experience many of the same difficulties and finance gaps of new businesses in gaining access to capital at affordable rates. Internal market imperfections occur when capital is not lent due to the existence of safer alternative investment opportunities elsewhere, the higher transaction costs of dealing with SMEs, the perceived relative risks of untried businesses and the scale and location of the business. Early stage businesses may still lack the sufficient track record of delivering returns on investment required by financial institutions to provide investment capital (Mason and Harrison 2002). External market imperfections arise from capital flight to sectors of the economy able to deliver relatively higher rates of return, for example property during a business cycle upswing.

Indigenous policy instruments have focused on debt or borrowings and equity or financial ownership as the two principal means of financing, especially as early stage businesses are often unable to draw upon retained earnings. The appropriateness of debt depends upon interest rates, loan terms and duration. The loan is typically secured on the assets of the borrower. Collateral and the viability of the business plan are the key criteria. Equity can be internal from owners or employees or external from financial institutions. Each can vary in their influence upon the business, from regular and passive monitoring of financial returns to more active involvement. Indigenous interventions to shape access to finance can include improving SME business plans and relationships with financial institutions and individual investors through effective brokerage and network building, providing grants or subsidised loans with high social returns locally and regionally and establishing financial institutions, programmes and markets with specific local and regional remits such as development banks and micro-finance initiatives (Armstrong and Taylor 2000; Beer *et al.* 2003; Klagge and Martin 2005). Social enterprises have tapped into local credit unions and other alternative sources of typically localised capital (Lee *et al.* 2004). The difficulties experienced by micro-businesses and SMEs in accessing capital have led to legislative changes to address discriminatory lending practices by financial institutions towards individuals and businesses. Notably, this includes the Community Reinvestment Act in the United States which requires depository institutions to invest specified amounts of capital in under-served sectors, groups, regions and businesses (Leyshon and Thrift 1995).

Skills and workforce development issues can become acute for growing and expanding micro-businesses and SMEs (Bennett and McCoshan 1993). Providing specialist and targeted training is a key area, particularly for owner managers having to deal with increasingly complex and growing businesses. Another is addressing labour market failures, particularly relating to skills shortage areas where the market is failing to reproduce skills and is stimulating labour poaching and wage inflation (Pike *et al.* 2000). Public intervention to kick-start or support the supply-side of the labour market in specific occupations or skill groups is a key policy instrument. Other initiatives can address recruitment difficulties since SMEs can struggle to offer competitive salaries and career

Plate 5.2 ***Supporting high-tech business: a technology park in Dortmund, Germany***
Source: Photograph by David Charles

development prospects compared to larger employers. We discuss indigenous approaches to upgrading and developing labour in more detail below.

Sites, premises and infrastructures remain critical for growing and expanding businesses. The kind of external economies central to endogenous growth theory can be generated by putting growing businesses together under one roof. Horizontal interventions can support cooperation among businesses, for example disseminating a new technology, and vertical actions which spread information and knowledge, for example managerial advice or common interest training programmes (Enright and Ffowcs-Williams 2001). The close proximity of firms can increase tacit learning, knowledge sharing and other productive relations that collectively improve the competitiveness of participant businesses (Storper 1997). This kind of policy intervention is a physical manifestation of the cluster policy discussed in Chapter 3. Managed workspaces, for example, target relatively established firms seeking access to subsidised premises and shared services. Similar to business incubators, managed workspaces can be targeted at specific activities, social groups or places for indigenous local and regional development.

Growing from their role in supporting the establishment of new businesses, research and science parks can underpin existing business growth through innovation and technology support. While international in their usage as indigenous policy interventions, research and science parks have been used as broader local and regional development

Example 5.2 The Bangalore technology hub, India

Descriptions abound of the oldest and best known science parks, those located around Stanford University in the United States and the University of Cambridge in the United Kingdom. A lesser known example is the burgeoning information technology hub of Bangalore, India, where rapid growth in software exports since the late 1990s has been partly due to government-owned and operated Software Technology Parks. Known by some as the 'Silicon Valley of India', the region is home to six interlinked parks within the state of Karnataka (Parthasarathy 2004). The parks are fed by research staff and technology developments from two major Bangalore and Karnataka regional universities and have facilitated the production of semiconductors and micro-processing chips for major multinational companies including Intel and Texas Instruments. The growth of the parks can be attributed partly to the policy support they receive from the national Ministry of Information Technology, especially legislation that promotes software exports by offering fiscal incentives to companies to encourage them to invest in telecommunications and other infrastructure. The dependent nature of relations within high-tech sectors between the nascent Silicon Valley's, including Bangalore, and the original Silicon Valley in California in the United States may shape its future development trajectory (Parthasarathy 2004).

Source: Meheroo and Taylor (2003)

tools to transfer technology, attract FDI, support R&D and create employment (Castells and Hall 1994). Such projects are significant in early stage industrialising countries, for example in South East Asian nations such as India's 'Silicon Valley', the Bangalore 'technology hub' (Example 5.2).

Developing and upgrading labour

Shaping the capabilities and skills of people is the other main element of indigenous local and regional development policy. People in localities and regions are a key resource given their potential ability to upgrade their skills and qualifications through education, training and development (Bennett and McCoshan 1993; Campbell *et al.* 1998). Increasing the productivity of labour is central to increasing incomes and living standards (Cypher and Dietz 2004). Education and training can increase the capabilities and capacities of labour – or 'human capital' – to accommodate new technologies and to innovate. Knowledge and skills are central to local and regional economic competitiveness and policy interventions (Keep and Mayhew 1999). Developing and upgrading labour occurs through the formal education process, from the school to the university level, and through learning by doing in the workplace.

The challenge for local and regional development policy in developing and upgrading labour is that there are marked geographical variations in the extent to which populations in localities and regions are engaged in institutions to develop and upgrade their qualifications and skills. Participation in advanced or tertiary-level education has an uneven regional distribution, for example, with the Slovak Republic exhibiting the largest regional variation in enrolment in tertiary or advanced education with a coefficient of variation of 0.88 (Figure 5.1). Moving away from the Keynesian emphasis upon managing the demand-side of the economy discussed in Chapter 3, recent indigenous initiatives have focused upon the local and regional supply-side of markets and have tended to steer clear of interventions to shape labour demand. Recent indigenous approaches focus upon developing and upgrading labour as a supply-side resource for existing businesses in localities and regions as well as a potential stimulus to entrepreneurship and new business start-ups.

Economic adjustment within local and regional labour markets often depends upon the ability of labour to adapt to ongoing change in labour demand as restructuring forces changes in employment levels, renders skills and knowledge obsolete and requires new competences and skills to be learnt continuously. 'Lifelong learning' describes this approach to ongoing skills development (Coffield 1999). Developing and upgrading indigenous labour can tap into the underutilised resources of workers who possess outdated or undervalued skills and cannot contribute their full potential to economic activity and growth because their skills are not in demand (Metcalf 1995). Besides not being an asset in the growth process, underutilisation denies labour the dignity of paid work, poses reproduction costs to society through welfare transfer payments and other social costs, and renders individuals, households and communities vulnerable to poverty and social exclusion (Campbell *et al.* 1998).

In neo-classical and, latterly, in endogenous growth theory, labour is understood as 'human capital' (Becker 1962) – an asset in which investment can be made and returns expected. 'Human resources' is another related term used (Bennett and McCoshan 1993). Developing and upgrading labour focuses on improving formal qualifications, skill levels and work experience to the benefit of individuals and the broader local labour pool as well as individual organisations in contributing to local and regional economic activity and growth. The central aim is raising the capacity of individuals and groups to develop and to use new innovations and technology to improve productivity. Local and regional institutions occupy a central role in indigenous approaches to developing and upgrading labour, from the level of schools up to the growing importance of workforce intermediaries – such as business associations, labour unions and other agencies – that operate

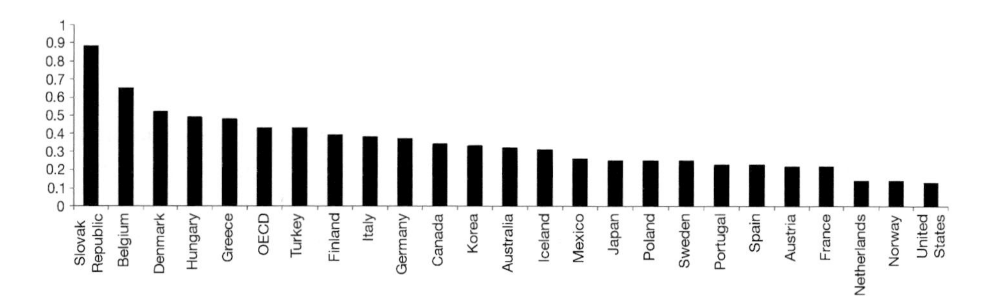

Figure 5.1 ***Regional variations in enrolment in tertiary education by country, 2001***

Source: Adapted from OECD (2005b: 146)

Note: Coefficient of variation, students per 100 population.

between state and market to develop workforce skills and provide support services (Bennett and McCoshan 1993; Benner 2003; Giloth 2003). In particular, local and regional employers in the private, public and voluntary sectors are interpreted as critical in understanding and articulating local and regional employment demands.

Numerous local and regional development policy approaches are used to promote labour market adjustment (Martin and Morrison 2003). Key issues concern demand- or supply-side orientations, public and/or private provision and the nature of interlinkages between industry and government, including educational institutions, which shape the adjustment capability of local and regional labour markets. Improved labour market information, for example, is a starting point for understanding local and regional labour market functioning, enhancing the amount of publicly available information on the local labour force, reducing information asymmetries, identifying skills gaps and revealing comparable skills capable of redeployment (Meadows 2001). Needs analysis and its connection with indigenous resources locally and regionally are often used as part of diagnostic techniques to better understand labour market change. Example 5.3 details such a framework and its utilisation for SME development in Germany.

Development and upgrading labour can have a broad ambit, including the promotion of business education at the school level within the public education system (Campbell *et al.* 1998). Raising aspirations and levels of educational attainment as well as encouraging entrepreneurialism as part of career development advice are examples. Too much business influence in schools may skew educational priorities, however, if students become too narrowly trained in accordance with the wishes of industry but without the adaptive ability think broadly, critically or creatively about problems outside the realm of commercial usefulness (Rodríguez-Pose 1998). Furthermore, business interests may have an economic rationale but socially perverse incentive to encourage educational institutions to create an overabundance of students educated in certain fields. This would depress employee earnings and intensify competition for scarce jobs. After two decades of school level entrepreneurship programmes in the United Kingdom, however, their effectiveness is weak (Greene 2002).

Strengthened further education systems have been sought to meet local and regional demand for vocational or specialist education and training, across the spectrum from general business studies to more specialised technical apprenticeships (Bennett and McCoshan 1993). In the context of the 'knowledge economy', the role of higher education and universities in local and regional development has received considerable recent attention developing and upgrading labour initiatives (Keep and Mayhew 1999), ranging from the need to retain graduates, particularly in lagging localities and regions, to the potential for advanced and postgraduate technical and managerial education, training and development.

Indigenous approaches use customised training programmes to improve the supply of skills where they are not being provided by the market and are constraining local and regional development. Such programmes may target groups experiencing chronic or temporary unemployment like women, youth, or older workers. Initiatives may seek to develop entirely new skill sets, readapt or modernise old skills, raise skills for entry-level work, or break the cycle of long-term unemployment through pre-employment training and intermediate labour markets (Belt 2003). Local governments in jurisdictions

Example 5.3 The competencies-based economies formation of enterprise for SME training, Germany

Training programmes for SMEs have been used extensively as local and regional development tools internationally. The German Foundation for International Development runs a comprehensive training programme called the 'competencies-based economies formation of enterprise' (CEFE) programme. It seeks to improve the performance of local SMEs through entrepreneurial skill building, self-analysis and encouragement of enterprising behaviour. The programme has been used in over sixty countries and works through four interlinked components that allow it to adapt and respond to local conditions. Each of the modules is interactive and participatory. Phase one draws in local residents with entrepreneurial interests who may be suited for participation in the programme, while at the same time letting the programme staff familiarise themselves with local cultural norms and the economic environment. Phase two assesses the needs of groups in the community – identified by their ethnicity, age, former occupation, educational background, the size of their business and so on – for targeted training in business development. In the third phase, complementary indigenous resources are identified by the programme managers to assist in the business development programme, for example local government or other donor agencies familiar to local businesses. In the final phase, candidates from the identified group are screened to select those business people with the strongest potential to contribute to the local economic fabric, and on whom resources will be most effectively spent. The designers and administrators of the training programme have found it especially effective for small businesses with one to ten employees, 98 per cent of whom claim that their income and turnover increased by at least 30 per cent after going through the CEFE training. The CEFE programme demonstrates two strengths of a human resources development programme. First, the degree of adaptability, as well as attention to local circumstances, is a key determinant of programme success. Second, human resources training is often best delivered as part of a larger business support scheme in which entrepreneurs are selected who will use the newly acquired skills directly within their existing enterprises. In this way, training can complement the existing indigenous strengths of the local economy. As we noted above, however, such programmes focus upon those often most able to help themselves and not the much harder to reach individuals and groups in need of support.

Source: Fisher and Reuber (2000)

with high populations of recent immigrants may, for example, establish language training programmes to improve access to employment (Meadows 2001). Concerns exist, however, about the effectiveness of policy and public subsidy to encourage individuals with low human capital to enter self-employment, especially in low enterprise areas (Acs and Storey 2004).

While Keynesian attempts to manage the demand-side of the labour market through active policy have received relatively less attention recently, they are an option for indigenous local and regional development. Employment maintenance schemes, for example, provide temporary subsidies to companies to keep workers employed at full wage levels until alternative employment and training can be set up rather than making social payments to workers who have been made redundant and risking their fall into unemployment and inactivity. By subsidising the cost of labour, these schemes temporarily postpone redundancy and preserve employment levels, maintaining the social networks of support and expectations of professional conduct that form among employees in a workplace (McQuaid 1996). Such programmes connect with longer-term solutions to worker redundancy including reskilling and retraining (Bosch 1992).

Area-based indigenous approaches to developing and upgrading labour focus upon local labour markets (Martin and Morrison 2003). Territorially based employment pacts, for example, organise agreements between local employers and labour exchanges, often supported by public funding, to orient recruitment towards the local labour pool. This stimulates local demand and, in the longer term, can underpin a specialised local labour market and raise aspirations among local residents to participate in education and training. Once in employment, access to further training and development can be supported by local and regional institutions. More recently, however, labour market interventions, for example the New Deal employment subsidy in the United Kingdom, are often connected to broader reform of the welfare system and the targeted experiments with 'workfare' initiatives that link social policy to economic outcomes and promote supply-side 'employability' (Peck 2001; McQuaid *et al.* 2005).

Conclusion

Shaped by the conceptual understandings of 'development from below', 'grass-roots' and 'bottom-up' perspectives and endogenous growth from within localities and regions, indigenous approaches contain inherent flexibility and adaptability allowing their specific design to address particular local and regional circumstances. The context-sensitivity of indigenous development strategies in connecting to local and regional assets and resources are strengths but may require high degrees of policy adaptation and learning to meet local and regional development needs. Establishing new businesses, growing and sustaining existing businesses and developing and upgrading labour necessitates detailed knowledge of the places in which policy is deployed. Capacity building and empowerment is integral too, particularly for the effective coordination and integration of policies within comprehensive development frameworks by institutions of government and governance working across geographical scales within a multilevel system. Indigenous approaches may require lower-level but sustained and longer-term funding to support long-term strategies by local and regional institutions. The potential of indigenous approaches for connecting to the kinds of exogenous or externally oriented approaches to local and regional development are addressed in Chapter 6.

Indigenous approaches are not without their drawbacks. Such strategies are often relatively slow to yield substantive rewards. In quantitative terms, growing micro, small and medium-sized businesses may not create substantial numbers of new jobs in short time periods, certainly in comparison with the often high levels of claimed employment creation by new inward investment projects. However, as we shall discuss in the next chapter, jobs created in the local and regional indigenous sector might prove more sustainable in economic, social and environmental terms over time. Developing and upgrading labour is a long-term and ongoing task, sometimes resulting in enhanced labour mobility and a 'brain drain' effect through the outflow of the skilled and the qualified from lagging localities and regions. Indigenous development may be smaller scale and perhaps less likely to effect a transformation in the development trajectory of a locality or region, for example in contrast to the ways in which large FDI projects may implant new facilities, occupations and industries in a place. In qualitative terms, the culture changes required to stimulate higher levels of entrepreneurialism and raise educational aspirations may take generations to achieve. Deciding upon the quality of specific forms of indigenous development to prioritise for policy intervention raises thorny questions that require value judgements – what types are encouraged and what forms are discouraged or ignored (Acs and Storey 2004). These kinds of development dilemmas are discussed further in the conclusions in Chapter 8.

An indigenous route to development may even be effectively closed-off in localities and regions that lack any discernible assets and resources amenable to development, rendering them dependent upon exogenous approaches. As a place, if the main resource is abundant and cheap labour then development options may be somewhat limited. Creating competitive advantage in the manner of Porter's theory through increasing the sophistication and value-added content of economic activities in the locality and region may then prove a somewhat challenging task. Upgrading indigenous economic structures faces the moving target of keeping up with the cumulative growth and developing advantages of more prosperous localities and regions. Echoing the potential problems of cluster policy and exogenous approaches discussed in the next chapter, the relative merits of specialisation and diversification remain an issue for indigenous development. Economic specialisation can yield external economies but may breed risk and dependence upon a narrow economic base. Diversification reduces reliance and spreads risk but may not benefit from dynamic agglomeration economies. The slower pace and more incremental nature of indigenous development may be more manageable and firmly rooted in place but it is also potentially less glamorous and affords fewer of the kinds of spectacular advances sometimes necessary to attract political support. Overall, indigenous approaches may be necessary and helpful but on their own they are not sufficient for local and regional development. The next chapter addresses their foil: exogenous or externally oriented approaches to development in localities and regions.

Further reading

On indigenous approaches, see Cooke, P. and Morgan, K. (1998) *The Associational Economy: Firms, Regions and Innovation*. Oxford: Oxford University Press; Crouch, C., le Galès, P., Trigilia, C. and Voelzkow, H. (eds) (2001) *Local Production Systems in Europe: Rise or Demise?* Oxford: Oxford University Press; Stöhr, W.B. (ed.) (1990) *Global Challenge and Local Response: Initiatives for Economic Regeneration in Contemporary Europe*. London: The United Nations University, Mansell.

For a review of the entrepreneurship literature, see Acs, Z. and Storey, D. (2004) 'Introduction: entrepreneurship and economic development', *Regional Studies* 38(8): 871–877.

On context-specific policy, see Storper, M. (1997) *The Regional World: Territorial Development in a Global Economy*. London: Guilford.

For a review of SME support policy, see Armstrong, H. and Taylor, J. (2000) *Regional Economics and Policy* (3rd edn). Oxford: Blackwell.

ATTRACTING AND EMBEDDING EXOGENOUS RESOURCES

6

Introduction

The fortunes of local and regional economies are crucially dependent upon their abilities to attract and embed exogenous resources. In practice, this has often meant attempts to attract the investment of transnational corporations and to exploit their potential benefits for local and regional economies. This chapter describes the growth of the TNCs and their changing form, especially the emergence of global production networks and the policies and instruments seeking to attract and embed FDI for local and regional development. It also charts the more recent concern with attracting particular occupational groups and 'creative professionals'. The decisions of TNCs to invest, reinvest or divest, and the phenomenon of territorial competition, have the power to shape local and regional development and to determine geographical patterns of prosperity and disadvantage.

The rise of TNCs was an important feature of the second half of the twentieth century. TNCs have been described as the 'movers and shapers' of the global economy (Dicken 2003). Local and regional development agencies have at various times placed a great deal of effort in attracting mobile investors, especially in manufacturing, through local and regional policy (e.g. Amin *et al.* 1994). Traditionally, TNCs have been viewed as providers of, often, large numbers of jobs, especially for disadvantaged regions. More recently, researchers have emphasised their role as bearers of new technology, innovative management practices and stimuli to local suppliers. Thus, the decisions of TNCs can have a great bearing on the development prospects of local and regional economies. As well as the potentially positive benefits accruing from investments by TNCs, the sudden withdrawal of large externally owned firms from local economies can have devastating social and economic effects. As we discuss below, some researchers suggest that an over-reliance on TNC investment can lead to the emergence of 'branch plant economies', where the development prospects of localities and regions are shaped by their place in the hierarchical 'spatial division of labour' discussed in Chapter 3, with TNCs providing at best only semi-skilled routine jobs (Massey 1995). More broadly, the growing power of TNCs has led some commentators to suggest that they have become more powerful than governments in determining where and when investment takes place in the economy (Hymer 1972; Strange 1994). As we detailed in Chapters 3 and 5, despite the resurgence of the small firm as a provider of employment, wealth and the subject of local and regional development, the evolving TNC continues to dominate the economic landscape (Harrison 1994; Amin and Tomaney 1995b).

This chapter aims to examine the evidence about the impact of TNC investment in local and regional economies. Theoretical and empirical developments have altered perceptions about the potential contribution of TNCs to local and regional development. At the same time, as we suggested in Chapter 2, questions of what type of local and regional development and for whom have dogged discussions about the role of inward investment in lagging regions. For instance, TNC investments in lagging regions have helped to reshape gender relations in some regions by specifically seeking to tap unused pools of female labour. The impacts of such developments have proved controversial: at one and the same time they can both provide jobs for women who may have been previously excluded from the workforce, while at the same time confining them to 'low-skilled' occupations (Massey 1995; Braunstein 2003). In the past, researchers tended to take a largely critical view of the impacts of foreign direct investment on host economies, suggesting that they generated 'dependent development' by creating 'cathedrals in the desert', which were only weakly embedded in the local economy (e.g. Amin 1985). These types of criticism continue to be levelled at some TNC investments in lagging regions, but lately some researchers have suggested that the changing nature of TNCs throws up a different range of local and regional development possibilities (Amin *et al.* 1994; Henderson *et al.* 2002). This has led to efforts to discover ways of 'embedding' TNCs in host economies often through local and regional development policies. At the same time, the growing tradability of services, along with more efficient systems of information and communication technologies, has led to the emergence of mobile service investment (Marshall and Wood 1995). Finally, the alleged shift to a knowledge economy has raised new challenges for local and regional development and attracting and embedding exogenous resources, including new technologies and managerial skills. Under this scenario, crucial to local and regional development, is the attraction of the specific occupations and professional groups that make up the so-called 'creative class' that underpin the knowledge economy (Florida 2002b).

The economic role of TNCs

There was a growth in foreign direct investment from the end of the 1960s, which accelerated in the 1980s and 1990s (Figure 6.1). TNCs account for two-thirds of global exports, of which a significant share is intra-firm trade (UNCTAD 2004). Thus, movements of goods, services and investment *within* firms can have large impacts on local and regional economies. TNCs make direct investments either by greenfield investments or through processes of merger and acquisition (M&A). The relative importance of these two modes of entry into local and regional economies varies over time and between places, but both have the potential to help reshape economies.

The overall figures for FDI growth obscure a highly uneven geography. Historically, it has been the case that developing countries have dominated flows of both inward and outward FDI (Ruigrok and van Tulder 1995; Boyer and Drache 1996; Hirst and Thompson 1999). Indeed, the United States and United Kingdom have dominated these

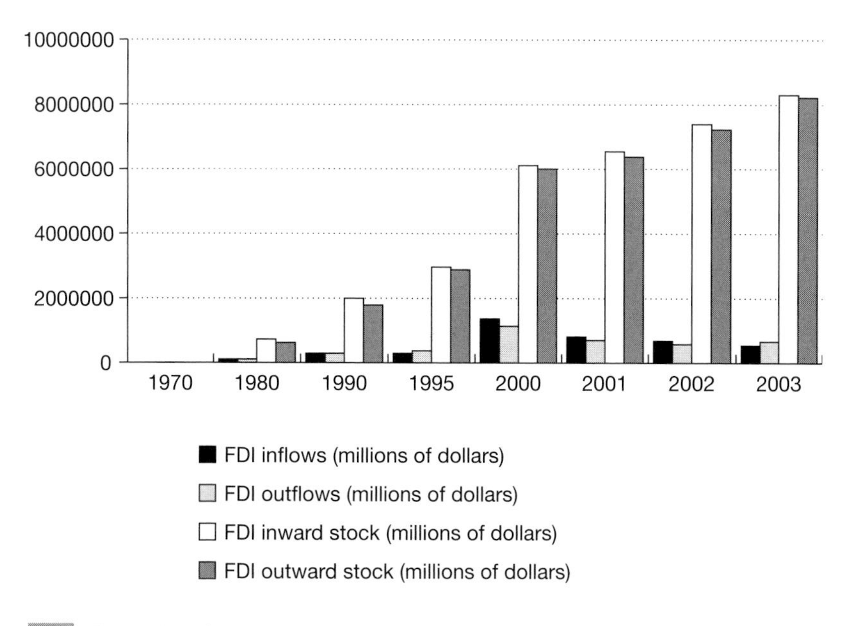

Figure 6.1 ***Global FDI indicators***

Source: Adapted from UNCTAD FDI Database

flows, although the last two decades of the twentieth century saw a sharp relative growth in outward FDI from other countries, notably Germany, France and Japan. Japan, especially, made an important contribution to the rapid growth of overall FDI in the late 1980s and 1990s reflecting the 'endaka' phenomenon of the appreciating Yen. This period also saw the beginning of outward FDI from developing and newly industrialised countries. These flows were dominated by a handful of countries, notably, before 1997, Hong Kong, along with Singapore, Taiwan, South Korea, Malaysia, Brazil and, more recently, China. Most FDI, perhaps contrary to the popular conception, as well as originating in developed countries, goes to developed countries. Thus, overall, FDI has played a bigger role in shaping local and regional development in the developed world than it has in the developing world. The United States and Europe attract most FDI, but Japan which invests heavily abroad attracts relatively little FDI itself. These trends partly account for the large trade surplus accumulated by Japan and trade deficit in the United States sustained by overseas investors in the 1980s and 1990s. Among developing countries, though, China emerged at the end of the twentieth century as an important recipient of FDI (Example 6.1). Overall, however, FDI flows decreased at the turn of the twenty-first century and the long period of FDI growth slowed (UNCTAD 2004).

As well as being geographically uneven, flows of FDI vary between sectors. Historically, FDI was very important in the primary sector, reflecting the internationalised character of sectors such as mineral mining and oil and gas production, which have been important in developing countries. Here, FDI provided the capital necessary to exploit natural resources. The growth of FDI from the 1960s and, especially, the large increase in the late 1980s and 1990s reflected the internationalisation of manufacturing activities, both through greenfield investments and M&A activity. Especially from the

Example 6.1 New locations for FDI

China's recent industrial ascendancy has made it a magnet for FDI, as this analysis from the OECD shows:

> China became the world's largest recipient of foreign direct investment (FDI) for the first time in 2002. Yet while this is impressive for a country that began reaccepting foreign investment only recently, this jump is amid declining FDI inflows to other countries. In terms of FDI per capita, China still ranks relatively low but has still been surpassed by several OECD member countries. According to OECD statistics, in 2000 the United States, Germany and France, among others, all received more in foreign investment than China's US$38.4 billion. In 2002, however, FDI to most OECD nations fell sharply, particularly in the US. Meanwhile, FDI into China rose to a record US$52.7 billion. Nearly half of cumulative realised FDI in China is listed as having originated in Hong Kong, China, though this includes an uncounted amount of FDI from the overseas Chinese diaspora, Chinese Taipei and from within China itself, via 'round-tripping' in Hong Kong to take advantage of fiscal incentives offered to non-mainland investors. In terms of investment from OECD countries, China's ranking has recently fallen. In 1995, China was in second place, but by 2000 it had dropped to fourth. Also, recalling the country's enormous population, FDI inflows in per capita terms remain far lower in China than in all OECD member countries, except Turkey. The composition of Chinese FDI inflows is also different; since the 1990s, most of the FDI among OECD countries has been due to mergers and acquisitions, which is negligible in China's FDI inflows.

Source: 'China ahead in foreign direct investment',
OECD Observer, no. 237, May 2003

1980s these developments were driven by the liberalisation of trade policies and the emergence of trading blocs such as the Single European Market in the European Union and NAFTA. TNCs were especially important in a number of key sectors, notably high technology sectors (e.g. pharmaceuticals and information and communication technology industries); the automotive industry and mass production industries such as tobacco, soft drinks and processed food. Within developed countries especially, there has been a decline in flows of FDI in lower-cost, lower-skill manufacturing activities, which has had significant implications for some local and regional economies. More recently, however, there has been a marked growth in FDI in the service sector, especially in financial services, marketing and distribution and teleservices, reflecting the general growth of services in the economy and the growth of information and communication technologies (ICTs) in their marketing and provision (UNCTAD 2004).

This brief overview of trends shows that flows of inward investment have grown in importance. But the ability of localities and regions to attract and embed FDI is variable and this has important implications for the prospects for local and regional development and policy, which are described below. However, the impact of FDI on host economies is highly uneven, often reflecting the different national origins and various destinations and diverse sectors of particular investments and different modes of entry into host economies. This suggests we should be wary about the value of general claims about the impact of FDI on local and regional economies. It points to the need for theoretical approaches to understanding FDI and its local and regional impact which allow for a range of motivations, practices and outcomes.

The theory of the TNC

The theory of transnational corporations has evolved over time, often reflecting attempts to explain shifts in the practices of MNCs. In the 1960s, Stephen Hymer was perhaps the first to develop a view of the international economy that went beyond the neo-classical trade theory introduced in Chapter 3 by focusing on the role of the firm and its organisation (Hymer 1979). Hymer's work proved both seminal and far sighted, anticipating many themes in the debate about globalisation that developed years later, accurately predicting the growth of a globally integrated economy in which TNCs would play a central role. Hymer's theory attempted to explain the large increase in US outward FDI that occurred from the 1960s onwards. He attributed this growth to the emergence of the large, multidivisional firm and improved communications technology, on the one hand, and the growth of European and Japanese competition on the other. For Hymer, the main motivation for firms to internationalise was to gain better access to foreign markets. A wave of US investment into Europe was a notable feature of this period, although Japan remained blocked to foreign investment, with Hymer (1972: 122) predicting it would remain 'a source of tension to oligoplistic equilibrium'.

Hymer's focus on the firm and its organisation was a breakthrough in the theory of the multinational enterprise. As we saw in the product life cycle theory in Chapter 3, a further element of theory was provided by Vernon (1966), who emphasised the importance of technological change in the internationalisation process. Again, attempting to develop a theory of the internationalisation of US industry, Vernon argued that firms were likely to introduce new products in their home market. The stimulus to product innovation was strong in a high income economy such as the United States. As overseas demand for products developed this would be served in the first instance by exports but, as products 'matured' and their production was 'standardised', firms tended to set up plants overseas in order either to reduce production and distribution costs or to gain better access to markets. In the first instance these plants would be located in other high-income countries, where demand would be high for the products in question. However, as the product became more standardised, firms were likely to open plants in developing countries. The emphasis here was on 'the timing of innovation, the effects of scale economies, and the roles of ignorance and uncertainty in influencing trade patterns' (Vernon 1966: 190).

John Dunning (1988) developed an eclectic theory which drew together a number of insights to develop a framework for understanding why firms choose to engage in international production. According to this approach, there are three features of firms that engage in international production. First, firms possess ownership advantages, notably size advantages – or scale economies – which allow them better access to finance or technology which they can transfer between locations (O). Second, they seek to exploit location specific assets such as markets or resources (L). Finally, in the face of imperfect markets, firms choose to internalise (I) activities in order to reduce the uncertainties of international activity. Dunning (1988) called this approach the 'eclectic paradigm' and others have called it the OLI approach (see also Loewendahl 2001; McCann and Mudambi 2004).

The work of Hymer, Vernon and Dunning has proved highly influential in the study of TNC activity, partly because it has spawned a wide range of research on international business and partly because the growth of TNCs has become central to broader debates about globalisation and local and regional development (for reviews see Hayter 1997; Held *et al.* 1999; Dicken 2003). Although principally driven by a desire to explain the activities of firms, these classic theoretical treatments all had geographical implications. In general, they implied that the local and regional impact of FDI would create a hierarchy of locations and that within this hierarchy certain regions were likely to be confined to subordinate roles in the economy. In particular, the slowdown of growth in the advanced world from the end of the 1970s, it was assumed, would force the development of 'a new international division of labour'. This would mean that 'the survival of more and more companies can only be assured through the relocation of production to new industrial sites, where labour-power is cheap to buy, abundant and well-disciplined; in short, through the international reorganisation of production' (Fröbel *et al.* 1980: 15; see also Massey 1995). Hymer, for instance, predicted that:

> a regime of North Atlantic Multinational Corporations would tend to produce a hierarchical division of labor between geographical regions corresponding to the vertical division of labor within the firm. It would tend to centralize decision-making occupations in a few key cities in the advanced countries, surrounded by a number of regional sub-capitals, and confine the rest of the world to lower levels of activity and income, i.e., to the status of towns and villages in a new Imperial system. Income, status, authority and consumption patterns would radiate from these centers along a declining curve, and the existing pattern of inequality and dependency would be perpetuated.
>
> (Hymer 1972: 114)

Early waves of international investment tended to focus on secondary manufacture associated with final assembly operations. Market size and cost factors were important factors in the attraction of FDI during this period with multidivisional firms typically organised on a *geographical* market basis. In addition, many of these branch plants were often dependent for material and technical inputs from their parent companies and had very little autonomy, with centralised, hierarchical management control the order of the day. In this perspective, the internationalisation of production would restrict some local and regional economies to the role of 'branch plants economies' which could become heavily dependent on decisions made by externally controlled firms (Government of

Canada 1972; Firn 1975; Telesis 1982). Such worries have been raised repeatedly (e.g. Amin 1985) and have also been raised in relation to the impact of mobile investment on Central and Eastern Europe (e.g Grabher 1994; Rainnie and Hardy 1996; Sokol 2001; Pavlinek 2004).

One common instrument used to attract FDI has been the export processing zone (EPZ). The creation of EPZs was widespread in the low- and middle-income countries in Latin America, the Caribbean, Asia and, to a lesser extent, Africa, from the 1970s onward. Their introduction generally signalled the beginning of a shift away from import substitution strategies. EPZs are advocated by the World Bank (1992) as a means of internationalising developing economies and stimulating the development of local industry. EPZs typically involve the provision of financial incentives to attract FDI, including tax exemptions, duty-free export and import and free repatriation of profits, the provision of infrastructure and exemption of labour laws. Despite their popularity many questions have been asked about the effectiveness of EPZs (Dunning 2000). Experiences with EPZs have been decidedly mixed. There are generally few examples of EPZs contributing to skill formation and the development of linkages with local industry. Even where these objectives have been partially realised, the offer of low-cost labour, generous concessions and enclaves with modern facilities have not always outweighed their costs, for instance in terms of foregone government revenue, or declining labour standards. The result is that the performance of EPZs has often been poor (Jauch 2002).

The changing nature of TNCs: towards global production networks?

During the 1990s, however, researchers noticed changes in the nature of the branch plant economy (e.g. Amin *et al.* 1994). Partly reflecting shifts in corporate organisation which saw product-based structures replace geographical market-based structures. Overseas plants began to assume responsibility for world product mandates (WPM) or continental product mandates (CPM) signifying an evolutionary upgrading of plants over time and holding out the possibility of regions altering their position in spatial divisions of labour. Foreign manufacturing units that were initially set up to produce standard products with semi-skilled labour, in time could acquire their own technical, managerial and marketing expertise. Engineering capabilities acquired to perform routine technical activities – service, maintenance and customisation of products to individual buyer needs – could evolve into R&D proper. Similarly, over time it was observed that some TNC affiliates increased the amount and range of their purchases from their host economy, thus increasing their positive local and regional economic impact. Researchers mooted a potential transition from 'branch plants' towards 'performance plants' with enhanced and potentially positive implications for local and regional development (Amin *et al.* 1994; Pike 1998). Table 6.1 summarises the potential implications of these changes for local and regional development. Local and regional development agencies both responded to and fomented these developments through policy support in order to maximise the development potential of FDI. By supporting entrepreneurial managers of local affiliates in the intra-firm competition for investment, local agencies could attract

Table 6.1 ***Dimensions of plant type and local and regional development implications***

	'Branch Plant'	*'Performance/Networked Branch Plant'*
Role and autonomy	External ownership and control; structured position and constrained autonomy; truncated and narrow functional structure involved in part-process production and/or assembly; cloned capacity and vertically integrated with limited nodes capable of external local linkage (e.g. suppliers, technology); state-policy subsidised establishment via automatic grants to broadly designated areas.	External ownership and control but possible enhanced strategic and operating autonomy as well as responsibility for performance increased within a 'flattened' hierarchical structure; wider functional structure involved in full process production tilted towards manufacturing rather than solely assembly; sole capacity with product (range), division or market mandate at the expense of rationalisation elsewhere; increased nodes capable of linkage (e.g. R&D with technology support, human resources with training); state-policy support for establishment on selective and regulated basis (e.g. job creation, local content).
Labour process	Labour-intensive, semi- and unskilled work; 'routinised' and specific tasks within refined technical division of labour; high volume production of low to medium technology products; standardised process technology; short-term, task-specific, on-the-job training integrated with production.	Capital and technology intensive, semi- and skilled work with increased need for diagnostic and cognitive skills; recombined job tasks and individual/team responsibility for performance; low to high technology and low to high volume production flexibility; flexible and re-programmable process technology; longer term, coordinated with investment, on- and off-the-job training.
Labour-management relations	Organised and unionised labour; job classifications, task assignments and work/supervision rules linked to seniority-based pay scales; formalised and collective negotiation and bargaining tied to employment contract; personnel management with administrative focus.	Business unionism; reduction and streamlining of grading, job titles and meritocratic salary structure; shift to company-based non-(traditional) union arenas, individualised negotiation and bargaining tied to 'enabling' agreements; human resource management techniques.
Labour market strategies	Employees considered interchangeable, replaceable and in need of constant supervision; limited screening and high labour turnover and absenteeism; reliance on external labour market.	Rigorous scrutiny and increased selectivity in recruitment; employees as human resources needing investment; teamworking to reduce labour turnover and identify employee with the goals of the company; development of core internal labour market and peripheral (part-time, temporary) segments.

Table 6.1 ***continued***

	'Branch Plant'	*'Performance/Networked Branch Plant'*
Supplier linkages	Limited since integration with broader corporate structures of production and supply chains; intra-firm linkages substituted for local ties; limited local supply chain knowledge and greater awareness of potential suppliers in headquarters region.	Outsourcing increase with JIT and synchronous suppliers; increased potential for local procurement and supplier agglomeration; first and second tier supply chain management; increased global sourcing and partnership relations; growth in dependence in the local supply network; geographically distributed production networks and JIT operated over (inter-) national distances.
Local economic development implications	Externally owned and controlled plants with limited decision-making powers locally ('dependent development', 'branch plant economy') vulnerable to closure or relocation ('footloose', 'runaway industries', 'hyper-mobile capital'); limited growth rates in employment and output; low technology and skills ('screwdriver plants'); few local linkages ('enclave development', 'dual economy', 'industrialisation without growth', 'cathedrals in the desert'); diversified industries not building upon or modernising existing regional industrial strengths; limited innovation potential and technology transfer from dedicated production processes and suppliers.	New concepts of externally owned and controlled plant with increased decision-making autonomy for strategic and operational issues, more rooted and anchored in the local economy ('embedded firm'), higher levels of technology and skills, higher innovative potential, more local linkages and increased technology transfer through research and technological development functions; supplier links upgrading process technology improvement and partnership development with suppliers; potential for the plant to be a 'propulsive local growth pole', 'vehicle/catalyst for local economic development' and capable of setting in train 'sustainable development'.

Source: Pike (1998: 886–887)

additional resources that could contribute to local and regional development, for instance by assisting the development of local supplier networks and R&D activities (Håkanson 1990; Amin *et al.* 1994; Young *et al.* 1994).

For some writers these changes were all closely connected and reflected structural transformations in the world economy:

> An important shift since the 1970s is a move from an international to a global economy. This global economy has three particularly important characteristics. First, industries increasingly function on an integrated world scale, through the medium of global corporate networks. Second, corporate power has continued to advance, so that the new industries are progressively oligopolistic, progressively cartelized. Third, today's global corporations have become more decentralized through increased 'hollowing out', new forms of sub-contracting, new types of joint ventures, strategic alliances and other new 'networked' forms of corporate organization.
>
> (Amin and Thrift 1999: 176; see also Loewendahl 2001; Ernst and Kim 2002)

The implications of these developments for local and regional economies were set out by Amin and Thrift:

> The majority of localities may need to abandon the illusion of self-sustaining growth and accept the constraints laid down by the process of increasingly globally integrated industrial development and growth. This may involve pursuing those interregional and international linkages (trade, technology, transfer, production) which will be of most benefit to the locality in question and upgrading the position of the locality within international corporate hierarchies by improvements to its skill, research, supply and infrastructure base in order to attract 'better quality' branch investments.
>
> (Amin and Thrift 1999: 182)

While such sharply drawn typologies can be readily criticised, various efforts to develop a complex and nuanced understanding of the flows and impacts of inward investment have focused on the emergence of 'global production networks' (Henderson *et al.* 2002; Coe *et al.* 2004; Phelps and Waley 2004). According to Henderson *et al.* (2002) global production networks have evolved in relation to a combination of liberalisation policies, the rapid take up of ICT and intensifying 'global' competition. This conceptualisation of international business recognises that firms, government and other actors from different societies, while responding to 'global' economic pressures, can have a variety of priorities in relation to profitability, growth and economic development, which are each necessary to understand in order to assess the production network's implications for firm and local and regional development. Global production networks comprise three principal elements: *value, power and embeddedness* (Henderson *et al.* 2002; Coe *et al.* 2004; Phelps and Waley 2004). Understanding the interplay of the relationships between these elements helps us to understand the potential policy interventions for attracting and embedding external investment for local and regional development.

Plate 6.1 ***Transnational corporations: global firms in Sydney, Australia***

Source: Photograph by David Charles

Value can be created through technological innovation, brand reputation, or special skill; it can be enhanced through technology transfer in connections between lead firms and their suppliers; and it can be captured for a host economy by means of the regulatory regime and the provision of financial incentives which place limits on profit repatriation, specify the local content of manufactured products or ensure decent labour standards. Power is exercised in three ways in global production networks. Corporate power accrues asymmetrically to firms and their affiliates, but in ways which can allow affiliates to exercise a degree of autonomy. Power is exercised by the national, regional or local state although in ways that are markedly different between countries. Also, power is exercised by international organisations such as the WTO or EU and power is also exercised by groups such as non-governmental organisations (NGOs) and trade unions which seek to influence the activities of international corporations. Finally, the global production network is embedded first in the set of connections that links firms, government and other actors which are often heavily influenced by history and the national contexts in which global firms originate. In addition, global production networks become embedded in places and this embeddedness can become an influencing or constraining factor in the local and regional development contribution of firms. In this perspective regional development is conceptualised as:

> a dynamic outcome of the complex interaction between territorialized relational networks and global production networks within the context of changing regional governance structures.
>
> (Coe *et al.* 2004: 469; see also Smith *et al.* 2002)

This suggests that the emergence of global production networks may raise new potentials for local and regional development.

The role of local and regional development agencies

National governments and local and regional development agencies have long been involved in providing location incentives to mobile investors. Indeed, the provision of such incentives was a central element of regional policies from the 1950s as we described in Chapter 4. During the 1960s and 1970s, there is evidence that investment incentives played a role in determining the geographical distribution of mobile firms within countries and in creating manufacturing employment in lagging regions (e.g. Moore and Rhodes 1986). But such policies have been criticised in part for providing incentives to firms for projects that would have occurred anyway – so-called 'deadweight' effects – generating wasteful inter-regional subsidy competition and contributing to the creation of branch plant economies by indiscriminately attracting plants which were weakly embedded in local economies and provided only semi-skilled forms of employment (Massey 1995). As we distinguished in Chapter 2, such policies helped to create development *in* the region, but not development *of* the region (Morgan and Sayer 1988). These criticisms and changing perceptions of the nature of FDI and its development potential altered approaches to the attraction and embedding of mobile investment; with regional and local development agencies generally adopting more proactive policies.

Young and Hood (1995) identify the challenges facing inward investment agencies seeking to maximise the local and regional development potential of FDI in the changing economic context. These include the need to target investment which can contribute towards the broader objectives of local and regional development, assessing the quality of investment projects (e.g. the type of jobs, occupations, functions, supply chains) and moving beyond narrow performance measures such as number of projects attracted and jobs created. The recent paucity of large-scale, greenfield manufacturing investment in some developed countries makes it necessary to generate expansionary investment from within the existing stock of investors in any given location and forestall rationalisations, leading to job loss and closure. It is also thought necessary to integrate inward investment promotion and broader local and regional development programmes, connecting exogenous and indigenous approaches. Attracting new types of mobile investment such as services, R&D and headquarters activities is also important. As we outlined in Chapter 3, the emergence of a knowledge economy places a premium on the development of regional innovation systems (Cooke and Morgan 1998). Young *et al.* (1994) advocate a strategy of 'developmental targeting', which amounts to:

> a process which identifies inward investment market segments which match the desired outputs from inward investment (in terms of employment, technology transfer, trade and balance of payments and linkage and spillover effects) to the competitive advantages of countries and regions.

This 'developmental' approach is adopted – in rhetoric at least – by many local and regional development agencies. As we discuss in more detail in Chapter 7, in Ireland a series of official reports signalled a move away from an indiscriminate attraction of foreign investment to a more targeted approach to particular industries and activities (such as R&D), a stronger emphasis on aftercare and more priority on promoting Irish-owned industry (e.g. Tomaney 1995; see also Fuller and Phelps 2004). The broad functions of inward investment agencies are set out in Table 6.2. Danson *et al.* (1998) observe that few inward investment agencies meet the ideal type. As we suggested in the discussion of mobilising indigenous resources in Chapter 5, successful regions though are generally able to tailor their activities to the concrete circumstances of their particular local and regional economies.

Exogenous or externally oriented strategies have the potential to connect with and complement the indigenous approaches discussed in Chapter 5. Explicit linkage programmes have been used to exploit the market opportunities for the supply of goods and services from local and regional businesses to inward investment plants, especially where autonomy has been decentralised and decision-making devolved to plant-level (see Amin *et al.* 1994 for an analysis). Such strategies can encompass supply chain initiatives by local and regional agencies focused upon technology, skills and training, and management. Such programmes can be deliberately oriented towards indigenous, locally and regionally owned companies. The knowledge transfer from externally owned plants can have a demonstration and upgrading effect upon indigenous firms, driving innovation through supply chains and enhancing their competitiveness, through initiatives such as auditing mechanisms, technology transfer and formalised relationships (Vale 2004).

Table 6.2 ***Functions of inward investment agencies***

Policy formulation	Liaison and dialogue with parent organisation Guidelines for inward investment policy Assessment of effectiveness of policy Integration with national and regional industrial policies Development of partnership scripts and protocols for joint working
Investment promotion and attraction	Marketing information and intelligence Marketing planning Marketing operations outside and inside the relevant area Management of overseas agents and offices
Investment approvals	Screening and evaluation of potential projects
Granting of incentives	Consideration of investment offers Incentive application advice and approvals (including direct financial incentives plus training grants, innovation grants, land and buildings, etc.)
Providing assistance	Assistance with public utilities (roads, water, electricity, sewerage, telecommunications) Facilities and site Training and recruitment Links with universities and research institutes Supply chain linkage and development
Monitoring and aftercare	Continuation of assistance post-launch Relationship management and liaison (including reinvestment projects, upgrading local suppliers, etc.)

Source: Adapted from Young *et al.* (1994)

Networks shaped by focal inward investors can amplify external economies, reducing transaction costs for network participants. Agencies have often used such policies to grow specialised local linkages in an attempt to anchor and embed externally owned plants in localities and regions. Yet, such initiatives can also lock-in highly localised suppliers into dependent relationships with externally owned plants. For the inward investor, local and regional assets and linkages can be mobilised to their advantage against other plants in other places within intra-corporate competitions for investment.

TNCs and regional economies in practice

Despite the claims of some commentators, Phelps and Waley (2004) suggest that much of the recent literature may have overstated the renewed local orientation of TNCs. They maintain that the extent of local linkages and the impact of local supplier initiatives must be set against trends towards the consolidation of procurement among major international partner supplier firms. The autonomy of local affiliates and the attempt to influence them by development agencies is limited by their dependence on decisions made in global production networks. There are few, if any, truly 'global' corporations, although many employ that rhetoric for marketing and other purposes. Most still have strong roots in their national economies. The most internationalised firms tend to originate in small countries with constrained domestic markets. Even among the largest firms most tend to locate the bulk of their production in their home economy, with key functions such as R&D only partially internationalised at best (Ruigrok and van Tulder 1995; UNCTAD 2004). Ruigrok and van Tulder (1995) contend that the nature of a firm's domestic bargaining arena (vis-à-vis the state and labour, etc.) is at the root of its internationalisation strategy. Thus, a company's internationalisation strategy is heavily influenced by the nature of its bargaining relations within its domestic industrial complex and, as we discussed in Chapter 4, its attendant economic, social and political structures.

Thus, Dicken (2003) warns against unambiguous, all embracing, evaluations of TNC impacts on host local and regional economies. The net costs and benefits will depend on the interaction between the attributes and functions of particular activities within their corporate system and the nature and characteristics of the host local and regional economy. The impact of these developments can be seen in contrasting case studies. Coe *et al.* (2004) have examined the impact of BMW's investment in the underdeveloped region of eastern Bavaria. BMW has traditionally concentrated its German investments in Munich and remains 47 per cent owned by the Quandt family from Bavaria. In the early 1980s, the company made a strategic decision to expand its production to eastern Bavaria, attracted by skilled labour and the willingness of the local union – in the face of opposition from its national headquarters – to accept new working practices. BMW invested €7 billion in the region, creating three plants, directly employing 35,000 people, with a further 20,000 employed in supplier firms. BMW's principal site in Regensburg was made available after the German government was forced to abandon the construction of a nuclear waste facility in Wackersdorf. The federal and state governments provided incentives and infrastructure to attract BMW to the region.

The new production system introduced by BMW required co-location of supplier firms and some of these global suppliers' branch plants have now become leading plants within their parent companies, setting benchmarks for other plants in the production network. Coe *et al.* (2004: 478) conclude: 'Eastern Bavaria's regional economy has, without doubt, benefited from globalizing processes linked to the region via BMW's production network'. It represents an example of value creation (through the new investment) and value enhancement (through the transfer of technology to supplier firms), facilitated by interactions and negotiations between the firm, government and labour representatives within the region and beyond, albeit in a context where BMW has considerable power by virtue of the scale of its presence.

By contrast, Phelps and Waley (2004) have examined 'one multinational company's attempt to disembed itself'. Specifically, they examine the linked investment and divestment strategies of Black and Decker in North East England (discussed in more detail in Chapter 7), northern Italy and China. Black and Decker captured a monopoly in the market for power tools in the 1960s, although by the 1990s this market had matured and the company's monopoly position was eroded by growing retailer power, diminishing brand strength and low-cost competition. The company rapidly internationalised in the 1960s from its base in Baltimore, Maryland, adopting a classic decentralised, multi-domestic organisation consisting of twenty-three wholly owned manufacturing operations around the world, giving a degree of autonomy to its local operations. Faced with the erosion of its monopoly, the firm began to consolidate its power tool operation and initiated a process of rationalisation, especially in Europe, with the intention of specialising in a narrow range of products.

> The autonomy of factories therefore began to decrease with the centralized allocation of new products to remaining production operations . . . [although] an element of affiliate management autonomy remained, promoting a degree of intra-corporate competition for the right to manufacture new products.
>
> (Phelps and Waley 2004: 200)

In 1998, Black and Decker announced plans to close four power tool plants, one of which would be in Europe. This generated an intense competition for survival between two plants, one in Spennymoor, North East England, and one at Molteno, near Milan, in northern Italy.

The Spennymoor plant operated in the distinctive context of the United Kingdom with its 'open door' policy to FDI, which involves 'minimal monitoring of or performance requirements placed on inward investors' and within a local economy heavily dependent on FDI (Phelps and Waley 2004: 200; see also Pike and Tomaney 1999). It is this lightly regulated context which makes both investment and divestment relatively straightforward and less costly than in more regulated economies in continental Europe. The plant was one of the company's largest and remained non-union in the most densely unionised region of the United Kingdom (Pike and Tomaney 1999). Faced with the threat of closure, though, local managers and development agencies sought to strengthen their bargaining position, by reducing production costs and by securing over £1 million in state regional policy support for new investment in the plant.

In a measure of the geographical interdependencies in international production networks, Spennymoor's success spelled disaster for the plant in Molteno in Lecco province in Lombardia. Northern Italy presented Black and Decker with a different socio-economic and institutional context. The plant was one of the few foreign-owned factories located in an industrial district dominated by small firms and subject to the more stringent Italian regulatory regime. Unlike the plant in Spennymoor, the Italian plant was unionised. A major difference between the cost structure of the two plants was that the relative prosperity of the northern Italian economy and the tighter labour market regulations meant that it was difficult for the Molteno plant to make use of temporary workers to cope with peaks and troughs in the production cycle. Overall, the Molteno plant had higher production costs than the UK plant and it was for this reason that the plant became marked for closure. The local unions and the management, together with the regional government, developed a case for retaining the plant, but when this looked forlorn, sought to maintain the site as a location for production. The more stringent legal requirements surrounding plant closure in Italy meant this local coalition compelled Black and Decker to run down the plant in ways which kept the facility and workforce intact, while new investors were successfully attracted to the site, partly through the provision of public investment incentives. Local and regional development institutions therefore played a critical role in mediating the processes of international economic restructuring with decisive implications for local and regional prosperity.

The rise of FDI in services raises the question of the degree to which these can become embedded in regional economies. Some economies have attracted a large amount of mobile service investment. Ireland, for instance, especially the greater Dublin region, has become a major centre for software production through the attraction of FDI (see Chapter 7). Employment in software grew from 7793 in 1991 to 31,500 in 2001, although it declined thereafter. The majority of this employment was in overseas firms, which generated over 90 per cent of the sector's revenue and a similar proportion of the sector's exports, with Irish-owned firms tending to be niche specialists. The attraction of a number of high-profile international investors including Microsoft helped to produce a pool of skilled software engineers which in turn helped attract further investments (National Software Directorate 2004). TNC affiliates of software firms in Ireland began in the main by manufacturing packaged software, but the role of affiliates evolved into the more complex task of 'localising' software for different markets, which requires a range of engineering skills, leading to an upgrading of the industry's skills profile. The development of software localisation activities has stimulated the growth of significant supplier industries in Ireland, notably in printing (of software manuals), turn-key services (notably packaging) and the out-sourcing of localisation activities themselves. Over time, TNCs have opted for longer-term supplier relationships offering development opportunities for the Irish economy. IDA Ireland, the development agency, strongly supported these upgrading efforts. This upgrading process means that 'Irish-based affiliates are becoming European corporate hubs, and now occupy more important roles within their corporate hierarchies than during previous rounds of inward investment' (White 2004: 252; see also Ó Riain 2000).

The Irish affiliate of the Canadian software firm Corel was widely cited during the 1990s as an example of successful plant upgrading, although, ultimately, this was insufficient to save it from closure. Corel located in Dublin in 1993, gradually expanding employment after demonstrating its ability to outperform its firm's headquarters in Canada in software localisation. By the end of the 1990s, the Irish affiliate employed 200 staff. Despite the profitability of the Irish operation, broader problems in the company led Corel to announce the closure of its Dublin office. White observes:

> The affiliate's superior performance or good relationships with local institutions proved inconsequential in determining the affiliate's future . . . Ultimately it was not local factors but rather factors internal to the corporation that determined the embeddedness of this particular affiliate.
>
> (White 2004: 254)

Despite the relative success of Ireland in attracting such 'high-quality' investment, its heavy dependence on decisions taken outside the country means that such employment is always at threat from corporate-driven rationalisation. Despite the sectoral differences and national contexts, the Corel and Black and Decker stories have important similarities and implications for local and regional development institution strategies aimed at attracting and embedding exogenous resources. The closure of Black and Decker's Italian plant as 'a centrally conceived parent-company strategy was no respecter of the accumulated status and network embeddedness of major affiliate operations' (Phelps and Waley 2004: 211). Indeed, the Black and Decker story had a further twist in 2001 and 2002 respectively, when the company shifted jobs from Spennymoor to Suzhou in China and to Usti in the Czech Republic in search of cheaper cost production locations. The threat of closure once again hung over the plant and was enough to extract further investment incentives from the UK state to 'safeguard' the remaining jobs (Phelps and Waley 2004). These issues not only confront individual firms, but also can sweep through entire regional economies when they become dependent on a concentration of externally owned branch plants in particular sectors, such as in 'Silicon Glen' in Scotland (Example 6.2).

Coe *et al.* (2004: 471) conclude that 'economies of scale and scope embedded within specific regions are only advantageous to those regions – insofar as such region-specific economies can complement the strategic needs of trans-local actors situated global production networks'. It is this process of 'strategic coupling' that determines the prospects for local and regional development emanating from the attraction and embedding of mobile investment.

Attracting and embedding occupations?

The challenges confronting those charged with attracting and embedding mobile investment in order to promote local and economic development remain profound, especially in lagging regions. The literature on mobile investment generally views the ability of regions to target specific types of mobile investment as important, yet difficult. Ann Markusen makes the case for targeting occupations in addition to industries based on acknowledgement of the 'decreasing commitment of both firms and workers to each other or to localities, due to integration and the ability to work from remote sites via

Example 6.2 Crisis in Silicon Glen

Scotland's Silicon Glen was an archetype 'branch plant economy', with a strong representation of computer hardware manufacturers. Scotland faced a crisis when these firms began to run-down employment and divest posing new challenges for development agencies, as the following report from *Business Week* shows:

> Clouds over Silicon Glen: Foreigners spawned a high-tech boom. Now they're pulling out
>
> A pall has descended over Scotland's once-vibrant Silicon Glen. On Aug. 1, Sanmina-SCI Corp., the Alabama-based maker of computer and electrical gear, announced it would lay off 750 workers at its plant in Irvine, Scotland. Sources at IBM's operation in Greenock, which employs 5,500 workers, say plans are under way to shift manufacturing out of Scotland as part of the company's recently announced restructuring. 'Five years ago, our staff spent all their time negotiating pay rises and better benefits for electronics workers,' says Danny Carrigan, the Scottish head of manufacturing trade union Amicus. 'Now they seem to spend all of their time dealing with layoffs.'
>
> Scottish officials began wooing foreign tech companies in the 1960s with regional government grants. A host of mostly American and Japanese multinationals took the bait, giving rise to a 50-mile high-tech corridor stretching from Edinburgh to Glasgow. By the early 1990s, electronics had displaced whisky as Scotland's top export.
>
> Now, the distress in the global technology industry is exposing the weakness of an economic policy built on winning foreign investment at the expense of developing homegrown industries. International heavyweights such as Motorola Inc. and NEC Corp. are shutting their Scottish plants and shifting production to Asia and Eastern Europe. Silicon Glen's assembly-line workers earn an average of $1,300 a month, approximately ten times more than their counterparts in China. Job losses at Silicon Glen over the past two years add up to some 15,000 – more than 20% of the area's workforce. 'Silicon Glen has been a major contributor to the Scottish economy, but we have become overly dependent upon a narrow base of foreign companies,' says Iain McTaggart, general manager of the Scottish Council for Development & Industry, a lobbying group in Glasgow . . . Scotland, home to 5 million people, depends more heavily on exports than does the rest of Britain. Electronics make up 51% of Scotland's overall exports. The plants in Silicon Glen shipped out $3 billion worth of goods in the first quarter of 2002 – down 16.3% from the same period last year . . . Scottish Enterprise, the government agency charged with promoting the electronics industry, is working with many of Scotland's most prestigious universities to spin-off new technologies and businesses.

Source: *Business Week*, 2 September 2002

the Internet' (2002: 1). Such an approach would suggest a subtle shift in local and regional development policy to one where the conditions are provided which attract particular types of workers, rather than a focus on particular economic activities or firms.

Richard Florida (2002a, 2002b) has gone further and argued that the future of local economies relies on attracting and retaining members of the 'creative class', comprising those who work in sectors such as technology, media and entertainment and finance and whose activities embody creativity, individuality and difference. Creativity, according to Florida, is the new motor of the economy. Creative types can be identified by their relaxed dress codes, flexible working arrangements and leisure activities focused on exercise and extreme sports and their preference for 'indigenous street level culture'. They are attracted to 'bohemian' environments that can provide the conditions which will support their creativity. Such cities are supposedly characterised by the '3 Ts': tolerance, talent and technology. Cities, in particular, according to Florida, must now compete to attract members of the 'creative class' by creating the environments in which they choose to live. The creative class values places for their authenticity and distinctiveness. Although influential – several US cities claim to have implemented policies aimed at attracting members of the creative class – Florida's analysis and prescription have been criticised. While Florida implies that many, if not all, cities can create the conditions for the growth of the creative class, in practice he seems to be describing conditions in cities which are at the leading edge of the knowledge-based economy, including 'world cities', and which may be consolidating their economic advantages. Peck (2005) has criticised such 'hipsterization' strategies and the analysis upon which they are based, suggesting that they involve little more than the celebration of the lifestyle preferences of high-income groups in ways which accelerate inequalities and intensify territorial competition with potentially wasteful side effects (see also Asheim and Clark 2001). Florida, himself, has stated that the emergence of the creative class appears to be linked with a growth in inequality in world cities (see Fainstein 2001; Buck *et al.* 2002; Hamnett 2003). Despite its weaknesses, Florida's work does draw attention to the importance of human capital in local and regional development, in general, and the attraction of mobile investment, in particular – as we noted in Chapters 3 and 5. Markusen's (2002) rather less breathless analysis points to the importance of wider labour pools in enhancing local attributes, not simply in satisfying the specific requirements of incoming firms. Activities which seem aimed at local markets may constitute regional assets with spillover effects on the productivity of other regional economic activities. This emphasises the importance of skilled labour 'because it increases the productivity and performance of a range of firms and industries, both indirectly and via in creating, attracting and retaining firms and thus jobs' (Markusen 2002: 7). Local and regional development institutions have only recently attempted to translate the ideas of attracting particular occupations – and even the 'creative class' – with as yet uncertain outcomes.

Conclusion

The attraction and embedding of exogenous resources in the form of mobile investments and occupations remains crucial to the prospects for local and regional development.

The policy instruments and measures of local and regional development institutions can be pivotal in securing, retaining and developing the potential as well as ameliorating the problems of potentially footloose forms of economic activity and growth. TNCs continue to be the movers and shapers of the world economy. Indeed, the growing trend towards deregulation of economic relations between and within countries is probably contributing to the strengthening of the position of mobile firms. Local and regional development prospects are influenced by the changing nature of the TNC and, especially, the emergence of global production networks. The emergence of such networks can have quite ambiguous implications for different regions. There is evidence that some localities and regions and their development agencies have proved adept at working with the grain of change in TNC organisation to maximise the local benefits of investments. Attracting particular occupations has emerged as a key element of exogenous-oriented development strategies too. Again, the role of local and regional development institutions and policy is critical in rooting such potentially footloose assets for local and regional development, even if the ambiguities and vagaries of territorial competition render any approach to externally oriented development a difficult and uncertain strategy.

The evidence discussed here confirms the themes explored in Chapter 4 and points to the importance of effective local and regional government and governance as a mechanism for dealing with the threats and opportunities posed by exogenous development strategies. Effective local and regional policy which is oriented towards 'development' and 'network' forms of intervention can potentially help to ensure footloose and mobile assets contribute to local and regional development – albeit for perhaps ever more fleeting time periods. At the same time, the role of national and international authorities in limiting the wasteful effects of the kinds of unregulated territorial competition described in Chapter 1 is increasingly critical too. The effectiveness of the multilevel governance system which was examined in more detail in Chapter 4 then looks even more important when considering exogenous approaches to local and regional development. The final part of the book – Part IV: Integrated Approaches – draws upon the introductory context, the discussion of what kind of local and regional development and for whom, the frameworks of understanding and policy interventions to consider our case studies of local and regional development in practice.

Further reading

On the relationship between TNCs and local and regional development, see Amin, A., Bradley, D., Howells, J., Tomaney, J. and Gentle, C. (1994) 'Regional incentives and the quality of mobile investment in the less favoured regions of the EC', *Progress in Planning* 41(1): 1–122; Dicken, P. (2003) *Global Shift: Reshaping the Global Economic Map in the 21st Century* (4th edn). London: Sage; Henderson, J., Dicken, O., Hess, M., Coe, N. and Yeung, H.W-C. (2002) 'Global production networks and the analysis of economic development', *Review of International Political Economy* 9(3): 436–464.

For a review of the roles of RDAs, see Danson, M., Halkier, H. and Cameron, G. (2000) *Governance, Institutional Change and Regional Development*. London: Ashgate.

PART IV

Integrated approaches

7 LOCAL AND REGIONAL DEVELOPMENT IN PRACTICE

Introduction

This chapter brings together the concerns of the book and considers integrated approaches to local and regional development in practice. It addresses a series of case studies in which the global context, models and theories of local and regional development, institutions of government and governance and approaches to indigenous and exogenous development policy have been played out in particular places in the quest for local and regional prosperity and well-being. Emphasising the point that local and regional development is a global concern made in Chapter 2, the choice of case studies is deliberately international drawing upon the experiences from across three continents globally: Europe, Central and North America and East Asia.

Each case study is organised around the key themes addressed in the book. These comprise, first, the kinds of local and regional development models and strategies. Second, the concepts and theories used to understand and interpret the local and regional development issues and to inform policy and institutional interventions. Finally, the specific development strategies and policy initiatives, the extent to which they have been successful and their future challenges. The cases are different kinds of territories in size, prosperity, development trajectory and government and governance context. Each reflects diverse experiences and particular characteristics, legacies and predicaments.

Yet, each case study shares common issues that are revealed by an analytical approach informed by the main themes in the book. Each case faces shared challenges of economic adjustment and change, particularly addressing the problems of decline and/or establishing new or recreating the conditions for growth that may prove more enduring and sustainable. Each place faces the task of analysing their local and regional development predicament and constructing feasible models and strategies to shape their future development trajectory. A common issue is the management of interventions seeking to balance indigenous and exogenous assets and resources, particularly between manufacturing and service activities. Each locality and region pursues its development in particular national institutional and political contexts of government and governance within an increasingly multilayered polity. Addressing the enduring problems of socio-spatial inequality engrained in uneven development connects the experience of our case studies.

We begin our analysis with the North East of England. This region of the United Kingdom was among the first to industrialise during the nineteenth century. The North East is now struggling with the legacy of the long-run economic decline of its traditional industries, branch plant economy tradition, weak service sector and limited local and regional institutions. Ontario – another early industrialising region and province within Canada – is attempting to effect a transition towards a learning region in the context of continental economic integration, increasing competition between North American region-states and multilevel government and governance in the Canadian federal system. Silicon Valley is a region of California in the United States that has experienced enduring growth from its post-war industrialisation based upon maintaining its leading position in high-tech growth industries and capacity for innovation and learning. Unequal social and spatial development and the challenge of sustaining the Silicon Valley model dominate its local and regional development concerns.

Following rapid industrialisation, Busan has experienced relative decline in the shadow of Seoul's accelerating growth and the sharpening of regional inequality in South Korea. A new commitment to 'balanced national development' and the profound decentralisation of the institutions of government and governance is seeking to address the local and regional development gaps. Ireland emerged from a period of economic crisis to experience very high economic growth rates from the late 1980s – the Celtic Tiger phenomenon – regulated by national social partnership. Sharpening social and spatial inequality has raised questions about the sustainability of this local and regional development trajectory. In Spain, Seville has a long history as a major city of Andalusia but its growth trajectory has been marked by economic and social under-performance since the 1980s. An ambitious development strategy to transform Seville into a 'technocity' – the 'California of Europe' – on the back of high-tech-led development has floundered and reinforced uneven social and spatial development. In Mexico, the State of Jalisco's industrialisation had provided a diversified industrial base prior to faltering in the 1980s. In the context of continental liberalisation and integration within NAFTA, development has revived through the support of indigenous industries more embedded in the local and regional economy.

North East England: coping with industrial decline

North East England is a classical example of an industrial region which experienced large-scale social and economic change in the final decades of the twentieth century. Situated in northern England and adjacent to the Border with Scotland (Figure 7.1), the region's growth from the middle of the eighteenth century to the end of the nineteenth was linked to the industries of the steam age. By the end of the twentieth century, scarcely anything was left of these industries. From the 1930s onward, the North East became a 'policy laboratory' where successive regional interventions were tested, but none of which have arrested the long-run *relative* decline of the region; indeed, that decline accelerated in the last decades of the twentieth century (Hudson 1989; Robinson 1989, 2002; HM Treasury and Department of Trade and Industry (DTI) 2001).

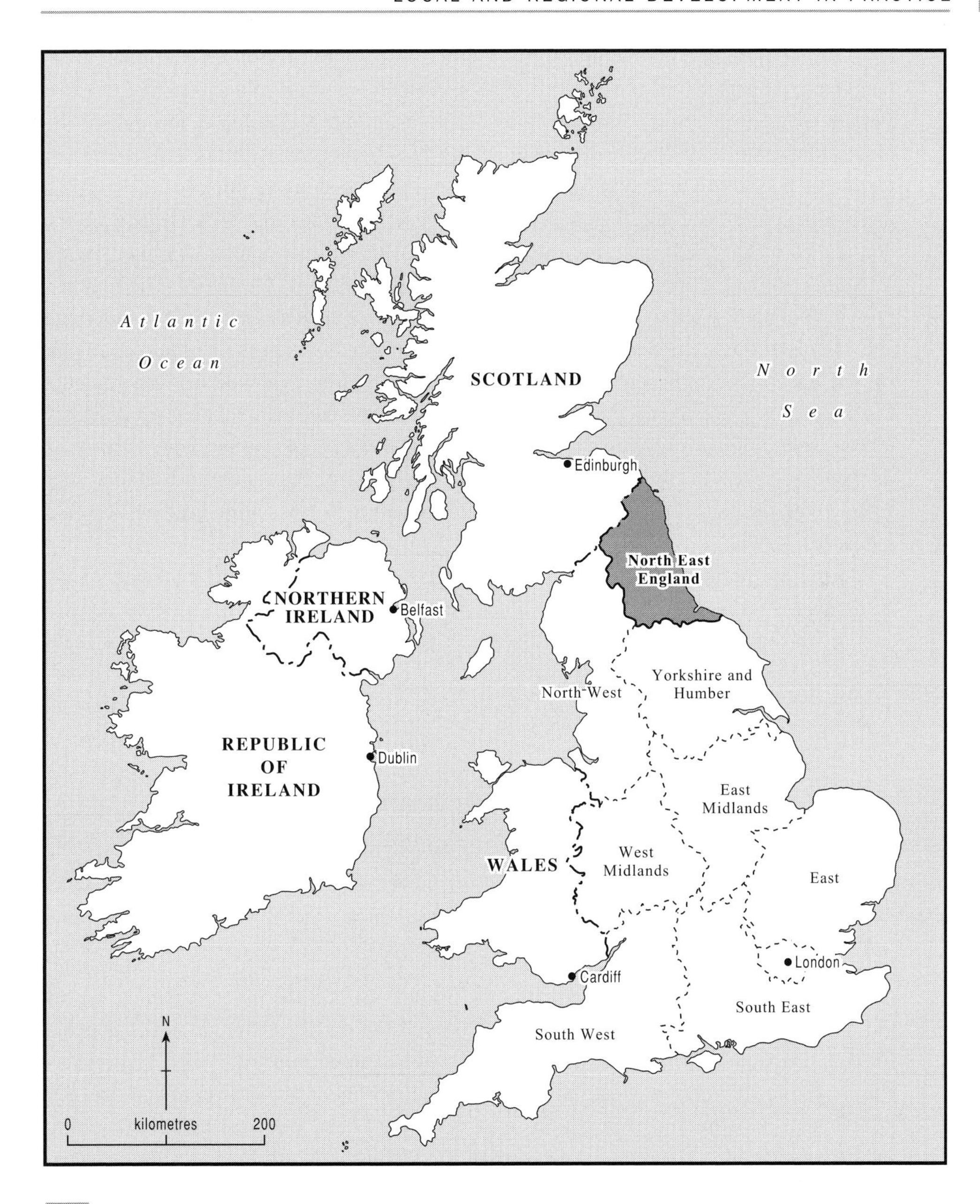

Figure 7.1 ***North East England, United Kingdom***

Historical growth and legacy

The city of Newcastle upon Tyne was an important trading centre during the Middle Ages, but the North East region's rapid growth from the eighteenth century resulted from the development of 'carboniferous capitalism'. This closely interlocked cluster of industries comprised coalmining, iron and (later) steelmaking, shipbuilding, heavy engineering including the manufacture of railway locomotives, bridges and armaments, underpinned by a local banking sector, grew serving the markets of the expanding British Empire. Indeed, the North East was a leader in steam age technologies throughout the nineteenth century. In the first decade of the twentieth century, fully one-quarter of the global output of the shipbuilding industry was produced on the banks of the region's three principal rivers, the Tyne, Wear and Tees (Hudson 1989). The legacy of the dominance of the traditional and often indigenous industries imparted a strong degree of path dependency upon the region's subsequent development trajectory. The region's growing population led to the expansion of urban centres such as Newcastle, Sunderland and Middlesbrough and a proliferation of mining communities in the northern coalfield. By the end of the nineteenth century, however, there was evidence that the region's key industries were losing their competitiveness in the face of declining productivity levels, technological changes and new international forms of competition. Control of the banking sector in the North East began to move out of the region to London, integrating with the capital markets in the City and loosening its links with local and regional industries.

During the period between the First and Second World Wars, the weakness of the North East's industrial base became apparent. Within the context of a global depression, collapsing markets for coal and ships in particular led to the emergence of mass unemployment and social conflict. The North East became defined as a 'problem region'. The British state eventually responded to the crisis with modest experiments in regional policy involving the provision of new factory space and incentives for firms to locate in the region. In parallel, local and regional institutions of capital, labour, the local state and civil society began to form regional organisations to represent their interests. At the same time, a debate began concerning the appropriate forms of government for the region, which focused on the need to move beyond a highly localised and fragmented form of local government (Tomaney 2002).

During the inter-war period, the technologies and industries of the post-steam age – with the important exception of bulk chemicals in Teesside – tended to develop in other regions of the United Kingdom, notably the South East and the West Midlands. It was the onset of the Second World War which proved the region's saviour as the demand for coal, ships and armaments increased and was sustained into the 1950s. The ownership structure of industry changed during this period under an interventionist national government. Coalmining was nationalised in 1947. Later, the steel and shipbuilding industries were also taken into state control, rendering the North East a 'state managed region' (Hudson 1989). Waves of merger and acquisition activities saw control of local industries shift out of the region, usually to London, southern England or beyond (Marshall 1978; Smith 1985; Pike 2006). From the 1960s onward, though, the pace of restructuring quickened. Coalmining contracted first in the 1960s and then following the defeat of the miners' strike in 1985. Tens of thousands of jobs were lost in the sector. Similar processes affected steelmaking, shipbuilding and engineering in the last quarter of the twentieth century often as a result of the interaction of market forces and state policies. State ownership, although originally conceived as a means of safeguarding employment, had become a mechanism of retrenchment and restructuring. Privatisation of the basic industries, as part of the 'free-market' policies pursued by the Conservative governments of the 1980s and 1990s, signalled the final rundown of these sectors and firms, often devastating local communities (e.g. Hudson 1989; Robinson 1989; Tomaney *et al.* 1999; Tomaney 2003).

The effects of this contraction in the North East were offset by two processes. First, during the 1960s and 1970s, successive national governments, following the Keynesian macroeconomic approach to national demand-management outlined in Chapter 3, operated an extensive regional policy aimed at tackling the problems of lagging regions like the North East. In general terms, this policy sought the geographical redistribution of growth through both restrictions on development in fast growing areas and incentives for firms to invest in designated 'development areas'. With relatively high rates of unemployment swelling the labour pool and depressing wage levels, exogenous development was sought as the North East became a focus for mobile forms of manufacturing investment, especially for labour-intensive activities owned by UK and US companies. Alongside the restructuring of manufacturing, the region benefited from the general growth of service industries which emerged as an increasingly important provider of employment. The growth of services in the North East, when compared to other regions of the United Kingdom, rested disproportionately on the expansion of the public sector, often through state-directed relocations of civil service jobs, while business services, for instance, tended to be under-represented (Marshall 1982; Robinson 1989). Shaped by the urban entrepreneurialism discussed in Chapter 4, the physical regeneration of some

Plate 7.1 ***Decay in an old industrial region: housing stock abandonment in Easington in North East England***

Source: Photograph by Michele Allan

parts of the region, notably the Newcastle–Gateshead quayside area, was a significant development at the turn of the twenty-first century. Property development and culture-led regeneration in this urban core helped to alter the external perception and image of the North East or Tyneside, but such developments tended to divert attention from the chronic underlying weakness and social inequalities entrenched in the regional economy (Robinson 2002; Byrne and Wharton 2004).

The North East experienced important economic changes during the twentieth century. However, while in absolute terms its wealth has grown and, in 2005, its unemployment rate was substantially lower than it had been in the 1980s, its place at the bottom of the economic and social hierarchy of UK regions was contested only by Northern Ireland. The region has failed to keep pace with the growth in economic prosperity in the English regional context (Figure 7.2). The North East had the lowest income per head, contained the largest proportion of communities with multiple forms of deprivation, the lowest rates of employment, the lowest levels of educational attainment, the lowest rates of entrepreneurship and, still, the highest rate of unemployment. For the last thirty years of the twentieth century, the region lost population (HM Treasury and DTI 2001).

Foreign direct investment and the branch plant economy

Reflecting the outcome of previous exogenous-based reindustrialisation strategies, patterns of merger and acquisition and the constrained nature of the growth of the service sector, the North East can be viewed as a classic 'branch plant economy'. Despite the shrinkage of the manufacturing base, the North East continues to derive a higher proportion of its GDP than most UK regions from manufacturing output with employment concentrated in externally owned firms (Jones and Wren 2004). Bedevilled by the issues

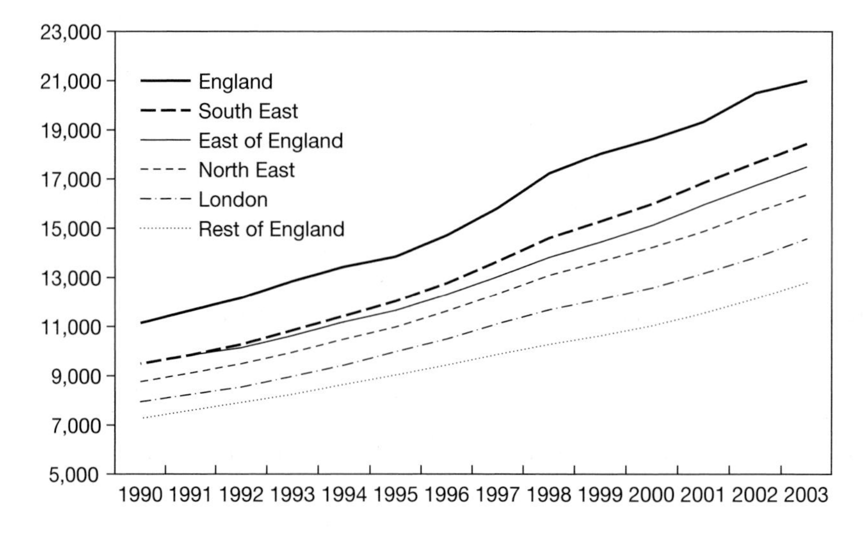

Figure 7.2 ***GVA per head, English regions, 1990–2003***

Source: Office for National Statistics

of attracting and embedding exogenous resources outlined in Chapter 6, inward investment has been the main industrial policy instrument for local and regional development in the North East for decades, with the aim of creating jobs and reducing unemployment. The provision of financial incentives as part of regional policy has been central to this approach. Equally important has been the distinctive 'open door' policy to FDI of successive national governments, involving minimal monitoring of the performance requirements placed on inward investors and the evolution of a light regulatory framework which makes both investment and divestment relatively easier and less costly than in comparative economies in continental Europe. While making the region attractive to major international investors such as Black and Decker, Nissan, Siemens and Fujitsu, it renders much of the region's manufacturing employment vulnerable to rationalisation. Large investors, such as Siemens, opened and closed major factories within a short space of time when confronted with sharp and unexpected deteriorations in product market conditions (see Chapter 6; see also Pike and Tomaney 1999; Loewendahl 2001; Dawley 2003; Phelps and Waley 2004).

Openness to FDI was the hallmark of the free-market policies pioneered by successive Conservative governments between 1979 and 1997. However, echoing the concerns about balancing and linking indigenous and exogenous local and regional development interventions, 'the United Kingdom government's market-led philosophy and belief in the unequivocal benefits of inward investment resulted in an industrial policy which made no attempt to link FDI to the competitiveness of indigenous industry' (Loewendahl 2001: 335). Surveys of local linkages showed that despite the processes of externalisation, branch plants in the North East were poorly integrated into the regional economy, reflecting the deep functional specialisation and integration of the region's manufacturing industry into an increasingly international spatial division of labour (Phelps 1993; Loewendahl 2001). Similarly, there is little evidence that FDI stimulated the development of private sector R&D in the region (Loewendahl 2001), which continues to perform poorly relative to the rest of the United Kingdom (HM Treasury and DTI 2001).

Service sector weakness

The continued relative importance of manufacturing draws attention to the enduring weakness and partial transition towards a service-dominated economy in the North East. Business services are unevenly distributed throughout the United Kingdom, but are especially under-represented in the North East. In part, this is a function of the branch plant character of the manufacturing sector. The absence of locally headquartered and strategic decision-making functions restricted the opportunities for the growth of business services in the region (Marshall 1982). The growth of financial services, which generally favoured the South East of England but which also fuelled the development of provinicial centres such as Leeds, Manchester and Edinburgh, occurred at a lesser rate in Newcastle (Gentle and Marshall 1992). The North East proved adept at attracting investments in call centres in the late 1990s. Almost 50,000 workers – over 4 per cent of the regional workforce, twice the national average – were employed in call centres in the North East in 2003 (DTI 2004). Call centre employment, however, replicated many of the features of the manufacturing branch plant economy providing mainly routine jobs and being only loosely rooted in the region (Richardson *et al.* 2000). The North East's disproportionate dependence on public sector employment remained. From the 1970s onwards, successive governments decentralised civil service 'back office' jobs to the North East, while generally retaining higher-level occupations in London and the south. Moreover, as we noted in Chapter 4, waves of privatisation, 'agencification' and restructuring in the public sector institutions of government and governance tended to impact disproportionately on the North East as routine jobs were automated or contracted out (Marshall *et al.* 2005).

Strengthening local and regional institutions?

Periodically since the 1930s, political debate in the North East has questioned whether the structures of local and regional governance are suited to the task of promoting the region's development. The notion that the North East's governance structures were ill-suited to promoting the adaptation of the region's economy and managing land-use planning has been a central and recurrent theme. In the 1960s and 1970s, the expansion of Keynesian redistributive regional policy was accompanied by the creation of new regional planning structures. These bodies though were appointed by and remained accountable to national government and had little executive authority. In the late 1970s, the first efforts were made to develop a coherent regional plan, but these were halted with the election of the Thatcher government in 1979. Thus, in common with other parts of England, the region's governance was characterised by fragmented and periodically restructured local government and a growing tier of regional organisations that were largely accountable to central government. While this witnessed the establishment of one of the first regional development organisations in England, Northern Development Company, its initial remit was to intensify the strategy of exogenous development based on attracting FDI. Another feature of the region during the last half of the twentieth century was the political dominance of the Labour Party at the local and national parliamentary level. This provided a degree of continuity, but, unlike the Communist influence in Emilia Romagna discussed in Chapter 4, this did not provide an ingredient for the emergence of a successful 'regional productivity coalition' (see also Tomaney 2002, 2005).

The idea that strengthening regional institutions was a necessary ingredient in efforts to tackle inter-regional inequality and the plight of lagging regions like the North East was taken up by the New Labour government in 1997. The new administration instituted a programme of devolution to establish the Scottish Parliament, National Assembly for Wales and legislative assembly in Northern Ireland. There were multifarious political and cultural factors that led to the adoption of this programme (Tomaney 2000). In England, the focus was increasing coordination and effectiveness of the plethora of government agencies already operating at the regional level and the creation of priority regional development agencies (RDAs) in every part of England to lead the economic development agenda. The RDAs are agencies of central government, rather than of the regions themselves, although together with regional chambers they did provide space for new actors to engage in regional policy-making (O'Brien *et al.* 2004).

Echoing the new regionalist agenda of raising regional and thereby national economic performance, RDAs were instituted in prosperous as well as lagging regions, breaking with the post-war regional policy tradition of discriminatory and selective institutional support for assisted areas (House of Commons 2003; Fothergill 2005). In the shadow of a relatively more powerful Scottish Parliament (Pike 2002a), demand for a devolved political settlement for the region grew among the North East's political class. A proposal to create an elected regional assembly in the North East, albeit with very modest powers, was rejected in a referendum in 2004. This left the region without institutions which could become the focus of 'development coalitions' (Keating *et al.* 2003) or, still less, for the basis of a 'developmental regional state' (Ó Riain 2004) which – as we discussed in Chapter 4 – are associated with successful local and regional development elsewhere in Europe.

The North East faced profound challenges at the start of the twenty-first century. It is an economically weak and politically marginal region in the UK context, with little evidence that the development gap between it and the rest of the United Kingdom was closing. Moreover, the perils of strategies based on exogenous development discussed in Chapter 6 meant that the forms of competitiveness that had underpinned its development through the attraction of FDI in the second half of the twentieth century were being eroded as cost-sensitive and often labour-intensive manufacturing and increasingly service-oriented investments were made in new locations such as Central and Eastern Europe, China, India and other parts of Asia. The limits of the national strategy of the previous decades were revealed:

> The key historic location advantage of the UK has not been its technological or skill infrastructure, but rather institutional openness to FDI and new work practices and cost advantages connected with low relative labour costs and the availability of government incentives. While FDI in the UK is likely to be associated with organisation innovation, as MNCs like Nissan introduce new work practices and supplier relations, flexibility and cost-oriented investment motivations are unlikely in themselves to lead to technological innovation and integration with local industry.
>
> (Loewendahl 2001: 337)

The region has been left with a constrained indigenous industrial base, little in the way of technological and skill-based advantages upon which to innovate and promote adjustment, dwindling and increasingly selective national and EU local and regional policy support, and poorly configured institutions of local and regional government and governance. North East England's local and regional development prospects appear relatively bleak.

Ontario: economic adjustment, continental integration and the limits to learning in a North American region-state

Ontario is the industrial heartland of Canada. The province is the economic motor of its national economy, accounting for around 40 per cent of Canada's total GDP and employment (Wolfe and Gertler 2001). Historically, Canada's national development strategy and its privileged position within the British Empire shaped Ontario's local and regional development trajectory. Latterly, the relationship with the United States has become economically and politically dominant, especially given its geographical proximity (Figure 7.3). The particular form of post-war growth and prosperity experienced by Ontario has left the province with a legacy of structural issues. These include a branch plant economy due to high levels of foreign-ownership, under-investment in R&D and a wage advantage for labour with comparable skills and productivity relative to the United States due mainly to state-provided social benefits, including public healthcare. In government terms, Ontario is a province within the Canadian federation.

From industrial heartland to learning region?

From the early 1980s, Ontario experienced a sustained period of economic restructuring triggered by the kinds of challenges we introduced in Chapter 1, globalisation, technological change and intensified competition, each of which process has been accelerated by economic integration with the United States (Wolfe and Gertler 2001). The early 1990s recession was the deepest since the 1930s Great Depression, worsened by tight macroeconomic policy, high interest rates and currency appreciation, a new Federal value added tax and a US business cycle downswing. Ontario's output contracted by 7.8 per cent and 320,000 jobs were lost, almost two-thirds in manufacturing, unemployment rose above 10 per cent and investment collapsed (Wolfe and Creutzberg 2003). Amidst

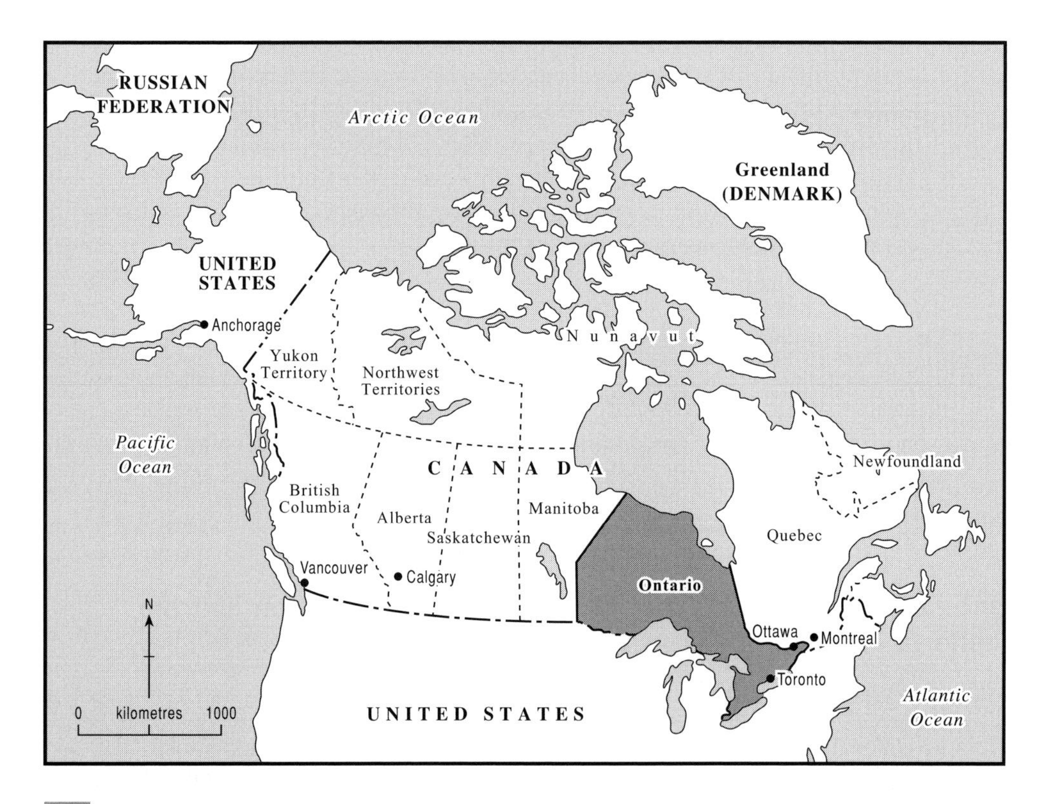

Figure 7.3 ***Ontario, Canada***

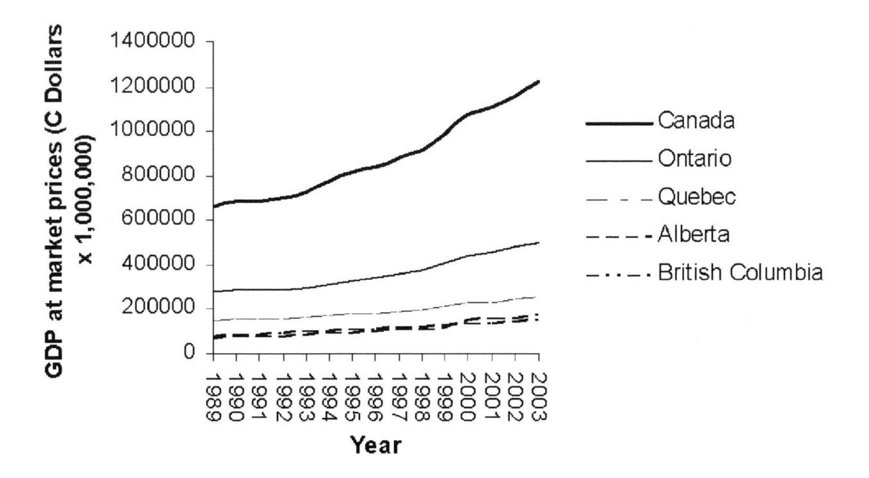

Figure 7.4 ***GDP at market prices (income-based) for selected provinces and Canada, 1989–2003***

Source: Calculated from Statistics Canada data

a modest recovery in the late 1990s and early 2000s, Ontario's growth has remained sluggish (Figure 7.4) and, while employment is growing in the upturn, unemployment remains volatile (Figure 7.5).

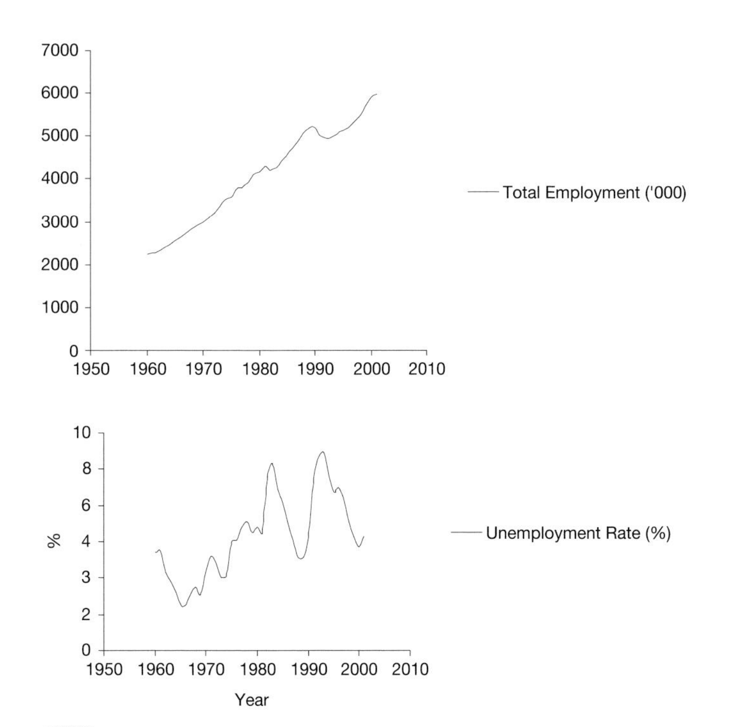

Figure 7.5 ***Total employment ('000) and unemployment rate (%), Ontario, 1960–2001***

Source: Calculated from Wolfe and Creutzberg (2003: 73)

Despite deindustrialisation and the structural shift towards services revealed in the high proportion of service employment (Table 7.1), Ontario remains a manufacturing heartland producing over 50 per cent of national manufacturing GDP in Canada (Wolfe and Gertler 2001). Nearly half of all jobs in the province are in manufacturing. Although the productivity gap with the United States remains, it has narrowed amidst lower employment levels. During the late 1990s economic recovery, high-tech sectors experienced high output and productivity growth while the overall number of plants and jobs fell (Wolfe and Gertler 2001). Table 7.2 illustrates that automotive assembly and parts dominate manufacturing, principally 'The Big Three' (Ford, GM and Daimler-Chrysler) and Japanese transplants (Toyota, Honda, Suzuki-GM), and the province is the second largest producer in North America after the US state of Michigan.

Electrical and electronics products, especially telecommunications equipment, remain important too. Leading high-tech firms have been attracted (e.g. Silicon Valley-based Cisco Systems) to tap into highly qualified, productive and, compared to the United

Table 7.1 ***Sectoral composition of employment, Ontario, selected years, 1955–2001***

Year	*Primary industry*	*Secondary industry*	*Tertiary industry*
1955	13.7	38.3	48.0
1961	9.6	33.3	57.1
1971	6.1	33.5	60.4
1975	4.7	30.8	64.5
1981	4.6	30.2	65.2
1991	3.3	18.0	71.7
1996	2.9	17.4	73.6
2001	2.0	18.2	73.1

Source: Adapted from Wolfe and Creutzberg (2003: 74)

Table 7.2 ***Top ten manufacturing industries in Ontario, 1999***

Industry	*Value of shipments and other revenue (CDN$ m)*	*% Top ten*	*% All industries*
Transportation equipment	98,637.3	46.6	38.4
Food and beverage	21,867.2	10.3	8.5
Electrical and electronic	20,277.7	9.6	7.9
Chemical	17,753.5	8.4	6.9
Primary metals	14,164.0	6.7	5.5
Fabricated metal	12,549.2	5.9	4.9
Machinery	8,889.1	4.2	3.5
Rubber and plastics	6,350.3	3.0	2.5
Refined petroleum and coal	5,935.5	2.8	2.3
Paper	5,052.2	2.4	2.0
Total	211,476.0	100.0	82.3
All manufacturing industries	257,033.3		

Source: Adapted from Wolfe and Creutzberg (2003: 75)

States, relatively lower wage labour. Supported by a dense research infrastructure, mirroring many of the attributes of a strong regional innovation system, information technology industries in Ontario include 8,000 firms and employ over 300,000 in the 'Technology Triangle' around Cambridge, Kitchener, Waterloo and Guelph and 'Silicon Valley North' in the Ottawa-Carleton region (Leibovitz 2003; Wolfe and Creutzberg 2003). Ontario, especially Toronto, is attractive for high-skilled labour from within Canada and internationally. In terms of attracting the exogenous resources potentially offered by the 'creative class' discussed in Chapter 6, recent research compares Canadian cities respectably on creativity indices with city-regions in North America (Gertler *et al.* 2002).

Alongside concentrations of foreign ownership, Ontario has a strong indigenous sector of home-grown international companies with headquarters and key R&D functions in the province, for example in telecommunications and aerospace (e.g. Nortel Networks, Bombardier). Indigenous banking is important too. Toronto is a key financial centre nationally and third after New York and San Francisco in North America, although its relatively small size, autonomy and national orientation has meant limited scale economies and an inability to promote indigenous industrial adjustment (Wolfe and Creutzberg 2003).

Local and regional development in Ontario has focused upon the challenge of adjusting to deindustrialisation and promoting the kind of innovation and shift to a 'learning economy' capable of sustainable development discussed in Chapter 3 (Gertler 1995). The aim has been 'to promote the transition of the Ontario economy towards those sectors and firms with the capacity to generate higher wage, higher value-added and environmentally sustainable jobs' (Wolfe and Gertler 2001: 585). Changing from the historical focus upon top-down, hierarchical, nationally-centred policy and 'hard' infrastructure in the post-war period of the kind introduced in Chapter 1, recent local and regional development policy has sought to build a more decentralised regional innovation system based upon intersectoral cooperation, trustful relations and social learning (Wolfe 2002). Mobilising indigenous and attracting and embedding exogenous resources are evident. Provincial government policy has invested long term in post-secondary education since the 1960s and constructed a strengthened research base and links with industry, upgrading existing manufacturing and, during the early 1990s, developing clusters.

Changes since the 1990s in Ontario's economy have been profound but uneven. Evidence exists to support the role of policy in improving the quality, sophistication, timeliness and innovation record of economic activities in the province (Wolfe and Gertler 2001). Elements of a regional innovation system exist alongside a fragmented policy context sometimes lacking coordination. Influenced by the dominant neo-liberalism of their close neighbour the United States, in some assessments Canada's individualistic and anti-cooperative culture has militated against private sector support and participation in the high-trust relations and partnership-based social organisation central to the more associative forms of governance described in Chapter 4 (Leibovitz 2003). Challenges remain for local and regional development in Ontario as it grapples with a pattern of growth that is more productive but generates fewer and specific sorts of jobs.

Continental integration and the emergence of a North American region-state

During the 1950s and 1960s, Canada's National Plan embodied Keynesian demand management and opened the economy to free trade and the foreign investment-led modernisation of its resource and manufacturing sectors (Wolfe and Creutzberg 2003). Post-war growth and prosperity disproportionately favoured Ontario through Federal import substitution industrialisation, patent legislation and preferred trading links in the Commonwealth remnants of the British Empire. This 'Golden Age' reached its limits with the crisis of Fordism we discussed in Chapter 3 and in the changing economic context of local and regional development. Stagflation, internationalisation and overcapacity intensified competition from the 1970s.

Crisis stimulated debate between competing visions of the state's role in the economy and industrial policy in the 1970s and 1980s, mirroring the tension between neo-classical and Keynesian interpretations of local and regional development discussed in Chapter 3. Trade-led adjustment struggled with a more interventionist approach to developing the 'technological sovereignty' of Canadian industry, inspired by the experiences of post-war France and Japan (Wolfe and Creutzberg 2003). Amidst the neo-liberal turn in the international political economy, the Royal Commission on the Economic Union and Development Prospects for Canada (1985) rejected interventionism for a market-oriented approach due to the openness of the economy, federal structure and fragmented social structures incapable of fostering consensus. Canada subsequently sought free trade and integration with the United States, tight monetary policy and labour market reforms.

Trade-led adjustment accelerated the global and especially US integration of Canada and Ontario in an emergent North American economy, following the Auto Pact in the late 1960s, the Free Trade Agreement (FTA) in 1989 and its successor the North American Free Trade Agreement with Mexico in 1994. Trade and investment flows were liberalised and cross-border production networks expanded as indigenous Canadian firms sought expansion into the larger and more lucrative US market (Wolfe and Creutzberg 2003). Exports and imports in key sectors are dominated by trade with the United States. Ontario's exports to the United States account for around 45 per cent of its GDP (Courchene 2001). Cross-border rationalisation has proceeded through mergers and acquisitions, forging links between Canadian and US and global firms (Wolfe and Gertler 2001). As a problem generated by exogenous development, low wage and low value-added activities have been readily outsourced to Southern US states and Mexico in the context of external control and foreign-ownership.

As part of North American economic integration, the 'East–West' national Canadian economy has been supplanted by a series of 'North–South', cross-border 'regional' economies between Canada and the United States (Courchene 2001). Ontario has reoriented itself from a provincial economic heartland and focal point for the trans-Canadian economy to a North American region-state, building upon its close geographical proximity to major US markets in the Great Lakes. Central to this transformation is Greater Toronto's evolution from a provincial capital with significant international reach to a global city-region. Ontario and Greater Toronto are seeking to build their broader roles in the North American and international context while preserving their national positions within Canada.

Multilevel government and governance in the Canadian federal system

Connecting to our discussion in Chapter 4, within Canada's multilayered system of government and governance, the federal level remains important for local and regional development. Indeed, the late 1990s economic recovery owed much to federal macroeconomic reforms and stability (Courchene 2001). Devolution to the provincial level has proceeded unevenly but has included post-secondary education and additional research funding (Wolfe and Gertler 2001). In Ontario, provincial–municipal relations between regional and local government were profoundly reorganised by the incoming Conservative Administration in 1996 as it sought to enhance the province's attractiveness in a North American context and wrest back power from public sector unions. 'Hard' services (e.g. property, infrastructure) were shifted to the municipalities and 'soft' services (e.g. education, health and welfare) to the provincial level (Courchene 2001). Intergovernmental coordination remains thorny and tensions continue between the Federal and provincial levels. Despite devolution to the provinces, no single level within a multilayered system controls all the policy instruments needed to implement cohesive regional industrial strategies (Wolfe and Creutzberg 2003).

In the context of the economic challenges of the 1990s, political change has complicated an already fragmented institutional structure in Ontario. Conservative governments ruled from 1943 to 1985, focused upon attracting investment and controlling public expenditure and debt (Wolfe and Creutzberg 2003). The 1980s economic crisis heralded a period of political swings and instability. Four different governments involving all three major parties ruled from the mid-1980s. In the early 1990s, the social democratic New Democratic Party's (NDP) local and regional development policy innovations were overshadowed by the near 'fiscalamity' of the doubling of the province's indebtedness and triggering of successive credit down-ratings (Courchene 2001; Wolfe 2002). The market-oriented Progressive Conservatives were elected in 1995 and 1999 with dramatic consequences for local and regional development policy. As part of their 'Common Sense Revolution' of public expenditure reductions, tax cuts and deregulation to stimulate growth, the administration terminated the NDP's cluster development strategy (Wolfe 2002). Following the publication of the Ontario Jobs and Investment Board's (1999) *A Roadmap to Prosperity*, only in the late 1990s was there a belated recognition of the role of public sector spending programmes and tax incentives to renew the infrastructure, encourage innovation and build the knowledge-economy for local and regional development (Wolfe and Gertler 2001).

Ontario's movement towards a 'learning region' remains uneven despite the evidence of investment in research, skills and education. Collaboration and networking continue to be stymied by the individualistic and anti-cooperative character of Ontario's industrial culture, its weak governance and coordinating capacity and state traditions, the neoliberal agenda of the current provincial government and continued under-investment in R&D (Atkinson and Coleman 1989; Wolfe and Gertler 2001). In the context of continental integration, key industries are being 'hollowed out' in both domestic and foreign-owned sectors. Enterprise remains relatively weak in the Canadian and especially US context limiting the effectiveness of the kinds of interventions in support of indigenous development identified in Chapter 5. Toronto ranked third behind Vancouver and Montreal for venture capital financing in 2001 (Wolfe and Creutzberg 2003).

Inter-territorial competition among North American city-regions to attract and retain investment, jobs and high-skilled labour has intensified, prompting renewed place marketing for Ontario. Public debate has focused upon Canada's productivity shortfall, the falling Canadian dollar and higher taxes relative to the United States (Institute for Competitiveness and Prosperity 2005). As Courchene (2001: 163) notes: 'Ontario probably takes due notice of tax rates in [the Canadian Provinces of] British Columbia and Nova Scotia. But it is far more concerned about tax rates in Michigan, Ohio, and New York'. The Canadian 'social envelope' is under pressure as part of moves towards dismantling the welfare state in a deregulatory 'race to the bottom' against US states, reinforced by the agenda of Ontario's Progressive Conservatives of stronger incentives for entrepreneurs, sound fiscal management and public expenditure restraint. The arena of inter-territorial competition may yet be expanding further through North, Central and South America in the potential transcontinental Free Trade Area of the Americas.

As we discussed in Chapter 4, the scope and capacity of the public realm and its institutions are in question. In this context, Ontario's governors have had to confront the local and regional development challenges generated by the diseconomies of growth in its major cities, the urbanisation of its social problems and the suburbanisation of its tax base (Courchene 2001). As evidence of the growing importance of sustainable development detailed in Chapters 2 and 3, the establishment of 'Smart Growth' panels are a tentative response across the province. Critically, however, while Canadian cities are usually politically weak creatures of their provincial governments (Courchene 2001), Greater Toronto's emergent role as a North American city-region has stoked up demands for further self-determination within Ontario province and the Canadian Confederation (Jacobs *et al.* 2000). In the context of regional tensions in province–Federal relations, Ontario faces its local and regional development challenges in a more flexible, multilateral, negotiated and associative – including state and non-state organisations – multilevel system of government and governance (Wolfe and Creutzberg 2003). The extent to which this much more complex institutional architecture promotes or inhibits local and regional development in Ontario remains to be seen.

Silicon Valley: regional adjustment, innovation and learning

Silicon Valley is a place of recurrent interest for local and regional development. As a geographical concentration of internationally competitive high-tech economic activities, Silicon Valley is an icon of successful growth and a powerful symbol of US industrial leadership and entrepreneurial spirit (Walker 1995). From the 1940s, Silicon Valley has forged and retained a role at the forefront of the fifth Kondratiev or long wave of economic growth as the crucible of innovation and technological change in electronics in the information and communication technology industries. Historically, Silicon Valley centred upon Santa Clara County in the San Francisco Bay Area, Northern California (Figure 7.6). The Valley has since extended geographically into Alameda, San Mateo and Santa Cruz Counties and encompassed emergent multimedia activities in San

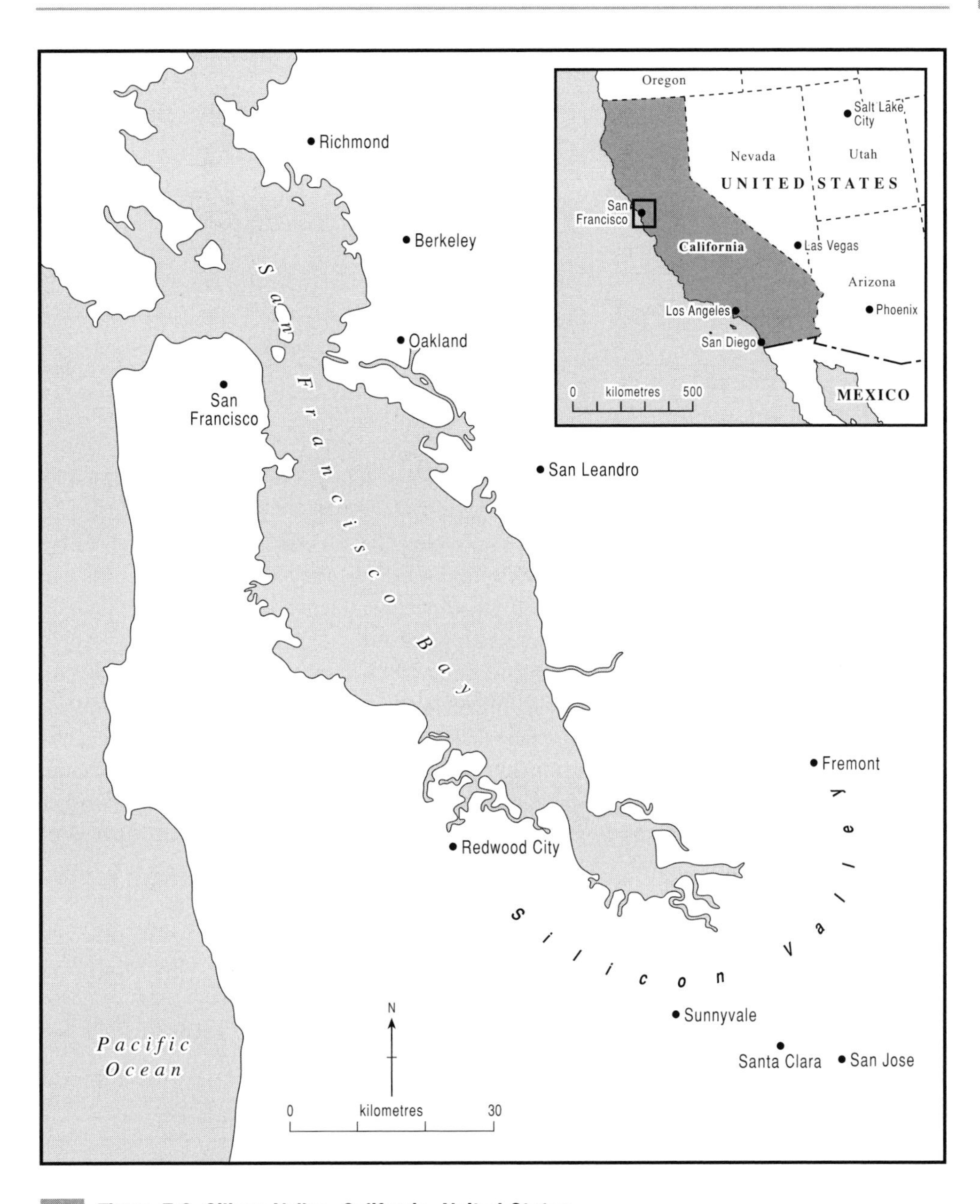

Figure 7.6 ***Silicon Valley, California, United States***

Francisco and software firms in the East Bay and Marin County (Saxenian 1995). The economy of Silicon Valley is dissonant with its territorial governance, spanning several authorities with limited political or cultural sensibility (Keating 1998). In line with economic change and linking to our discussion of the institutions of government and governance in Chapter 4, Table 7.3 shows how Silicon Valley's governance has evolved through distinct stages marked by elite networks at state and Federal level and an increasing focus upon local and regional issues and public–private partnership (Henton 2001).

Table 7.3 ***Economic eras and their form of regional governance in Silicon Valley, 1950s***

Economic era	*Regional governance*
Semiconductors (1959–1975)	Laissez-faire
	■ Semiconductor Industry Association
Personal computers (1975–1985)	Business-led
	■ Santa Clara Manufacturing Group
Software (1985–1995)	Collaboration
	■ Joint Venture: Silicon Valley Network
Internet (1995–)	Networking
	■ Silicon Valley Civic Action Network

Source: Adapted from Henton (2001: 396)

Following a downturn in the late 1990s and early 2000s, Silicon Valley increased its stock of fast growing firms in 2004 for the first time since 2000 and accounted for 10 per cent of all patents filed nationwide in the United States (Joint Venture: Silicon Valley Network 2005). The Valley's diversified mix of sectors, specialisations, firm sizes and indigenous and exogenous ownership structures are highly internationalised within global export markets, including well-established firms (e.g. Hewlett-Packard, Apple Computer) and rapid growth start-ups from the 1980s (e.g. Sun Microsystems, Silicon Graphics). The majority of firms in Silicon Valley are SMEs, however, occupying specialised niches in intra-industry trade and representing the dynamism of indigenous development described in Chapter 5. Key functions of headquarters, R&D and design as well as production reside in the Valley as an example of regional industrial rather than functional specialisation within the spatial division of labour. Since the 1960s, routine manufacturing has been outsourced to newly industrialising countries (e.g. Hong Kong, the Philippines) and lower-wage regions of the United States (e.g. Texas, New Mexico) (Scott 1988). Employment in the core technology companies is over 350,000 and, as Table 7.4 illustrates, labour productivity is more than twice the national average (Bay Area Economic Forum 2004). New firm formation rates are high, new business starts added 166,200 firms while deaths subtracted 125,000 firms between 1990 and

Table 7.4 ***Output per capita, selected US regions, 2002***

Region	*Output per capita ($ Thousands)*
Bay Area, CA	63.4
Boise, ID	48.5
Austin, TX	46.7
New York	47.4
Boston, MA	44.0
Seattle, WA	41.2
Houston, TX	40.5
Los Angeles, CA	32.6
United States	32.7

Source: Adapted from Bureau of Economic Analysis (BEA) cited in Bay Area Economic Forum (2004: 6)

2002 (Joint Venture: Silicon Valley Network 2005). New start-ups typically attract over one-third of total US national venture capital investment funds, worth over $7 billion in 2004. Silicon Valley's economic vitality underpins relatively high growth rates. At $52,000 in 2004, per capita income levels are significantly above the US national average ($32,800) albeit skewed by a select group of individuals in receipt of very high incomes.

Explaining Silicon Valley

Conventional explanations of Silicon Valley focused upon the external economies explained in Chapter 3 (Scott 1988). Production has disintegrated vertically and localised to enable flexible adaptation to market uncertainty and rapid technological change (Castells 1989). Our other frameworks of understanding from Chapter 3, geographical economics (Krugman 1991) and clusters (Porter 1990), interpret Silicon Valley as a classic agglomeration economy, benefiting from the cumulatively self-reinforcing processes of increasing returns and localised technological spillovers. Other analyses see an archetypal industrial district (Piore and Sabel 1984) or the coincidence of technology cycles and regional growth (Hall and Markusen 1985).

Contrasting the divergent performance of Silicon Valley and its competitor high-tech regional complex 'Route 128' in Massachusetts, Saxenian (1994) challenged conventional accounts. External economies assume clearly defined boundaries between the inside and outside of the firm. Drawing upon the socio-economic approaches detailed in Chapter 3, Saxenian's (1994: 1) network approach argues that this focus neglects the 'complex and historically-evolved relations between the internal organization of firms and their connections to one another and the social structures and institutions of a particular locality'. Firms are embedded in networks of social and institutional relationships that shape and, in turn, are shaped by their actions. Over time, such networks evolve distinctive industrial systems embedded in places:

> Silicon Valley . . . has pioneered a decentralized industrial system in which firms specialize and compete intensely, while collaborating in informal and formal ways with one another and with local institutions, like universities, to learn about fast changing markets and technologies. The successes of Silicon Valley firms thus depend as much on being a part of local social and technical networks as on their own individual activities.
>
> (Saxenian 1995: 2)

Such networks spread the costs of new technology development, reduce time to market and enable reciprocal innovation:

> Silicon Valley is far more than an agglomeration of individual firms, skilled workers, capital and technology. Rather it is a technical community that promotes collective learning and flexible adjustment among specialist producers of complex related technologies. The region's dense social networks and open labour markets encourage experimentation and entrepreneurship.
>
> (Saxenian 1995: 12)

In Saxenian's interpretation, the foundations of Silicon Valley's local and regional development trajectory were laid by leading individuals, such as Frank Terman, Dean of Engineering at Stanford University, in the 1940s. His vision for a 'technical community' revolved around the kinds of competitive and collaborative relationships between individuals, firms and institutions envisaged by socio-economic approaches and was founded upon porous institutional boundaries, continuous experimentation, information exchange and innovation through strong and local university–industry links (Saxenian 1995). Such developments are evidence of the approaches to indigenous local and regional development approaches detailed in Chapter 5.

Rather than heroic individual entrepreneurialism, federal government established the military–industrial–academic complex that stimulated Silicon Valley's early development and periodic, defence-related expansions (Leslie 1993). Stanford University, in particular, has been a key locus for education, research, skills and finance. Given its relative growth and prosperity, local and regional policy has been negligible. Over time, path dependency has locked-in virtuous networks of social and productive relations, institutionalised both informally and formally, that underpin the dynamism of Silicon Valley. The social and productive networks are geographically embedded – they are the 'relational assets' forged in and through Silicon Valley as a place (Storper 1997) – and are capable of adaptation and renewal over time. It is these social and productive networks rather than specific individuals, firms or technologies that provide the adjustment mechanisms underpinning the evolution and resilience of Silicon Valley. Figure 7.7 describes how Silicon Valley has evolved and led each techno-economic paradigm in the post-war electronics industry.

Competition and cooperation characterise Silicon Valley's social and productive networks. Competitive rivalry compels firms to define and defend markets, intensifying innovative activity (Saxenian 1995). Entrepreneurial 'heroes' pioneer new firms and

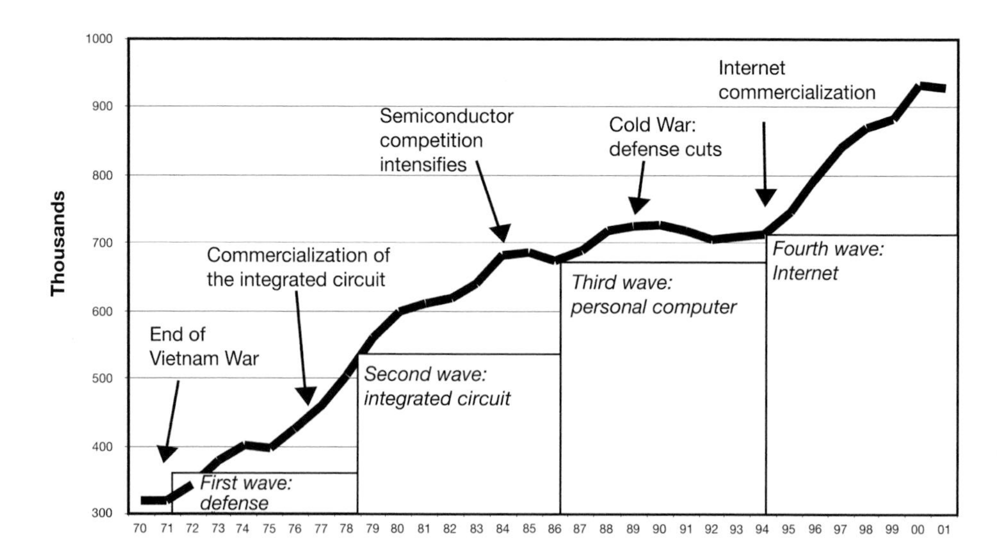

Figure 7.7 ***Employment and technological eras in Silicon Valley, 1970–2001***

Source: Henton (2001: 7)

technologies. High risks are matched by potentially high rewards. Silicon Valley's rich social, technical and institutional infrastructure of specialist goods and service suppliers provides a uniquely supportive context. A shared commitment to technological advance underpins an open and reciprocal culture. Informal dialogue and collaboration pools leading-edge knowledge about fast changing technologies in competitive markets among industry participants. Many new ventures fail but this is tolerated as necessary experimentation – a rite of passage and an opportunity for the types of collective learning deemed significant in the innovation, knowledge and learning approaches in Chapter 3. Although the proliferation of start-ups and multiple businesses competing to develop similar technologies has been considered wasteful (Florida and Kenney 1990), many more technological and organisational alternatives are explored than could be pursued within a large firm hierarchy or region with less fluid social and industrial structures (Saxenian 1994).

As a key element of Chapter 5's indigenous approaches, the labour market is central to the enduring economic dynamism of Silicon Valley. Social and professional networks endure beyond the firm and transcend sectoral and occupational boundaries. An open market exists for particular occupations and groups, facilitating job search by individuals and skills matching by existing and new firms. Supported by geographical proximity, high rates of labour mobility between firms are considered normal. Multi-firm career paths underpin adaptability and reinforce the culture of continuous change, experimentation and innovation in Silicon Valley (Saxenian 1995). Commitments to profession and technological advance rather than firms make engineers receptive agents of the Schumpeterian 'creative destruction' in long wave theories.

In addition to indigenous skill formation, Silicon Valley is a magnet for attracting and embedding exogenous assets and resources such as industry functions and talent from within the United States and internationally, attracted by leading-edge technology, dynamism and potential financial rewards. High-skilled immigration, particularly from the Pacific Rim, has further stimulated renewal. Entrepreneurial immigrants have established new businesses, forging relations between Silicon Valley and emergent high-tech complexes in Bangalore, India, and Hsinchu, Taiwan. In 1996, 1,786 firms with $12.6 billion in sales and 46,000 employees were run by Indian or Chinese executives (Saxenian 1999). While their scarce technical skills are valued and external connections are being built, immigrant entrepreneurs have also faced discrimination in capital and labour markets and have created social enclaves and networks for mutual support.

Unequal growth and the 'Two Valleys'

High growth rates in Silicon Valley have confounded neo-classical explanations of local and regional development. Silicon Valley has historically been a relatively high-cost location. High levels of innovation, skills and productivity offset high factor costs in a 'high road' local and regional development model (Walker 1995). Historically, amidst periods of accelerated industrialisation in the post-war period, the diseconomies and negative externalities expected by the external economies approach have been accommodated in Silicon Valley. Geographical expansion and skilled immigration have helped, especially international relationships with the nascent 'Silicon Valleys' outside the United States (Parthasarathy 2004). Sustainability has become a critical issue, however. The Bay remains among the fastest growing urban areas in the United States, stoking inflation in land, housing and labour markets and generating congestion and extended commuting, particularly for the peripheral workforce from San Jose, Gilroy and East Bay. While relatively high, Table 7.5 shows how median incomes have not kept pace with rising housing prices. Environmental damage is evident too, caused by the production of toxic waste, particularly among the myriad of manufacturing subcontractors (Pellow and Park 2002). The recent character of local and regional development has led to calls for more sustainable development and 'liveable communities' in Silicon Valley (Collaborative Economics 1998).

Growth has been distributed unequally between social groups and places in Silicon Valley. For Siegel (1998), 'Two Valleys' exist divided between the relatively prosperous, inner 'core' suburbs of white professional and skilled workers and the relatively disadvantaged, South and East Asian, Black and Hispanic 'peripheral' hinterlands of semi- or unskilled labour. This dualism may not adequately capture the complexity of occupational segregation and geographical separation along class, gender and racial lines, however.

While a source of regional adjustment and vitality, labour market flexibility and segmentation are key sources of social and spatial inequality in Silicon Valley. Benner (2002) identifies three trends: non-standard employment growth (e.g. temporary and independent contracting), high job turnover rates and inter-firm mobility and increased skills obsolescence. The upper echelons of the core labour market are characterised by high skills and rewards, often accompanied by high risks. Peripheral labour market segments exhibit functional and numerical flexibility. Class, gender and racial divisions are marked (Siegel 1998). Despite government and non-profit training programmes, for example, non-white workers constituted just over a quarter of managers and professionals but two-thirds of operators, labourers and service workers in the top thirty-three firms in 1996 (Benner 2002). Weak employment protection and poor health and safety records characterise peripheral subcontract firms. Collaboration's darker side is evident too. Firms have sanctioned whistle blowers, prevented government regulation and

Table 7.5 ***Median household income, median home prices and purchasing power, selected US regions, 2002***

Region	*Median household income ($000)*	*Median home price ($000)*	*Purchasing power (income as % of home price)*
Bay Area, CA	72	441	16.3
Boston, MA	62	411	15.1
Los Angeles, CA	45	284	15.8
Seattle, WA	49	256	19.1
New York	44	309	14.4
Austin, TX	52	157	33.1
Houston, TX	48	132	36.1
Boise, ID	46	130	35.1
US average	43	166	25.9

Source: Adapted from Population Demographics cited in Bay Area Economic Forum (2004: 13)

withheld 'commercially sensitive' safety information (Pellow and Park 2002). Labour organisation enjoys limited success in this traditionally non-union and fragmented industry, except among subcontracted janitors (Siegel 1998). As firms outsource 'human resource' functions to specialist service providers, the insecure and volatile labour market in Silicon Valley means employees increasingly depend upon intermediaries (e.g. professional associations, unions) for benefits and retraining (Benner 2002).

Importing and sustaining the Silicon Valley model?

The Silicon Valley experience has proved attractive to policy-makers seeking to replicate its local and regional development benefits. Recurrent attempts to distil the essence of a 'Silicon Valley model' have shaped policy. Approaches based upon a mix of indigenous and exogenous approaches in a supposed formula of research universities, science parks, skilled labour and venture capital in an amenable environment have often foundered (Saxenian 1989). The ingredients have often lacked the supporting context of historical Federal infrastructure investment, social and productive network relations central to Silicon Valley's continued dynamism and ability to adjust, innovate and learn in demanding, fast changing and competitive markets.

While it has repeatedly confounded critics that foresaw its demise like other regionally specialised industrial complexes such as Detroit and cars and Pittsburgh and steel in the United States, questions exist concerning Silicon Valley's continued viability and sustainable development. Some fear Silicon Valley may have outgrown the open, competitive and cooperative culture central to its adaptability (Saxenian 1995). The hubris of individual greed and increased litigation against former employees and/or suspected imitators are threatening to weaken the collective social and productive networks upon which Silicon Valley depends (Saxenian 1995). The state-led social investment of the post-war period that powered the early development and infrastructure of Silicon Valley (e.g. universities) has not been adequately renewed amid the tax cuts, reduced spending and state-level budget deficits of the 1980s and 1990s (Bay Area Economic Forum 2004; Walker 1995). Record layoffs and stagnating wages accompanied the boom and bust of the 'dot.com' bubble of the late 1990s. Jobless growth marks the recent recovery (Mullins 2005). Internal contradictions of unsustainable growth accompanied by labour market insecurity and inequality may challenge the economic and social cohesion of the Valley as concern grows about unemployment, low wages, commuting times, job competition, housing costs and health care. Whether Silicon Valley can regain its industrial leadership or see it pass irreversibly to the emergent centres in the Pacific Rim is unresolved (Walker 1995). Silicon Valley may yet evolve into a lesser node in an internationalising network potentially eroding the local and regional development benefits of its growth trajectory.

Busan: industrialisation, regional inequality and balanced national development

Busan, South Korea's second largest city, is located in the far south-east of the peninsula and emerged as an important industrial centre during the country's rapid rise to industrial prominence from the 1960s (Figure 7.8). Busan's growth was fuelled by the development of the footwear, textiles and shipbuilding industries. South Korea's remarkable economic ascent occurred following its devastation in a civil war during the 1950s in which virtually all major cities were destroyed and millions of people were killed. A rapid process of industrialisation and urbanisation in the 1960s and 1970s meant that Korea's per capita income rose from 15 per cent of the OECD average in 1970 to 70

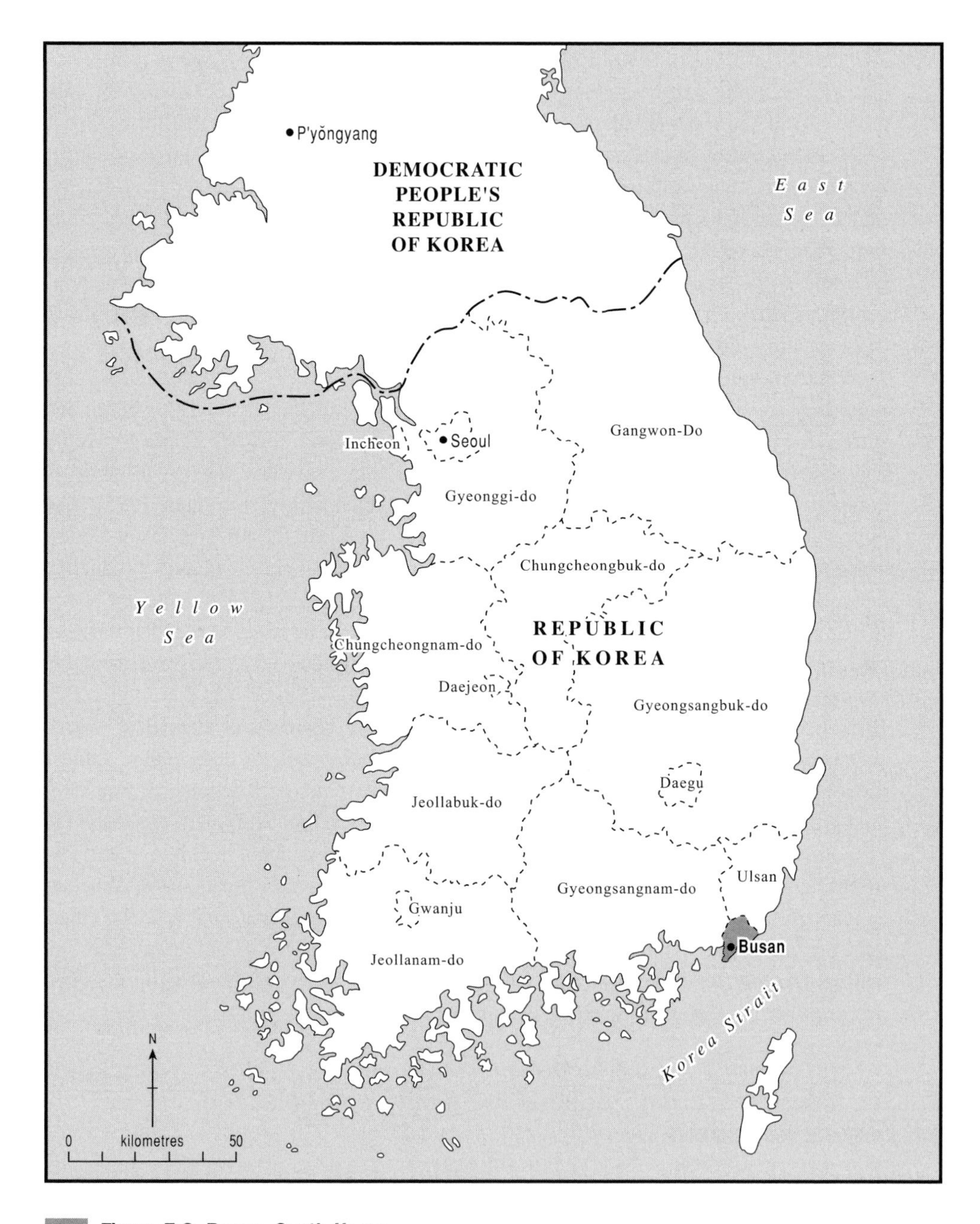

Figure 7.8 *Busan, South Korea*

per cent in 2005 (OECD 2005a) and Korea was the eleventh largest economy in the world (Cumings 2005).

The developmental state and national economic growth

Under the successive authoritarian governments of General Park Chung-hee in the 1960s and 1970s Korean industrialisation followed the import substitution model, a pattern of development that was repeated across East Asia during this period (see Chapter 2; see also World Bank 1995). The early Park Chung-hee governments promoted labour-intensive industries, notably footwear and textiles. In the 1970s, the Korean government shifted to promoting export industries and launched the 'heavy and chemical industry drive', focusing initially on petrochemicals and automobiles, as well as steelmaking, engineering, shipbuilding and, later, consumer electronics. The role of the government during this period focused on protecting the domestic market, tightly controlling the banking system in order to provide cheap finance for industry and ensuring a repressive labour relations regime. These conditions facilitated the development of the *chaebol*, the family-owned industrial groups, which were later to become global brands such as Daewoo, Hyundai, LG and Samsung (Amsden 1992; Evans 1989; Cumings 2005). The close relationship between government, industry and finance led Korea to be given the sobriquet 'developmental state', that is a system in which the state's primary role is to facilitate industrial growth (see Johnson 1999; Wade 2003), but in the Korean case was also associated with political corruption (Cumings 2005).

The highly centralised nature of the authoritarian and military governments of this period meant that industrial policies were not concerned with achieving territorially balanced national development. Despite its ethnic homogeneity, however, regionalism has deep historical roots in Korean political culture, north and south, and the pro-democracy movements of the 1970s and 1980s partly had their foundations in strong regional identities. As industrial location policies developed from the end of the 1960s, regional partisanship played a part in its designations, often favouring Seoul and cities such as Busan and Ulsan in Gyeongsang province, at the expense of Gwangju and the south-western Jeolla region (Cumings 2005; see also Kim and Kim 1992; Kim 2003).

The growth of the Korean economy continued more or less unabated, surviving the oil price rises of the 1970s, until the economic crisis that swept through the Asian economies in 1997 (see Korea National Commission for UNESCO 2001). It is widely agreed that, especially in the case of South Korea, the crisis was not an outcome of poor macroeconomic fundamentals, but of the rapid deregulation of the financial system under pressure from the United States, leading to unstable asset speculation (Stiglitz 2002). Although the Korean economy recovered surprisingly quickly from this setback, by the beginning of the twenty-first century, Korea's development path was at a crossroads. On the one hand, Korea came under pressure from institutions, such as the IMF, to open its economy more directly to global market forces and FDI. On the other hand, new democratic governments, notably under presidents Kim Dae-jung and Roh Moo-hyun, committed themselves to a 'participatory economy' aimed at tackling inequality and injustice (see Kim 1996). In particular, the government of President Roh Moo-hyun, which took office in 2003, committed itself simultaneously to reducing the power of the *chaebol*, attracting FDI through the creation of three 'free economic zones' in Incheon, Gwangyang and Busan-Jinhae, strengthening subnational government and a far-reaching regional policy aimed at achieving 'balanced national development' and closing the widening gap between the Greater Seoul region and the rest of the country. The new regional policy included placing limits on industrial development in the Greater Seoul region; moving the capital city out of Seoul and relocating over 170 government agencies to the regions of Korea, in the process reducing the proportion of public organisations located in the capital region from 85 per cent in 2004 to 35 per cent by 2012.

The new policy did not go uncontested. The legislation, passed in 2004, to create a new administrative 'capital' in Chungcheong Province was declared unconstitutional by the Constitutional Court. The government then proposed the 'Special Act for the Construction of the Administrative City', which was enacted in 2005. In contrast to the 2004 plan, Seoul will remain as the capital and home to the legislative and judicial branches, as well as the president. According to the revised plan, forty-nine government agencies, including twelve ministries, will be relocated by 2014. The population of the new city is targeted at 500,000 by 2030. The potential regional development impacts of this move, in theory at least, are substantial (Lee *et al.* 2005). In addition to the new Administrative City, the government pressed ahead with its plans to move 176 public organisations outside of the capital region by 2012. Taken together these measures, nevertheless, represent one of the most radical attempts in the world to alter a national spatial economy.

Globalisation and the growth of regional inequalities: the challenge for Busan

The commitment to balanced national development presents a set of challenges and opportunities to Busan, which lies 500 km south-east of Seoul and has a long history as a port and as a trading centre with Japan that predates the era of industrialisation (Figure 7.8). Traditionally, it was also the capital of Gyeongsang province. However, reflecting the city's growing size and industrial importance, Busan was given the status of a separate metropolitan authority as far back as 1962, although its boundaries were extended subsequently. In 2005, the metropolitan region had a population of 3.75 million.

Although remaining an important industrial centre, Busan's relative position deteriorated during the 1990s, with its share of national GDP declining from 8 per cent in 1990 to 6 per cent in 2002. Busan's loss was Seoul's gain as the capital and the neighbouring Gyeonggi Province grew rapidly. This Greater Seoul region now contains over 40 per cent of the national population. The widening gap is mostly attributable to low labour productivity, which in turn reflects the sectoral composition of Busan's economy. Busan's relative prosperity in the 1970s was based on footwear and textile production and other traditional forms of manufacturing, which accounted for 25 per cent of Korea's exports. The decline of these industries reflects the growth of low-wage competition from elsewhere in Asia (Shin 2004), but the gap between Busan and Seoul widened after the 1997 crisis. Seoul's industrial mix of R&D-based firms, ICT industries, business services and media industries and its prospective role as a 'world city' prospered in the more liberal post-1997 environment (OECD 2005d). While Busan experienced some recovery, this was at the cost of a weakening export performance – its share of Korean exports fell from 10.4 per cent of the Korean total in 1990 to 3 per cent in 2005 – and a shift to the use of temporary forms of labour. Overall, Busan's performance relative to that of the capital region (Seoul, Incheon and Gyeonggi Province) deteriorated during the 1990s (Table 7.6).

It was the rise of Korea which presented new problems to regions such as North East England from the 1970s that we discussed earlier in this chapter, but Busan is now facing similar problems to old industrial regions in Europe and North America. The economy contains relatively few of the high-tech industries that characterise the knowledge-based economy and hardly any locally headquartered firms. Busan's port remains important – and one of the largest in the world – but faces growing competition from elsewhere in North East Asia, especially China where massive expansions in port capacity have taken place. There is more general evidence that firms are leaving Busan for neighbouring or remote regions and that new firms are growing at an insufficient rate to offset this, while footwear and textile production has relocated to the Philippines, Vietnam and China among others.

Recent developments, however, need to be put into context: much of the employment change in manufacturing reflects productivity gains and many firms have relocated to the wider region. Moreover, Busan remains a world centre for shipbuilding, along with neighbouring Ulsan. The most likely medium-term prognosis for Busan is that it will remain an important industrial region, but with manufacturing providing a smaller proportion of overall employment. The region needs to generate employment from other

Table 7.6 ***Weight of GDP by regions, South Korea, 1985–2000***

	1985	*1990*	*1995*	*2000*
Nationwide	100	100	100	100
Seoul	25.8	26.3	23.6	20.3
Busan	8.2	7.8	6.7	5.8
Daegu	3.9	4.1	3.9	3.2
Incheon	4.3	4.9	5.0	4.2
Gwangju	—	2.2	2.4	2.1
Daejeon	—	2.4	2.2	2.1
Ulsan	—	—	—	4.8
Gyeonggi-do	12.9	15.5	17.0	18.7
Gangwon-do	3.7	3.0	2.7	2.5
Chungcheongbuk-do	3.4	2.9	3.3	3.5
Chungcheongnam-do	6.1	3.5	4.1	4.3
Jeollabuk-do	4.1	3.5	3.6	3.2
Jeollanam-do	7.6	5.2	5.3	4.9
Gyeongsangbuk-do	7.4	6.8	6.7	6.3
Gyeongsangnam-do	11.6	10.9	12.2	6.6
Jeju-do	0.9	1.0	1.0	0.9

Source: KOSIS (Korean Statistical Information System) Database, cited in Lee (2004)

Note: Under the 1997 administrative area revamping, Ulsan was elevated to a metropolitan city, and taken out of Gyeongsangnam-do province.

sources. It has witnessed some growth in ICT-based business, although not on the scale of Seoul, while Pusan (Busan) International Film Festival (PIFF) is the largest of its kind in Asia and can be seen as evidence of successful place marketing and the possible catalyst for new media industries. Nevertheless, Busan faces considerable challenges in the context of intensifying competition in the Asia-Pacific region, where export markets in the early 2000s grew at about twice the global average (OECD 2004b).

The task facing Busan is to maintain its important port role while developing new and complementary economic specialisms. Critical to the region's strategy is the development of the Busan-Jinhae Free Economic Zone (FEZ), established by the Korean government and opened in 2003, a form of export processing zone of the type discussed in Chapter 6. The FEZ provides tax breaks for foreign investors, designed to increase flows of FDI, which remain low in Korea, in general, and Busan, in particular, despite the IMF-imposed reforms mentioned above. In the Busan context, the limits of such initiatives are magnified because another FEZ is located in Incheon, adjacent to the principal international airport and Seoul. FEZs could lead to increased 'fiscal wars' and territorial competition within Korea of the type discussed in Chapter 1 and which Incheon seems better placed to win.

Developing indigenous assets through a regional innovation system

The national and metropolitan authorities in Busan have sought to develop key sectors, although by the mid-2000s these were still defined rather broadly. At the same time, national and local actors were showing interest in the concept of a 'regional innovation system' (RIS). The Roh government's decision to decentralise major government agencies outside the capital region presented opportunities to develop the RIS. In a bid to forge linkages with its indigenous assets, Busan applied for the relocation of public institutions related to its port economy – the Korea Ocean Research and Development Institute (KORDI) and the Korea Institute for Maritime Fisheries Technology (KIMFT) – and related to its emerging industries – the Korea Film Archive (OECD 2004b).

Busan has twelve universities and several research institutes but its workforce has lower than average educational attainment and the region suffers from a 'brain drain' as graduates are attracted to the opportunities provided by Seoul. There are marked regional differences in R&D performance in Korea (Park 2000). Moreover, Busan lacks a well-developed technology transfer network that links the production and diffusion of knowledge. None of the top 100 Korean firms is located in Busan. Instead, HQ functions are heavily concentrated in Seoul and Gyeonggi province. In fact, there is evidence of growing geographical concentration of key functions: in 2004 Renault-Samsung Motors moved its technical centre to the capital region in pursuit of skilled labour. The large proportion of firms in Busan is SMEs but these have a poor record of innovation and Busan attracted only 0.6 per cent of Korean venture capital in 2003 (OECD 2004b).

Given the importance and difficulty of mobilising indigenous potential discussed in Chapter 5, improving the dynamism of the regional innovation system is a priority for national, regional and local institutions in Busan. Central government required all regions in Korea to produce 'regional innovations plans'. Busan developed its plan, which led to the launch of a Regional Innovation Committee in 2004. The Committee comprises fifty-six representatives from the metropolitan government, business, universities, research institutes and civil society but is loosely institutionalised and has a limited direct role. Its role is to monitor regional innovation policy, coordinating and networking and making proposals to central government with the aim of promoting specialisation in the R&D effort. In this field, and in others, actions aimed at promoting developments in Busan quickly spillover to a wider region which includes Gyeongnam province and Ulsan. The question of how local and regional development is governed has thus emerged as an important issue in Korea (OECD 2004b).

The government and governance of local and regional development in the shadow of Seoul

Korea is a unitary country with a history of centralised government and limited local autonomy. The legal foundations of local government were established in 1948, but limited local elections began only in the 1980s. In effect, local authorities were instruments of central government. Local authorities began to gain power in the 1980s and elected local government became the norm in the 1990s. But local governance remains underdeveloped:

> First, the historical legacy of central control has left excessive authority at the central level. In particular, the education budget is largely outside the control of local authorities. Moreover, weak self-governance and a lack of capability at the local level have limited the effective use of the power that they do have. Second, the severe imbalance in the financial resources of central and local governments requires local authorities to rely on large transfers from the central government, including earmarked grants. Third, there is a large variance in the fiscal independence of local governments, with local own-source revenue accounting for 95 per cent of spending in Seoul compared to only 17 per cent in lower-level cities.
>
> (OECD 2005a)

The system of subnational government in Korea has evolved over time to reflect the degree of urbanisation and functional specialisation. Busan is designated as a metropolitan city under the Korean system. In 1995, metropolitan cities were granted the powers of provincial governments. This means that Busan's city limits usefully match the functional region and allow effective land-use planning, but there are problems in effectively coordinating relationships with the neighbouring province (Gyeongnam) in some policy areas (Shin 2000), while the metropolitan region might be too large a scale to engage citizens in local governance (OECD 2004b).

Decentralisation policies, such as those announced by the Roh government in 2003, present an opportunity to address some of the challenges faced by Busan. But they also raise new issues, such as how to raise political capacity at the local level in order that the new powers can be effectively used. The recent struggle for democracy in Korea means that the country has a well-organised civil society and, echoing the concerns about democratic renewal raised in Chapter 4, a challenge is to find ways of involving civil society more actively in local governance (see Kang 2003). Busan has experimented with 'local autonomy committees' at the eup, myeon and dong level, reflecting the 'participatory government' principles of President Roh, although they remain relatively minor players in the governance system (OECD 2004b).

The story of Busan's recent relative decline is, in part, also the story of Seoul's accelerating relative growth. Seoul's ascendancy, in large measure, is a product of globalisation and the advantages this bestows on (potential) 'world cities'. Tackling the widening regional divide is central to the strategy of 'balanced national development', but its success is likely to require enduring action by both national and regional governments.

Ireland: the 'Celtic Tiger', social partnership and socio-spatial inequality

The recent experience of rapid growth and economic transformation in Ireland has attracted worldwide attention and led to the sobriquet 'Celtic Tiger' (Sweeney 1997; MacSharry and White 2000). The high national growth rates have raised fundamental questions for local and regional development. Traditionally, the problems of the Irish economy were analysed in terms of 'dependency theory', with Ireland compared to 'Third World' economies and defined, perhaps contentiously, as 'Western Europe's only colony' (Kirby 2004; see also Crotty 1986; Munck 1993). Ireland is a relatively small country of 3.9 million people, with some 1.1 million of these concentrated in the Greater Dublin area. Given the size and openness of the Irish economy in the context of its peripheral position at the Atlantic edge of the European Union (Figure 7.9), though, O'Donnell (1993) conceptualises the Irish economy in terms of some of the regional development theories examined in Chapter 3. As such, the Irish experience – and the factors that underpin it – have attracted the attention of regions around the world. Despite being a nation state, in some important respects Ireland is more like a regional economy than a national one, particularly due to its high dependence on international trade and the historic role of labour migration in shaping its development (O'Donnell 1993; O'Donnell and Walsh 1995; Krugman 1997; Smith 2004). Observing the Celtic Tiger phenomenon at its height, Krugman noted that 'Ireland's boom has many elements of a typical regional economic take-off, together with some interesting dynamics related to the uncertainties of foreign investment' (Krugman 1997: 48). O'Donnell and Walsh (1995: 223) conclude, then, that Ireland is 'both a region and a state' within the EU.

During the 1990s, Ireland was one of the fastest growing economies in the developed world (Table 7.7) and, by some measures such as export trade (Table 7.8), one of the most globalised. Ireland's feat during this period was all the more remarkable when measured against its historic performance over the previous 150 years. From the mid-nineteenth century, Ireland performed poorly. Following the agrarian crisis of the 1840s, Ireland's population halved and thereafter continued to be characterised by high levels

Figure 7.9 ***Ireland***

of emigration. In the period after Ireland achieved independence from Britain, it pursued a policy of agrarian autarky and limited industrialisation behind the protectionist wall of very high trade tariff barriers (Lee 1989). However, during the 1960s, Ireland began a gradual process of opening up its economy, signing the Anglo-Irish Free Trade Agreement, acceding to the General Agreement on Tariffs and Trade (GATT) in 1967 and joining the EEC in 1973 (Lee 1989; Kennedy 1992; O'Malley 1992).

Table 7.7 ***Annual average GDP growth rate, Ireland, 1995–2003***

Country	*%*
Ireland	8.1
United States	3.3
Spain	3.3
United Kingdom	2.7
Netherlands	2.5
France	2.1
Denmark	2.1
Belgium	2.0
Italy	1.5
Germany	1.2

Source: Adapted from OECD Productivity Database, September 2004, cited in IDA Ireland (2004)

Table 7.8 ***Destination of exports, Ireland, 2004***

United Kingdom	17%
Rest of European Union	41%
United States and Canada	22%
Rest of world	19%
Unclassified	1%

Source: Adapted from External Trade, Central Statistics Office, October 2004, cited in IDA Ireland (2004)

Liberalisation and economic crisis

Apeing the exogenous growth strategies discussed in Chapter 6, the opening of the Irish economy was intended to force its modernisation by reorienting indigenous firms towards export activity and attracting foreign investment. Liberalisation occurred during a period of sustained growth for the world economy and, especially, the internationalisation of US manufacturing capital. A new economic governance institution, the Industrial Development Authority (IDA), was established to promote Ireland as a destination for foreign investment and the Irish state offered strong fiscal and financial incentives principally to foreign as well as indigenous firms (Tomaney 1995; Bradley 2005).

Ireland experienced strong economic growth and rising living standards during the 1960s and a rapid shift between the stages of modernisation from an agrarian to an industrial economy. However, Irish-owned indigenous industry contributed comparatively little to this growth and experienced decline following the removal of tariff protection. Ireland came to rely heavily on attracting successive rounds of foreign investment in order to maintain employment levels. Moreover, questions were raised about the long-term contribution of TNCs to Irish development. Echoing the problems of exogenous development identified in Chapter 6, Ireland seemed mainly able to attract only routine production activities and lower and semi-skilled occupations, rather than R&D and marketing activities which provide more skilled jobs and greater local and regional development potential (National Economic and Social Council (NESC) 1982; Culliton

1992; O'Malley 1992; Tomaney 1995). Indeed, Ireland became an important focus during this period for research about the problems of branch plant development and the policy debates, rehearsed in Chapters 5 and 6, focused on how Ireland could develop stronger indigenous enterprise. Concerns about the performance of the Irish economy became grave during the 1980s. Mjøset (1992) attributed this poor performance to weaknesses in the 'national system of innovation' which failed to improve productivity and living standards and generated cycles of emigration contributing to a vicious circle of decline. Successive Irish governments expanded public expenditure and borrowing during the 1970s and the onset of recession in the 1980s led to worsening balance of payments and public accounts and a rapid rise in unemployment. By the mid-1980s, Ireland faced a deep economic crisis.

The rise of the 'Celtic Tiger' and national social partnership

It is in this historical context that Ireland's more recent economic performance has effected economic, social, political and cultural transformation. Isolating individual causal factors is problematic but any account of the Celtic Tiger needs to acknowledge a number of issues. Ireland's accession to the European Union meant that it was integrated into a continental European market and, with the reform of the European Union Structural Funds, the historically low levels of prosperity meant the whole country was entitled to the highest levels of European regional policy funding for most of the 1990s. Since joining the EEC in 1973, Ireland received over €17 billion in EU Structural and Cohesion Funds support up to the end of 2003. Under the programming period 2000–2006, Ireland will receive a further €3.35 billion from the Structural Funds. The Cohesion Fund contributed €586 million to Ireland during the period 2000–2003 (Matthews 1994; Walsh 1995). Significantly for local and regional development, the Irish government used these resources judiciously, partly to improve the physical infrastructure but also to support shifts in industrial policy.

Important, if gradual, shifts in Irish industrial policy occurred from the 1980s onwards often supported by EU regional policy funds (Tomaney 1995). Connecting with the themes of institution-building, indigenous development and context-specific policy developed in earlier chapters, O'Donnell summarises these changes:

> There was increased emphasis on developing indigenous enterprises, and separate agencies for FDI and indigenous development were created. Industrial policy became more selective and demanding. The development agencies worked closely with firms in devising and implementing company development strategies. The National Linkage Programme was created to increase the number and capability of Irish sub-suppliers to MNCs. Science and technology policy was reorganised and new sector-specific agencies were created. The approach to inward investment became more selective, targeting leading firms in the high-growth, high-technology sectors: computers, pharmaceuticals, medical equipment and software.
>
> (O'Donnell 2004: 54; see also Sweeney 1997; Ó Riain 2000, 2004)

Plate 7.2 ***Global connections: the International Financial Services Centre, Dublin, Ireland***
Source: Photograph by David Charles

In an early precursor of the importance of the economics of knowledge and learning, an additional feature of the Irish economy was the commitment of the Irish state to investment in education, especially at further and higher levels. By the 1990s, Ireland had one of the highest proportions of graduates in its working population in the world. Echoing the transition in the government and governance of local and regional development, Ó Riain (2004) argues that these changes signalled Ireland's transformation into a 'developmental network state', which attempts to nurture localised production and innovation networks within global investment flows by shaping the character of its various local connections to global technology and business networks: 'This is made possible by the multiple embeddedness of state agencies in professional-led networks of innovation and in international capital, as well as by the state's networked organizational structure' (Ó Riain 2004: 5; see also Ó Riain 2000).

Ireland continued to attract FDI which was mainly concerned with exporting to EU markets. Critical to drawing such investment was Ireland's very low rate of corporate tax – 12.5 per cent in 2004 compared to rates of 30 per cent or over in the United Kingdom, Belgium, France, the Netherlands and Spain (IDA Ireland 2004) – along with relatively low and internationally competitive salary costs for labour of comparable skills and productivity (Figure 7.10). The sources of inward investment were highly concentrated in sectoral terms and in terms of country of origin. Fortuitously, the emergence of the 'Celtic Tiger' coincided with a boom in the United States economy. As Table 7.9

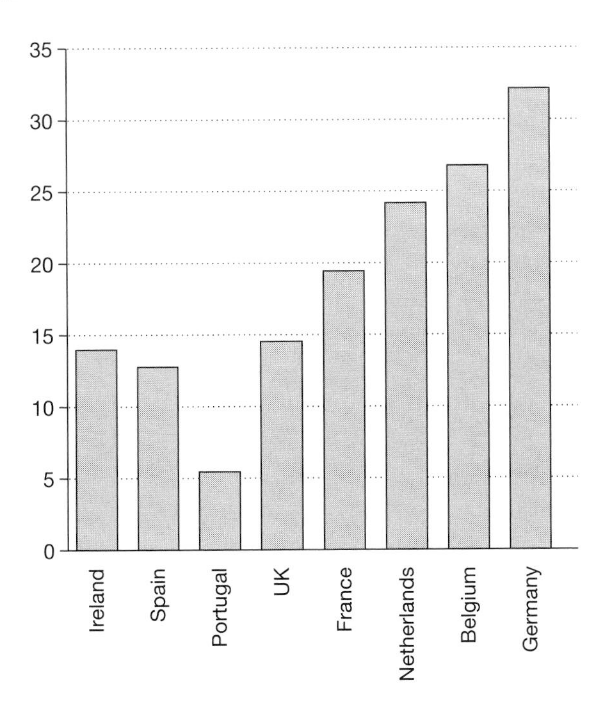

Figure 7.10 ***Cost of payroll, selected countries, 1995***

Source: IDA Ireland (2004)

Note: Hourly compensation including additional costs (in US$), 1995.

illustrates, there were 1,025 foreign-owned companies in Ireland in 2003, employing over 127,000 people, but some 70 per cent of these were in US-owned firms.

Ireland's institutions of government and governance were transformed during the emergence of the Celtic Tiger. Confronted with the severe fiscal and economic crisis of the mid-1980s, the Irish government and the principal social partners (business, unions and farmers) negotiated a 'Programme for National Recovery' which ran from 1987 to 1990. This proved to be the first of five agreements that lasted over a decade and amounted to a new and innovative form of negotiated social and economic governance (O'Donnell and Thomas 1998; O'Donnell 2004). The content of the agreements produced by this national social partnership shifted over time. The initial agreement was concerned with enlisting trade union support for efforts to correct Ireland's parlous public finances and to achieve macroeconomic stability; later agreements were more concerned with questions of employment creation, productivity, combating social exclusion and promoting local and regional development and, more latterly, extending partnership to the workplace level.

Although these features of Ireland's recent development are widely agreed, the origins and consequences of the Celtic Tiger have been strongly contested (see O'Hearn 1998, 2000; Allen 2000; Kirby 2002, 2004 for highly critical interpretations of the Celtic Tiger, albeit from different theoretical starting points). Among other things, critics stress the heavy dependence of the Irish economy upon a small number of large US plants

Table 7.9 ***Foreign investment in Ireland, 2003***

Country of origin	*Number of companies*	*Total employment*
Australia	9	286
Austria	2	244
Belgium	12	890
Bermuda	8	308
Canada	16	1,033
China	1	7
Denmark	9	2,015
Finland	4	420
France	39	2,181
Germany	149	11,394
Greece	1	48
Israel	1	75
Italy	24	625
Japan	32	2,461
South Korea	9	283
Liechtenstein	1	11
Luxembourg	4	175
Netherlands	39	2,602
South Africa	4	39
Spain	3	16
Sweden	17	2,229
Switzerland	25	2,635
Taiwan	2	186
Turkey	5	26
United Kingdom	118	8,086
United States	489	89,158
Other non-European	2	75
Overseas	1,025	127,578
Ireland*	29	1,415
Total	1,054	128,993

Source: *IDA Employment Survey 2003*

Note: * Accounted for by Irish financial services companies.

(O'Hearn 1998; Murphy 2000b). Additionally, in a dimension of the impacts of TNCs discussed in Chapter 6, some commentators claim that growth rates have been exaggerated as a result of the effects of transfer pricing. Specifically

> because of Ireland's low rate of corporation tax on manufactured goods and financial services (10%) it is in the interests of the MNCs to attribute very high levels of output to their Irish-based plants. In this way, growth that is for the most part produced by workers in the United States is attributed by corporate accountants to Irish workers for tax reasons.
>
> (Murphy 2000b: 4; see also Shirlow 1995)

O'Hearn (1998) concludes that Ireland's dependent position in the international economy has been reinforced because the Celtic Tiger has not generated sustainable indigenous local and regional development.

The emergence of the Celtic Tiger was associated with a growth of income inequality, partly linked to the pattern of tax cuts that some see as fuelling the Celtic Tiger (e.g. Kirby 2002, 2004). The proportion of Irish people at risk of poverty, after pensions and social transfer payments were taken into account, was 21 per cent in 2003. This was one of the highest rates in the European Union. The effect of pensions and social transfers on reducing the at-risk-of-poverty rate was low in Ireland compared with other EU countries. In 2001, social protection expenditure in Ireland was 15.3 per cent of GDP. This was half of the rate in Sweden and the lowest of the then fifteen EU countries (Central Statistics Office (CSO) 2005). Murphy (2000) argues the key features that underpinned its growth included low taxation and a liberal regulatory framework, although major tax cuts, in fact, followed the take-off of growth (Sweeney 2004). More worryingly, increased inequality may be linked to the very transformations that have generated economic growth. Ó Riain (2004: 5) notes: 'These multiple state-society alliances led to the uneven internationalization of society and growing inequality, generating political tensions with which the fragmented state structure cannot deal effectively'. Thus, Kirby (2002, 2004) concludes that Ireland falls short of being a 'developmental state', but instead remains trapped in a cycle of underdevelopment. Allen (2000) goes further, suggesting that the Celtic Tiger story merely obscures enduring class inequalities.

There is something in all of these criticisms, but there is also a danger of underplaying the major changes that have occurred in Ireland, particularly in terms of the balance and interrelation between exogenous and indigenous development. While the Irish economy remains dominated by TNCs and has had to cope with some high-profile closures – such as the cases discussed in Chapter 6 – there is evidence that the interaction of TNCs and local firms has had some positive impacts on indigenous companies (Coe 1997; O'Malley and O'Gorman 2001). Some foreign-owned plants have upgraded their activities over time through the localisation of R&D or increased local supplier linkages (Amin and Tomaney 1998). Moreover, while transfer pricing is undoubtedly a feature of the Irish economy, there is sufficient evidence of real growth in terms of employment, output and exports. In general, there is evidence that indigenous industry improved its performance, notably during the 1990s in sectors such as software (O'Malley 1998; Ó Riain 2000; O'Malley and O'Gorman 2001). Although difficult to specify precisely, the role of industrial policy in these developments seems important, with the Irish state and its governance institutions proving adept at providing the kinds of territorial assets that attract the sorts of TNCs that will contribute to development. Ireland may provide an example of a somewhat successful 'strategic coupling' between domestic and foreign-owned firms of the kind described in Chapter 6 (Tomaney 1995; O'Malley 1998).

Social and spatial inequality

It is the growth of income and spatial inequality in Ireland that remains the most profound local and regional development challenge facing policy-makers, especially as this growth seems implicit in the strategies developed by the 'network developmental state'.

Ireland has weak and rather confusing structures of local and regional government and governance (Morgenroth 2000). Local authorities are the main service providers at the county and city level. At the regional level, eight regional authorities (NUTS III level) coordinate local authority activities and play a monitoring role in relation to the use of EU Structural Funds. The country is further designated into two NUTS II regions and two regional assemblies are responsible for managing the regional programmes of the National Development Plan. An amendment to the Constitution of Ireland in 1999 gave clear constitutional status to local government for the first time and made it a mandatory requirement for local elections to be held every five years. There are no direct elections to the two regional levels, instead members are nominated to these levels by their local authorities.

Geographical inequalities find their chief expression in the expanding economic and social disparities between Dublin and regions such as the west and south-east. Civic voices, such as the Catholic bishops, have given political voice to concerns about the widening local and regional development gap within Ireland (e.g. Siggins 2005; see also Byrne 2004; O'Brien 2004). Generating and feeding off strong agglomeration economies, recent growth in sectors such as software has concentrated in the greater Dublin region. Irish policy-makers have only just begun to address the consequences of such geographical disparities by establishing a National Spatial Strategy designed to achieve 'balanced regional development' in ways which relieve congestion in Dublin and accelerate the development of less prosperous parts of Ireland; initiating a radical decentralisation of civil service jobs from Dublin to outlying parts of the country and recasting social policy around the notion of a 'developmental welfare state' (see Government of Ireland 2002).

Some commentators have raised doubts about whether the Irish government's local and regional development initiatives will be sufficient to offset the growing dominance of Dublin. McDonald (2005) has argued that Ireland 'is drifting towards a city-state, with Dublin as its only real powerhouse'. Official projections suggest the Greater Dublin area will have a population of 2.1 million in 2021, accounting for 40.7 per cent of the forecast population of the country as a whole. Dublin's growth has generated calls for new structures of government in the form of a Greater Dublin Authority to aid integrated planning across the city-region (Newman 2004). At the same time, voices in peripheral parts of Ireland have drawn upon 'new regionalist' arguments and called for improved regional governance capacity in order better to develop their own and more appropriate local and regional development strategies (O'Toole 2004).

The Celtic Tiger, Irish national social partnership and the network developmental state have helped to reorder Ireland's relationship with the global economy, but they have not entirely resolved the development problem:

> A crucial terrain of the global political economy will be these networked developmental strategies and class compromises, state and social institutions, and patterns of solidarity and inequality that emerge around them.
>
> (Ó Riain 2004: 5)

The Irish example reflects the tension of unequal growth for local and regional development and the importance of political choices, conflicts and compromises; albeit within the constraints of history and the external economic and political context. Fintan O'Toole (2003), the Irish journalist and political commentator, suggests:

> If the Irish boom is misunderstood as the product of neo-conservative economics, the agenda for sustaining prosperity is obvious: more tax cuts, more privatisation, a weaker State, an expansion of the ethos in which Irish people are to be understood as consumers rather than as citizens. If the boom is understood for what it was – a complex product of left-of-centre values which has not ended the spectacle of social squalor even while removing the excuse for it – the agenda is equally clear. Without a strong, active, imaginative public sphere in which all citizens have the capacity to participate, we will look back on the boom as a time of unfulfilled promises.
>
> (O'Toole 2003: 168–169)

Given its dramatic experiences during the 1990s and enduring development challenges, it is likely that Ireland will continue to attract the attention of those concerned with local and regional development in the future.

Seville: high-tech-led development in the 'California of Europe'?

In Andalusia in south-western Spain (Figure 7.11), the city of Seville's development strategy yielded much lower returns than originally expected. In the mid-1980s, the city of Seville – with the explicit support of the regional government of Andalusia and the Spanish state – embarked on an ambitious development strategy aimed at making the city not only the shop-window of a modern and dynamic Spain, but, inspired by the potential of technology-led development discussed in Chapter 3 and in particular the experience of Silicon Valley considered in this chapter, also the 'California of Europe' (Castells and Hall 1994: 198). Yet twenty years on, there was little evidence that the strategy served to dynamise the city and its surrounding hinterland, let alone transformed the city into the technological hub of Europe. The bases of Seville's development strategy are described and contrasted with its economic trajectory relative to the region of Andalusia and the rest of Spain, before highlighting the reasons for the failure of the development strategy.

Economic and social under-performance

Depending upon how metropolitan areas are measured, Seville is the third or fourth largest city in Spain. With a metropolitan population slightly over 1 million people, Seville is the capital of Andalusia, the largest Spanish region in population terms, and articulates a complex urban hierarchy that expands throughout south-western Spain. Seville has a rich history. Founded in pre-Roman times, it has served as an important urban centre for more than two millennia. Romans, Arabs and later the Spanish state made the city an important political, economic and trade hub. Seville was one of the

Figure 7.11 ***Seville, Spain***

largest and more economically dynamic centres in the world in late medieval and early modern times. Its status as a trade centre was enhanced by being designed as the Spanish gateway to its colonies in the Americas. Nevertheless, from the seventeenth century onwards, and coinciding with the move of the India fleet trading with the Spanish colonies in the Americas to Cadiz, the city started a long and constant decline that has lasted three centuries. The Industrial Revolution of the late nineteenth century bypassed the city and, with the exception of the tobacco monopoly, at the beginning of the twentieth century the city had a limited industrial base.

By the late twentieth century, Seville had numerous weaknesses that could jeopardise its economic future (Table 7.10). These included a lack of industrial development, serious technological backwardness with little investment in R&D, lack of scientists and researchers, and a dearth of competitive firms. The overall skills of the population were also relatively underdeveloped. Although the general level of education of the population had increased significantly during the 1980s, as a result of the Spanish government's efforts to improve education and skills (Rodríguez-Pose 1996), Seville still lagged behind most of the rest of Spain in general educational attainment, and especially in the number of graduates with a science and engineering background. The city also suffered from poor accessibility, with inadequate road and rail connections to other Andalusian cities and the rest of Spain, a port that could no longer cope with modern

Table 7.10 ***SWOT analysis for Seville, 1980s***

Strengths	*Opportunities*
Low cost of labour	Free mobility of factors (Single Market)
Excellent quality of life	Development of Latin America
Strong local and regional identity	Development of the Mediterranean axis
Strong local and regional government	Increase of foreign direct investment in Spain
	Expansion of quality tourism
Weaknesses	*Threats*
Lack of industrial development	Increasing competition from neighbouring cities
Technological backwardness	Emergence of Madrid as the main economic hub in Spain
Defective accessibility	Increasing competition in global tourism
Lack of adequate skills	
Low productivity	

Source: Authors' own research

ships, and a small and antiquated airport. Together these factors combined in a local economy characterised by low productivity relative to the rest of Spain. Not all local conditions were unfavourable, however. In the 1980s, Seville had a high quality of life, relatively low labour costs, and a strong local and regional identity, with strong local and regional governments, whose capacity to implement independent policies was among the highest in Spain. Its rich historical heritage also made it a city capable of attracting high-quality tourism.

In the context of deep-seated weaknesses and potentially positive indigenous endowments, Seville faced a series of challenges and opportunities. In 1986, Spain became a member of the European Union, when the prospect of the Single European Market and free mobility of capital, labour, goods and services from 1993 onwards was already on the horizon. Seville could expect to benefit from these flows of capital and goods. It is situated at the end of a potentially dynamic 'Mediterranean axis'. The expansion of high-quality tourism – beyond the traditional sun and sea – was also likely to favour a city known by its historical monuments. Given its economic weaknesses, EU membership also implied significant investments for lagging regions through EU regional policy. Andalusia and Seville were to benefit from the Structural Funds, especially after their reform in 1989. Seville had been the traditional Spanish gateway to Latin America and could play an important role, especially in the case of a potential economic recovery of the subcontinent following the economic crisis of the 1980s. The threats for Seville were, however, also significant. First, there was increasing competition from neighbouring cities, both for industry and tourism. Málaga, the second largest city in Andalusia, had developed healthy economic foundations based on tourism and experienced rapid growth. Córdoba and Granada could also count on rich historical heritages. In addition, exemplifying the self-reinforcing growth effects of Keynesian theories of cumulative causation, the national capital city of Madrid was emerging as the main economic hub in Spain and absorbing resources from cities and regions across Spain.

The 'technocity' development strategy

In 1982, Seville embarked on an ambitious local and regional development project aimed at redressing its secular economic decline. This project was articulated around the urban entrepreneurialism of the bid and award in 1983 by the International Exhibition Bureau (BIE in its French acronym) of the right to host a World Fair in 1992 to mark the five hundredth anniversary of the discovery of America. The World Fair project was combined with a longer-term development strategy called Cartuja '93, commissioned in 1988, and conceived by the eminent planners Manuel Castells and Peter Hall. Drawing directly from the theories of a fifth Kondratiev or long wave of techno-economic development, their vision was to create 'the largest technocity of Southern Europe' (Castells and Hall 1994: 193) by putting together 'an agglomeration of R&D centres and training institutions, excluding all manufacturing on-site, in some of the leading late twentieth-century technologies: computer software, microelectronics, telecommunications, new materials, biotechnology, and renewable energy' (Castells and Hall 1994: 194). The agglomeration economies generated by high-level functions and fast growing industries were interpreted as a sustainable local and regional economic structure. The site and buildings used for the Expo '92 were to be recycled to host the high-tech firms of the technological and scientific park at the heart of the Cartuja '93 project.

The strong high-tech emphasis of Seville's development project was in tune with the vision of a modern and dynamic city of the leaders behind the project. As Figure 7.12 shows, this vision was three-pronged. First, they envisaged Seville as a city that would

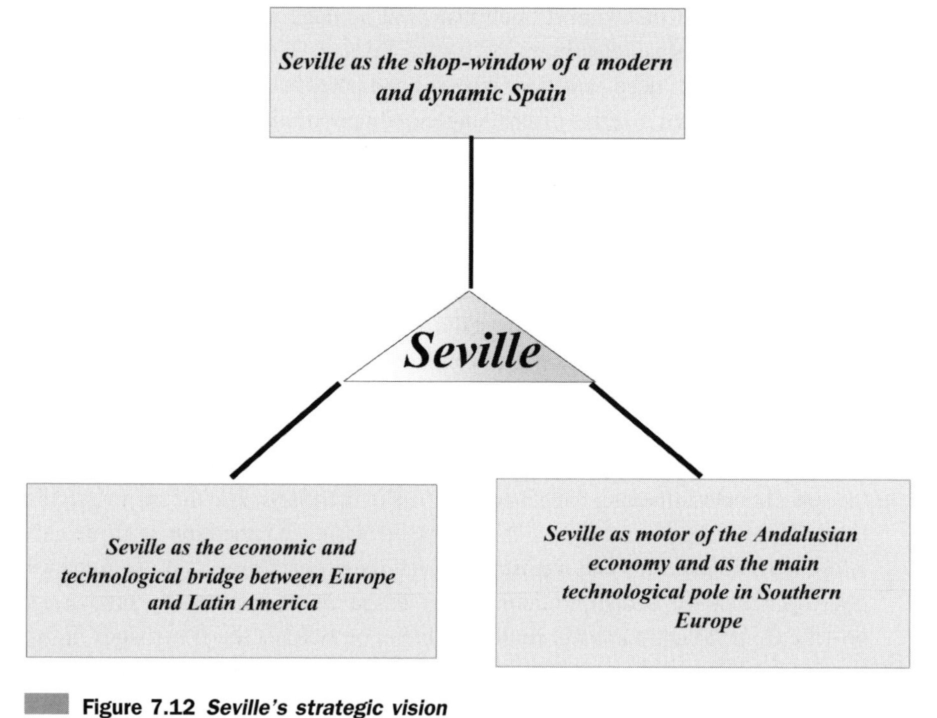

Figure 7.12 ***Seville's strategic vision***

Source: Authors' own research

act as the shop-window of a modern and dynamic Spain. Second, Seville would become the economic motor of the Andalusian regional economy and the main technological growth pole in Southern Europe. And finally, Seville would act as the technological bridge between Europe and Latin America. The general aim was to turn Seville and the rest of Andalusia into one of the economic motors of the Spanish national economy.

Linking the local and regional development strategy to a far-reaching and globally visible event such as a World Fair had important advantages. First, it contributed to generating a large social and political consensus around the event and the associated development strategy. All the key institutions were keen to be seen to take part in the development effort for the Expo '92 and the Cartuja '93 projects. This led to the horizontal and vertical coordination of government and governance institutions within the multilayered system. Horizontally, this involved the local government as the main driving force with the support of the Employers' Confederation, the local trade unions, the Chambers of Commerce, a series of lobbies, and the population of the city as a whole, through a series of initiatives to encourage civic engagement and popular participation. Vertically, linking the different scales of government and governance, in addition to local government, the Andalusian regional government, the Spanish government, and the European Union became keen contributors. The fact that, at the time, both the Spanish Prime Minister and the Deputy Prime Minister were from Seville certainly helped in securing this support.

Integrating both indigenous and exogenous approaches to local and regional development, the actual strategy for the Expo '92 and the Cartuja '93 projects was based on four pillars: investment in infrastructure, improvement of education and training, attraction of inward investment, and image promotion of the city and the region (Figure 7.13). The plan was to set up the basic conditions for the attraction and endogenous generation of high-tech industry, as a way to materialise the idea of the 'California of Europe'. In practice, the strategy tilted towards two of the pillars. Seville launched an aggressive image promotion campaign throughout the world. It aimed not only at promoting the Expo '92, but also at dissociating the image of Seville from that of a traditional tourist centre and creating that of a forward-looking city-region, capable of leading Southern Europe into the challenges of the twenty-first century. The second pillar prioritised was infrastructure development. A huge effort was made to transform the endowment of the whole city and of Andalusia in a period of barely six years. Not only was the island of Cartuja on the Guadalquivir river prepared and developed as the site of Expo '92, but also all accesses to the city and transport infrastructure within the city were improved. A new and modern airport and railway station were built; the ring-road around the city was finished; and the highways linking Seville with Madrid and with the other main cities in Andalusia completed. The icing on the cake was the first high-speed rail link in Spain between Seville and Madrid, inaugurated on 21 April 1992.

Not all the emphasis was on transport. Telecommunications infrastructure was also at the centre of the strategy. The buildings in the Cartuja Island site were designed as 'intelligent buildings' and fibre optic cables were used profusely throughout the city, making Seville at the time one of the most advanced cities in the world in telecommunications infrastructure. As the Spanish Prime Minister of the time, Felipe González, put it in a speech on 28 November 1990:

Figure 7.13 ***The pillars of Seville's development strategy***

Source: Seville Authority

> the Expo '92 gave us the opportunity to make an enormous effort in the development of infrastructure – highways, high-speed train, optic fibre, telecommunication systems – the excuse and the stimulus to do what, probably without an event of this nature, would have taken decades to do following the normal priorities.

And all this local and regional development came at a price. As Castells and Hall (1994: 197) underline: 'between 1985 and 1992 alone, some $10 billion (US) were spent on a variety of public works programs in Andalusia'.

The education and training pillar of the development strategy was, however, much weaker. Although substantive improvements in the general level of education were achieved in Seville and Spain during the 1980s, little was done in terms of generating the type of graduates needed to sustain the kind of technological growth pole envisaged by Castells and Hall. The expansion of higher education, including the creation of a new University, concentrated on humanities and social sciences. The imbalance between natural sciences and, above all, engineering and social sciences and the arts in the whole of Andalusia remained among the highest in Spain (Rodríguez-Pose 1996). With regard to the inward investment pillar, it was somehow expected that the supply-side approach linked to infrastructure upgrading would suffice to attract the relevant foreign firms. Replicating the problems of exogenous approaches in Chapter 6, the support to existing indigenous firms in the city-region was somewhat overlooked, on the basis that their existing sectoral specialisation and weak competitiveness hardly fitted the grand vision of Seville as the main technological pole in southern Europe.

Uneven local and regional development and the illusory 'technocity'

Despite the grand vision and huge effort committed to the local and regional development plan, the economic evolution of Seville in the years following Expo '92 has been lacklustre. Its GDP growth between 1992 and 2003 puts it in the middle of the Spanish provinces, mostly due to a significantly improved performance since 2000. In fact, with respect to the rest of Spain and Andalusia, Seville has experienced worse relative economic performance than in the years prior to Expo '92. As can be seen in Figure 7.14, which plots the growth of GDP in Seville and in Andalusia relative to that of Spain using a two-year average, while Seville was performing better than Andalusia and the

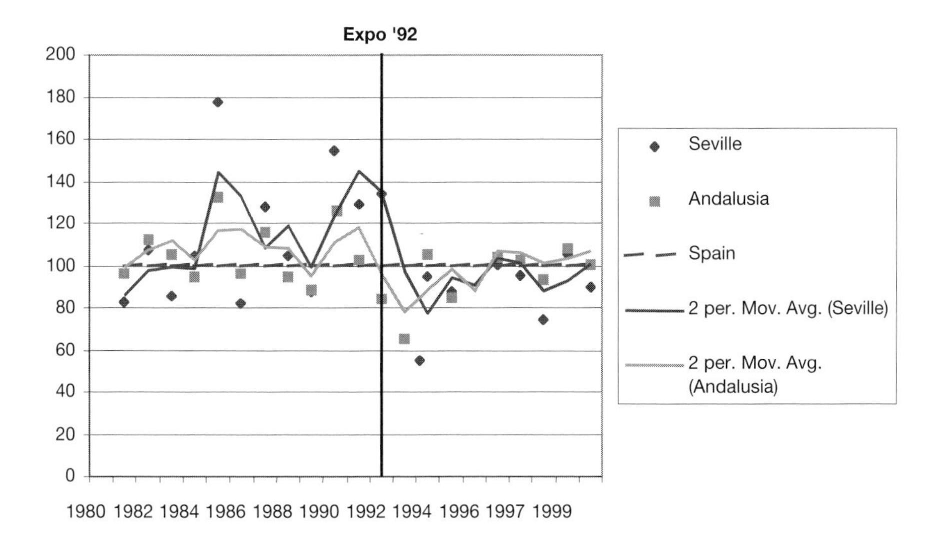

Figure 7.14 ***GDP in Seville, Andalusia and Spain, 1980–2001***

Source: Authors' own elaboration using INE data

rest of Spain throughout the 1980s and especially in the wake of the 1992 World Fair, since 1992 this economic dynamism has waned. The contrast is particularly poignant with respect to some of the provinces of eastern Andalusia, such as Almería and Málaga, that in the same period have been among the most dynamic in Spain. Given the relatively high demographic dynamism of Seville and its potential to boost economic growth, the relative decline in GDP per capita has been even worse.

Seville's unexpectedly weak economic performance is reproduced in other realms, such as unemployment and FDI. Seville´s unemployment has tended to grow relative to Spanish levels since 1992 (Figure 7.15). The whole of Andalusia, despite representing around 18 per cent of the Spanish population, managed to attract only around 2.6 per cent of total FDI between 2001 and 2004 (Spanish Foreign Investment Registry). There is also precious little evidence in the Cartuja Island site of the kinds of high-tech investment the project was supposed to attract.

Why has Seville's local and regional development not delivered? Why is there little sign in Seville of the ambitious 'technocity' the Cartuja '93 project was supposed to create? It is true that the infrastructure investment regenerated large parts of the city, but there is scarce evidence of the innovation pole that was supposed to become the economic motor of Andalusia. And although – despite serious problems in finding tenants until 1999 – the site of Cartuja '93 is now in full occupancy (Vázquez Barquero and Carrillo 2004), the majority of it is not occupied by the type of high-tech tenants originally planned. Leisure spaces (the theme park Isla Mágica), offices, and public

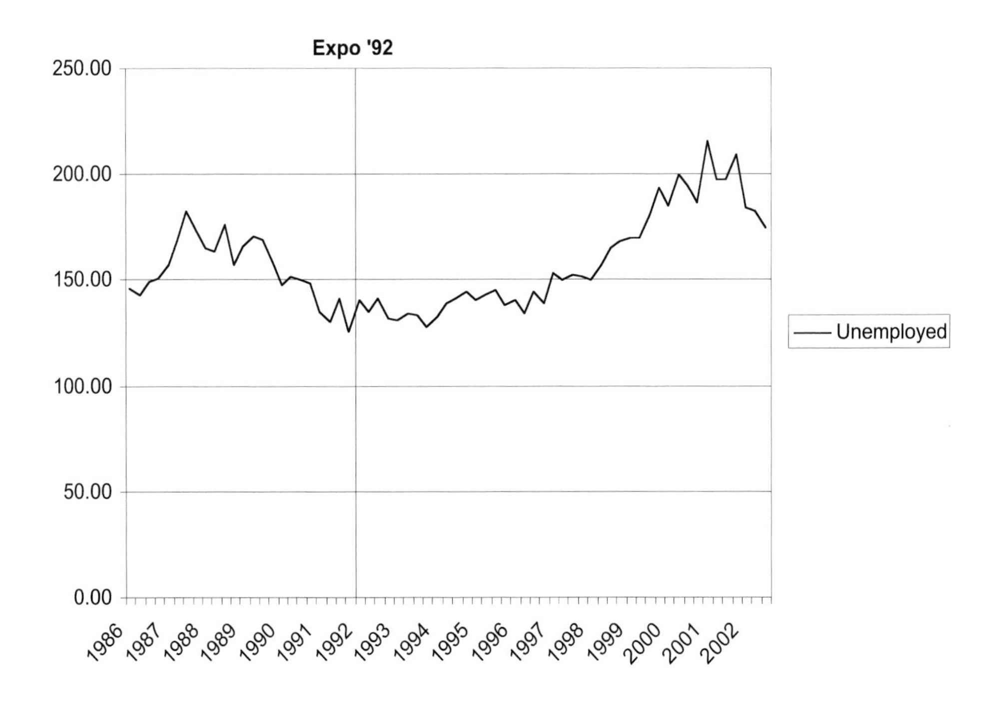

Figure 7.15 ***Unemployment in Seville and Spain, 1986–2002***

Source: Authors' own elaboration using INE data

sector and university buildings dominate the site. Indeed, in tune with Porter's cluster theory and Florida's notion of the creative class, the Cartuja Island is nowadays more of a media cluster than a real technology site. Few tenant firms conduct R&D and much of the R&D effort is associated with the presence of some public research centres.

Why did the vision of Seville as the technological motor of southern Europe not materialise? Some have put the blame on the economic downturn that affected the world economy after 1991–1992. The economic recession in the United States, Japan and Europe thwarted a large percentage of the potential high quality FDI that might have come from technologically advanced countries (Vázquez Barquero and Carrillo 2004). However, at least some of the blame lies in an overambitious vision that tried to transform the economy of the city radically, creating a technological pole *ex nihilo* and ignoring some of the harsh economic and social realities of the international economy and the city.

In addition, the strategy was applied in an unbalanced way. Reflecting the traditional, top-down and 'hard' assets approach to local and regional development introduced in Chapter 1, far too much weight was put on infrastructure. The advantages of this sort of strategy were clear to see. Infrastructure development was visible, swallowed an enormous amount of cash, and, above all, proved relatively easy to deliver and popular. In particular, it provided a lot of ribbon-cutting photo-opportunities for politicians in the run-up to the 1992 World Fair. As we noted in discussing the indigenous interventions in Chapter 5, developing human resources and local firms is longer term and less glamorous with fewer spectacular advances. By 1992, the human resources in Andalusia were ill prepared to assimilate, let alone generate, an advanced technological pole. Much of the expansion in higher education had taken place in humanities and social science, subjects that were less costly to teach and where teachers could be easily found, but that could hardly generate the sort of labour force that could attract high-tech firms or stimulate innovation. The indigenous assets of local and regional firms, whose day-to-day reality was a world away from the science and technology park of the vision, were left to fend on their own and not brought into the project.

This unbalanced approach left the whole strategy of the city dependent on image promotion and infrastructure provision. With a weak industrial fabric and inadequate human resources to fulfil the vision, the attraction of high-tech FDI proved almost impossible. Improvements in infrastructure may have contributed more in reinforcing alternative development poles, such as Madrid, with its much stronger economic base, than in guaranteeing the accessibility of Seville's output to the market. Many firms that used to service Seville and western Andalusia from the city have been forced to close or relocate to Madrid since the opening of the Madrid–Seville highway (Holl 2004). In addition, Seville's development strategy was further marred by inadequate institutional arrangements and tensions in the system of government and governance. The whole process was not immune to opportunistic defection. Many of the agents involved tended to defend their own particular interest at the expense of the collective good, reflecting shifts in the balance of power between agents. This contributed to blur objectives and to sudden changes in the steering of the project, well documented by Castells and Hall (1994). Seville's experience demonstrates how a large gap between a grand vision and the existing socio-economic base as well as the dominance of individual interests over collective interests contributed to the generation of an unbalanced local and regional development strategy, which yielded much lower results than originally intended.

Jalisco: liberalisation, continental integration and indigenous industrial revival

Struggling with the lack of competitiveness of its key industrial sectors, the state of Jalisco in Mexico has redressed its long-term economic decline through integrated, sustainable and indigenous local and regional development policies.

Industrialisation and diversification

Jalisco is located on the Pacific coast of Mexico (Figure 7.16) and has a population of over 6 million. With almost 4 million people living in its metropolitan area, Guadalajara, the capital of Jalisco, is the second largest city in Mexico and has historically been one of its three major manufacturing centres. Guadalajara's and Jalisco's industrial base took off through the early stages of economic transition in the 1930s, supported by the Mexican government's import substitution industrialisation strategy which lasted until the mid-1980s. Aimed at reducing Mexico's dependence on manufactured prod-

Figure 7.16 *Jalisco, Mexico*

ucts, the policy transformed Mexico into the second largest industrial economy in Latin America, behind Brazil. Mexican industry became fundamentally concentrated around three urban poles: Mexico City in the centre of the country, Monterrey in the north, and Guadalajara in the west. With a diversified manufacturing economy based on sectors such as food, tobacco, beverages, fabricated metal products, chemicals, rubber, plastics and textiles, the economy of Jalisco prospered under much of the import substitution industrialisation period. However, the lengthy period of economic protection fostered industries capable of benefiting from captive markets, but that had little incentive to respond to changes in market demand, innovate, apply new technology or increase productivity. Much of the industrial capacity in Jalisco evolved in ways that became increasingly vulnerable especially to international competition. In any case, much of the industrial tissue of the state was seriously vulnerable to competition.

By the end of the 1970s Jalisco's manufacturing base was beginning to show the first signs of exhaustion, and GDP per capita in the state started to grow at rates below the Mexican average. Industrial decline became even more evident during the 1980s as the Mexican economy became more open and liberalised from 1985. As Figure 7.17 shows, Jalisco started to grow significantly below the national average. Jalisco's industries struggled to adapt to the challenge of greater competition, leading the state into a slow but unrelenting process of economic decline. This trajectory contrasted strongly with the relative economic dynamism of the Mexican states bordering the United States.

Since the mid-1990s, however, Jalisco has managed to redress this process of economic decline and bring its economic performance back on a par and even marginally

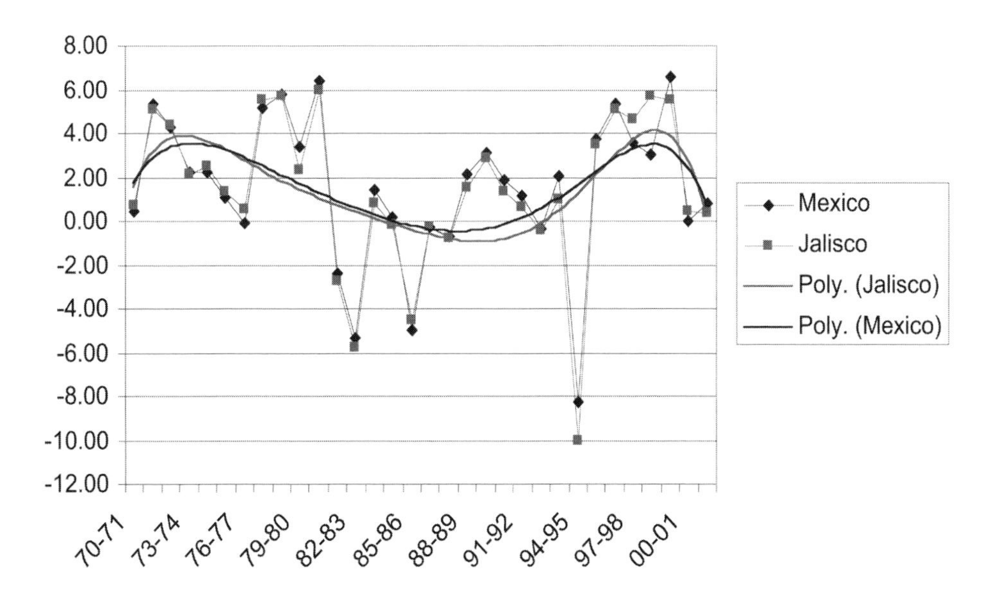

Figure 7.17 ***GDP per capita, Jalisco and Mexico, 1970–2001***

Source: Authors' own elaboration using INEGI data

exceed that of Mexico. Figure 7.17 reveals Jalisco's revitalised GDP, employment and investment growth that has been achieved without sacrificing the quality of employment and embarking on the kind of 'low road' strategy discussed in Chapter 2. One of the main drivers of this turnaround has been an effectively designed and implemented local and regional development strategy by the State Government, built around the principle of maximising the state's endogenous development potential. Drawing upon the kind of interventions discussed in Chapter 5, the strategy has contributed to the generation of almost 400,000 permanent jobs between 1995 and 2002, to raise the number of entrepreneurs by 17,000 in the same period, and significantly to raise the flows of investment into the state, both from other Mexican regions and from abroad (data from the State's Economic Promotion Secretariat). In addition, Jalisco has managed to curb its declining share of Mexican exports, which was the norm during much of the 1980s and early 1990s. Since the launching of the strategy in 1995, the participation of Jalisco's industries in Mexico's total exports (excluding oil) has risen continuously from 5.5 per cent of the total in 1995 to close to 10 per cent in 2001 (Woo Gómez 2002). The expansion in GDP, employment and exports has retained employment quality as most of the new jobs have salaries that range between two and five times the minimum wage (MW) level (Figure 7.18).

Liberalisation, continental integration and indigenous local and regional development

What explains the relative turnaround in Jalisco's economic development trajectory since 1995? Mexico has undergone profound changes since the mid-1980s. Import substitution industrialisation in place since the 1930s gave way to a relative opening of the economy through membership of GATT from 1985 and continental economic integration in NAFTA from 1994. Economic liberalisation went hand-in-hand with political change. After more than fifty years of rule by the Revolutionary Institutionalist Party

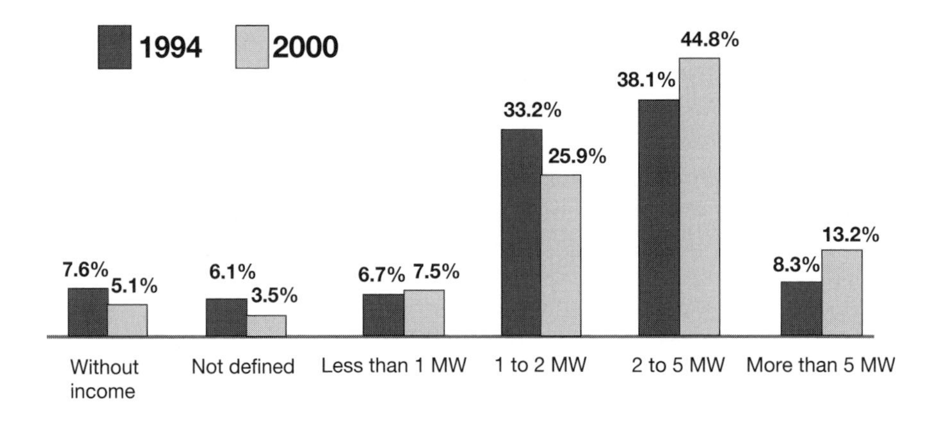

Figure 7.18 ***Salary structure in the Guadalajara Metropolitan Area, 1994–2000***

Source: Calculated from Government of the State of Jalisco (2001: 80)

(PRI), opposition governors began to be returned to office after the election of Ernesto Ruffo in Baja California in 1989 (Shirk 2000). The election of opposition governors often had implications for state, local and regional development strategies, as members of the opposition were keen to show their capacity to deliver sound economic policies (Rodríguez and Ward 1995; Díaz-Cayeros *et al.* 2001). In Jalisco, such an electoral change led to the approval of an ambitious development plan for the state. The election of Alberto Cárdenas Jiménez of the National Action Party (PAN) as Governor of Jalisco in 1995 brought a young and dynamic team into government. One of the priorities of the new state government was to set up a comprehensive development plan aimed not only at stemming Jalisco's economic decline, but also at addressing the serious economic problems that Mexico's 1994–1995 economic crisis – known as the 'Tequila effect' – had generated for the state.

The new strategy sought to mobilise and build upon indigenous assets by identifying key development sectors, and trying to minimise economic exposure by supporting a diversified range of well-established local sectors whose competitiveness had already been upgraded in response to national economic liberalisation (Government of the State of Jalisco 2001: 87–88; Woo Gómez 2002). A series of nested niche strategies were established to focus support on existing sectors deemed capable of competing in a more integrated and global economic context. The indigenous priority of the strategy did not imply disregarding the exogenous growth of attracting and embedding FDI in Jalisco.

Localising and adding value in agriculture and tourism

Among the strategies aimed at traditional indigenous sectors, agriculture took an important role. The government set up schemes to reinforce and promote links between agriculture and industry by strengthening existing production chains and trying to retain as much value-added as possible within the borders of the state (Government of the State of Jalisco 2001: 125). Traditional sectors which were amenable to local processing were prioritised to sidestep barriers that limit trade in agricultural sectors and minimise exposure to the volatile price fluctuations of commodified produce. These sectors included many natural medicine plants, the agave (the plant from which tequila is made) as well as green house produce mainly targeted at the US market. Sustainable local and regional development of agriculture was important and there was an early emphasis on organic food and better management of forest resources.

In the service sector, the main focus was tourism. In addition to the promotion of the main coastal resort of the region, Puerto Vallarta, the strategy targeted ecological and senior citizens rather than traditional tourists. The state encouraged rural tourism in haciendas and rural houses and health spas, often combined with ecotourism in inland areas, relatively isolated from the traditional coastal tourist destinations. Other traditional indigenous sectors were also promoted, such as local crafts, textiles, furniture, jewellery and shoes. The emphasis in these sectors was on redeveloping long-established crafts, training, improving the quality of materials and outputs, and assisting exports (Government of the State of Jalisco 2001: 131–134).

Indigenous industrial redevelopment and SME upgrading

While agriculture and tourism were important, the bulk of the local and regional development efforts in Jalisco focused on industry. In contrast to the high-risk exogenous-oriented strategies of trying to lure new sectors to the state through huge restructuring plans or FDI, Jalisco emphasised the traditional low technology sectors that made up the bulk of the local industrial fabric and that were already embedded in the region. Although this indigenous focus made for a somewhat unglamorous strategy, it turned out to be a strategy that worked.

Overall, the SME support strategy echoed the indigenous development approach in Chapter 5. Initiatives were based upon the provision of selective incentives for niche sectors, the creation of training and R&D facilities, and facilitating access to funds and financial support for entrepreneurs, frequently through venture capital funds. Targeted human resource measures were also used, such as the improvement of vocational training schemes or the concentration of resources for graduate and postgraduate training in selected areas. Support to the large number of local SMEs in the region took centre stage. A large network of support and research centres for SMEs – some partially financed by the private sector – was created to cater for the basic needs of SMEs and to improve production and competitiveness in the supported sectors. The network of support centres seek to help local SMEs to address the challenges of the new economy by promoting the use of IT in firms, by helping with software and advising SMEs how to engage in e-commerce. This effort has been combined with a series of measures to reduce red tape and encourage 'electronic government' in public institutions. The centres addressed SME needs including financial support, the promotion of an entrepreneurial culture, quality control, improvements to the packaging of products, technological support and advice on intellectual property, and access to information. Centres have also been established for advice on mergers and acquisition and on the creation of industry associations, for the promotion of innovation, design, management and marketing, and trade, as well as for the improvement of local crafts and for the development of essential infrastructure. Echoing the importance of strong regional government in leading and driving local and regional development strategies discussed in Chapter 4, the Government of the State of Jalisco, through its ministry of economic promotion, is at the apex of this intricate network of support centres.

While the fundamental focus of Jalisco's local and regional development strategy has been the SMEs which make the vast majority of firms and the largest proportion of employment in the state, larger firms and the exogenous approach of attracting and embedding FDI have not been overlooked. Attempts have been made to connect firms in sectors that could create links to existing networks of SMEs and thus help to embed production in the state. Steps have been taken in order to facilitate the shift from traditional *maquila* – assembly plants created under a scheme of tax exemption for the export of output – to second or third generation *maquila*. These are plants that increasingly substitute foreign inputs in the areas of technology, management or design with local inputs (Shaiken 1993). Much of the foreign direct investment the state government has sought to attract has been in sectors where production chains could be established with local firms and embedded for local and regional development. The state has business promotion centres in Mexico City and Los Angeles to support this selective and targeted strategy.

In addition, the state government prioritised investment in telecommunications and education infrastructure. Transport infrastructure investment was mainly aimed at preventing an increasing concentration of economic activity in the metropolitan region of Guadalajara and facilitating the kind of more balanced territorial development envisaged in Chapter 2 (Government of the State of Jalisco 2001). And, in contrast to Seville, the cost of a strategy not based on grand infrastructure policies or on huge incentives for the attraction of FDI turned out to be relatively low, as much of the strategy implied a re-prioritisation of existing state capital expenditure.

Although the change in economic fortunes cannot be solely attributed to the local and regional development strategy, Jalisco's experience demonstrates the potential impact of thoughtful design and the implementation of appropriate policies sensitive to the assets and needs of particular places. Using and upgrading existing indigenous resources has yielded local and regional development in Jalisco even in the context of increasing competition and continental economic integration. While the economic transformation has not been as radical or spectacular as Ireland, Jalisco's advances have been incremental and are likely to be more sustainable as a result. The almost two decades of entrenched economic decline has been halted and the seeds of a local and economic recovery have been planted. Whether Jalisco can sustain this growth trajectory in the context of further transcontinental economic integration throughout the Americas is the central issue for future development.

Conclusion

'Local and regional development in practice' uses the main themes of this book critically to analyse the experience of our international case examples. The local and regional development stories of North East England, Ontario, Silicon Valley, Busan, Ireland, Seville and Jalisco emphasise the importance of our central concerns. While working with particular histories, legacies and contexts, each place has faced the shared challenges of grappling with appropriate local and regional development models in a changing context. Common projects emerge whether managing industrial decline (North East England), attempting adjustment (Ontario, Jalisco), seeking to rebalance the national economy (Busan, Ireland), trying to sustain an existing trajectory (Silicon Valley) and embarking upon (Seville) or dealing with transformation (Ireland). Questions of what kind of local and regional development and for whom come to the fore. Each place has had to find its own way or synthesis of local and regional development in its own particular context. Similarly, the concepts and theories of local and regional development are central to deliberations about the diagnosis of current problems and the kinds of strategies, interventions and policies that might shape or influence them.

North East England is seeking to manage the decline of an industrial economy, an uneven transition to a service economy and the retreat of Keynesian state management. Ontario is attempting economic adjustment and finding the limits of the regional learning economy in the context of continental integration and its North American region-state status. Silicon Valley is trying to sustain its dynamic high-tech and knowledge economy-based local and regional development model in the face of growing pressures of social and spatial inequality and sustainability as well as international competition and rivalry in its leading industrial sectors. Busan is seeking to benefit from the stronger local and regional policy developed to rebalance the territorial development of the national economy and narrow the regional inequalities driven by the growth of the capital Seoul. Ireland too is wrestling with the economic and socio-spatial inequalities of its rapid growth and transformation as a FDI-oriented manufacturing and service-based economy. Seville is reflecting upon the relative failure and unevenness of its ambitious high-tech-led strategy and connection to the fifth Kondratiev or long wave to transform its local and regional development prospects. Jalisco is seeking to cement and embed the indigenous industrial renewal, built upon localisation, adding value and upgrading traditional activities, now integral to its local and regional development futures.

Institutions of government and governance were critical in each local and regional development experience. The continued fragmentation and weakness of its local and regional institutions mean North East England has struggled to address its plight. Ontario's economic adjustment strategy is influenced by the complexities of coordination and joint working in the multilevel federal system in Canada and the context of competition between North American region-states. Silicon Valley has hitherto relied upon limited institutional structures and business-led dynamism, although this may change given the magnitude of its sustainable development challenges. South Korea's developmental state has re-emerged with the intention of rebalancing the national economy and closing regional disparities, and local and regional institutions in Busan and other regions are seeking to gain from extensive decentralisation. Ireland's innovative national social partnership is pivotal in its national growth coalition but it faces serious challenges to address growing social and spatial inequality. Seville's experience reveals that the alignment of city, regional and national interests in transformation projects does not necessarily result in substantive local and regional development. Jalisco emphasises the potential of strong regional government in constructing an appropriate, context-sensitive local and regional development strategy as well as leading on its implementation.

Interventions and policies in the case studies connected directly with indigenous and exogenous approaches. North East England continued to suffer from the branch plant economy entrenched by its historical emphasis on exogenous approaches in manufacturing and services. Ontario remained vulnerable to external control and foreign-ownership especially in the context of NAFTA yet sought to grow and build its strong indigenous strengths to support the learning region strategy. Silicon Valley historically needed little policy intervention to support its innovative and dynamic networks beyond the federal university and defence expenditure that supported its development. The growth of labour market intermediaries and other institutions to address the Valley's sustainability questions may herald a turn to a greater degree of intervention than hitherto. Busan is attempting to reinvigorate its indigenous assets and promote adjustment with the help of stronger local and regional policy and the decentralisation of public institutions. Ireland remains explicitly conscious of the need to balance and connect exogenous and indigenous approaches in manufacturing and services in more inclusive ways to sustain its local and regional development trajectory. Seville's transformation strategy sought both indigenous and exogenous growth yet ultimately failed to deliver sustainable local and regional development. Jalisco prioritised indigenous development while exercising greater selection and targeting in its approach to exogenous resources.

The cases emphasised central elements of the frameworks of understanding in this book: the context-specific, path-dependent and locally and regionally sensitive nature of development trajectories and the role of policy. While subject to generalised quantitative assessment, the depth, character and sustainability of 'success' and 'failure' in development are often shaped by localities and regions. Ongoing adjustment and the search for sustainable local and regional development preoccupies the institutions of government and governance in each of our case studies. 'Local and regional development in practice' has provided an integrated and critical analysis of the experiences of the international case studies. The Conclusions that follow offer an opportunity to draw together the main themes of the book and to reflect upon their significance for local and regional development.

Further reading

For each of our case studies see the following:

North East England: Hudson, R. (1989) *Wrecking a Region*. London: Pion; Robinson, F. (2002) 'The North East: a journey through time', *City* 6(3): 317–334; Tomaney, J. (2002) 'The evolution of English regional governance', *Regional Studies* 36(7): 721–731.

Ontario: Courchene, T.J. (2001) 'Ontario as a North American region-state, Toronto as a global city-region: responding to the NAFTA challenge', in A.J. Scott (ed.) *Global City-Regions: Trends, Theory, Policy*. Oxford: Oxford University Press; Wolfe, D.A. (2002) 'Negotiating order: sectoral policies and social learning in Ontario', in M.S. Gertler and D.A. Wolfe (eds) *Innovation and Social Learning: Institutional Adaptation in an Era of Technological Change*. Basingstoke: Palgrave Macmillan; Wolfe, D.A. and Gertler, M. (2001) 'Globalization and economic restructuring in Ontario: from industrial heartland to learning region?', *European Planning Studies* 9(5): 575–592.

Silicon Valley: Benner, C. (2002) *Work in the New Economy: Flexible Labor Markets in Silicon Valley*. Malden, MA: Blackwell; Henton, D. (2001) 'Lessons from Silicon Valley: governance in a global city-region', in A.J. Scott (ed.) *Global City-Regions*. Oxford: Oxford University Press; Saxenian, A. (1994) *Regional Advantage: Culture and Competition in Silicon Valley and Route 128*. Cambridge, MA: Harvard University Press.

Busan: Cumings, B. (2005) *Korea's Place in the Sun: A Modern History* (2nd edn). New York: Norton; Lee, S. (2004) 'Economic change and regional development disparities in the 1990s', *Korea Journal* spring: 75–102; Shin D-H. (2000) 'Governing interregional conflicts: the planning approach to managing spillovers of extended metropolitan Pusan, Korea', *Environment and Planning A* 32: 507–518.

Ireland: O'Donnell, R. (2004) 'Ireland: social partnership and the "Celtic Tiger" economy', in J. Perraton and B. Clift (eds) *Where are National Capitalisms Now?* London: Palgrave Macmillan; Ó Riain, S. (2000) 'The flexible developmental state: globalization, information technology and the Celtic Tiger', *Politics and Society* 28(2): 157–193; Tomaney, J. (1995) 'Recent developments in Irish industrial policy', *European Planning Studies* 3(1): 99–113.

Seville: Castells, M. and Hall, P. (1994) *Technopoles of the World: The Making of 21st Century Industrial Complexes*. London: Routledge; Vázquez Barquero, A. (2003) *Endogenous Development: Networking, Innovation, Institutions and Cities*. London and New York: Routledge.

Jalisco: Giuliani, E., Pietrobelli, C. and Rabellotti, R. (2005) 'Upgrading in global value chains: lessons from Latin American clusters', *World Development* 33(4): 549–573; Rabellotti, R. (1999) 'Recovery of a Mexican cluster: devaluation bonanza or collective efficiency?', *World Development* 27(9): 1571–1585.

CONCLUSIONS

8

Introduction

The growing extent and importance of local and regional development has provided the central focus for this book. We have sought to provide an accessible, critical and integrated examination of contemporary local and regional development theory, institutions and policy. Part I introduced our initial starting points in considering local and regional development. Chapter 1 detailed the growing importance and significance of local and regional development for national, regional and local institutions of government and governance internationally in recent years. The challenging context of local and regional development was outlined encompassing a more complex, knowledge-intensive and 'globalised' capitalism marked by rapid economic, social, political and cultural change, territorial competition and concerns about its future economic, social and ecological sustainability. The need for alternative local and regional development strategies was outlined in the context of the changing institutions of government and governance and the evolution of public policy interventions from top-down, national and centralised approaches towards bottom-up, local and regional and decentralised and integrated forms. Geographical disparities and inequalities in prosperity and well-being frame questions about the aims, purpose and social justice of local and regional development.

Chapter 2 focused upon the fundamental questions of what local and regional development is, what it is for and, in normative terms, what it should be. The definition of local and regional development explored its meanings, historical evolution and geographies of space, place, territory and scale. Local and regional development has recently broadened to encompass economic and social as well as environmental, political and cultural dimensions. The sustainability of local and regional development is now paramount. Local and regional development is socially determined in the context of historically enduring themes, principles and values that vary between places and over time. The varieties, objects, subjects and social welfare dimensions of local and regional development often have socially and geographically uneven distributions of who and where benefits or loses from particular forms of local and regional development.

Part II established the frameworks for understanding local and regional development. Chapter 3 reviewed each of the main existing and emergent approaches, comprising neoclassical; Keynesian; theories of structural and temporal change (stage, cycle and wave theories; Marxist and radical political economy; transition theories); institutionalism and socio-economics; innovation, knowledge and learning; extended neo-classical theories (endogenous growth theory, geographical economics, competitive advantage and clusters); sustainable development and post-developmentalism. The discussion of each set of ideas focused upon their starting points, aspirations and assumptions; concepts, relationships, causal agents, mechanisms and processes; their relations to policy and their limitations in seeking to understand and explain local and regional development across space, in place and over time.

Chapter 4 discussed the inevitably intertwined relationships between institutions of government and governance and local and regional development. The chapter emphasised the changing nature of the state but continued importance of government in the context of the emergence of the broader institutional structures of governance, 'new regionalism' and the creation or strengthening of the capacities of local and regional development institutions. It also noted the enduring influence of the distinctive historical institutional legacies of particular national varieties of capitalism that shape the specific compromises and syntheses between growth, cohesion and sustainability that influence local and regional development. Discussion focused upon the increasingly multilevel systems of government and governance across a range of geographical scales for local and regional development and the growing importance and involvement of civil society, innovation and experimentation in democratising and encouraging participation in institutional and political structures. Understanding the institutions of government and governance is vital in framing the relative autonomy and degree of agency localities and regions have been able to exercise in reshaping existing and developing new approaches for local and regional development.

Drawing upon the emergent context, discussion of definitions, principles and values and frameworks of understanding, Part III addressed interventions and their instruments and policies of local and regional development. Themes reviewed comprised approaches seeking to harness both internal or indigenous and external or exogenous resources and forms of growth and development, the growing importance of context-sensitive rather than universal policy and the significance of policy learning and adaptation rather than policy transfer. Chapter 5 reviewed the tools aimed at mobilising local and regional economic potential and promoting indigenous and endogenous development within localities and regions. Building upon the bottom-up approach discussed in the introduction, the chapter detailed the instruments developed within programmes supporting the establishment of new businesses, the growth of existing businesses and upgrading and developing labour. Chapter 6 discussed approaches to understanding mobile investment and the ways in which local and regional institutions have attempted to attract and embed it to promote local and regional development. Institutions and policies seeking to attract and embed international firms, encourage reinvestment and form linkages to indigenous development strategies and dealing with divestment, and attracting and retaining specific occupations, were reviewed.

Part IV sought to pull the main themes of the book together and consider integrated approaches. Chapter 7 utilised the main themes, frameworks of understanding and interventions outlined in the book to analyse case studies of local and regional development in an explicitly international context drawing upon different places and territories from

East Asia, Europe, Central America and North America. The cases comprised North East England, UK; Ontario, Canada; Silicon Valley, USA; Busan, South Korea; Ireland; Seville, Spain; and Jalisco, Mexico. For each case, the analysis explored the common and particular definitions, principles, values and explanations of local and regional development, development strategies and policy approaches, the role of government and governance, their degree of success and future issues. Our analysis emphasised the often context-specific, path-dependent and locally and regionally sensitive nature of development trajectories and the role of policy learning in shaping institutional intervention for local and regional development. Rather than definitive and simplistic stories of 'success' or 'failure', the case studies suggest local and regional development is a question of the extent and nature of adjustment to both external and internal stimuli and constraints that evolve and change over time in particular localities and regions. Institutions of government and governance and social agency in localities and regions are critical in shaping the local and regional development trajectories of particular places.

This chapter has summarised the content of the book thus far. Yet, having reached this point, we are acutely aware that we have not spelled out our version of what kind of local and regional development and for whom. Such a discussion is imperative, as Glasmeier reminds us:

> in the absence of discussion on the goals and purposes of economic development policy, we will remain in a period of policy formulation which favours interventions targeted toward either reducing the costs of doing business or improving the competency of firms. Such emphases will ensure that theory is invoked to justify current practice, further diverting attention from the deeper underlying bases of economic deprivation.
>
> (Glasmeier 2000: 575)

We seek to draw upon some of the current concepts and theories of local and regional development detailed in Chapter 3 to outline our vision of what holistic, progressive and sustainable local and regional development might look like.

Holistic, progressive and sustainable local and regional development

Building upon the discussion about definitions, geographies, varieties, principles and values and distributional questions in Chapter 2, here we outline our version of holistic, progressive and sustainable local and regional development. Our definition of 'development' seeks the establishment of conditions and institutions that foster the realisation of the potential of the capacities and faculties of the human mind in people, communities and, in turn, in places (Williams 1983). In our view, the 'development' of localities and regions should be part of a more balanced, cohesive and sustainable project. Reducing the social and spatial disparities and inequalities between and within localities and regions is integral to this understanding of 'development'.

The holistic dimension

A 'holistic' approach sees close relations between the economic, social, political, ecological and cultural dimensions of local and regional development (Beer *et al.* 2003; Perrons 2004). The traditional priority of '"fixing the economy" as a prelude to, and as a platform for securing social well-being' (Morgan 2004: 883) is challenged. Instead, the holistic approach seeks to promote better awareness and balanced integration between the economic, social, political, ecological and cultural facets while acknowledging that trade-offs and conflicts may be involved (Haughton and Counsell 2004). Holistic thinking connects to the broadened notion of development as a wider and more rounded conception of well-being and quality of life discussed in Chapter 2. For Beauregard (1993), this includes a connection to the sphere of social reproduction within families and households and, in particular, the gender division of labour. This broader notion includes economic concerns – such as 'competitiveness', growth and productivity – but is not reducible to them. Indeed, the holistic approach attempts to move beyond the narrow economism and 'desiccated indicators' (Morgan 2004: 884) like GDP and income per head to develop new metrics that better capture the broader and more sustainable nature of local and regional development (Bristow 2005; Geddes and Newman 1999). Each of our case study analyses suggests local and regional development in practice wrestles with exactly these kinds of relationships and dilemmas, for example sharpened socio-spatial inequality threatening continued local and regional growth in Silicon Valley and Ireland as well as Ontario's learning region strategy to generate high-wage, high value-added and environmentally sustainable jobs. Holistic local and regional development attempts to integrate the concerns of economic efficiency and social welfare (Perrons 2004).

Critics of holistic thinking may question the practical feasibility of such an all-encompassing approach to local and regional development. Institutions and policies perhaps cannot attempt to intervene and shape such a wide and complex set of relationships for the good of localities and regions. The challenge to integrate the concerns of economic efficiency and social welfare may be formidable, particularly when economic priorities can dominate local and regional development. Yet, unless we begin to unpick dominant ideas of local and regional development and reveal the relations between broader notions of economic, social, political, ecological and cultural development, more balanced, cohesive and sustainable development of localities and regions may remain out of our reach.

The progressive dimension

A progressive approach to local and regional development is underpinned by a belief in the social injustice of uneven development and spatial disparities and inequalities. The unfairness of people's life chances and opportunities being shaped by where they live and their social context is integral to this principle. Progressive local and regional development is potentially holistic and recognises the relations between economic, social, ecological, political and cultural change. The roots of progressivism are in radical critiques and attempts at contesting and managing capitalism, a progressive value system and normative, often social-democratic and Left politics, and the aspiration for more geo-graphically even development in and across localities and regions over time (Massey 1993; Marquand 2004).

A progressive approach emphasises the role of the state together with other social institutions within civil society in tackling disadvantage, inequality and poverty in localities and regions. To varying degrees and in different ways, the experience of the case studies supports this analysis. In the position of a lagging regional economy, North East England is struggling with the shift away from an interventionist national state while Busan in South Korea may stand to benefit from an episode of strong and redistributive regional policy in the interests of more balanced national spatial development. While the narrow theoretical efficiency of competition and markets in allocating resources is recognised, it is explicitly argued, particularly in institutionalist and socio-economic approaches, that they are not free floating social phenomena but are underpinned by frameworks of institutions and conventions (Polanyi 1944). As Scott (1998: 102) argues: 'superior levels of long-run economic efficiency and performance are almost always attainable where certain forms of collective order and action are brought into play in combination with competition and markets' (see also Amsden 1992; Wade 2003). Markets, then, need to be tamed and regulated to ameliorate their tendency towards instability and the unequal economic, social and spatial outcomes that may undermine aspirations for more balanced, cohesive and sustainable local and regional development.

In opposition to a progressive approach are regressive forms of local and regional development that may encourage or entrench – through design or default – social injustice and local and regional disparities and inequalities. The promotion of economic liberalisation, state restructuring, welfare reform and territorial competition has been accompanied by regressive changes generating growing social and spatial inequality (Jessop 2002; Peck and Tickell 2002). Similarly, narrowly business-led or dominated structures of government and governance may seek to subordinate local and regional development solely to business interests and the economic concerns of competitiveness, growth, innovation and labour market flexibility. Holistic approaches may wither in such a cold climate. The experiences in our case studies echo such concerns. Silicon Valley and, to a lesser degree, Seville, revealed some of the contradictions and problems generated by network forms of governance in which business interests are powerful in promoting a particular kind of unbalanced local and regional development. Similarly, Ontario's experience illustrated how an ingrained individualistic and anti-cooperative business culture worked against private sector involvement and undermined experiments in the more collaborative and associative forms of governance considered necessary to construct a dynamic regional innovation system. Regressive forms of local and regional development may be characterised by the kinds of wasteful inter-territorial competition discussed in Chapter 1, zero-sum notions of places 'developing' at the expense of other places and an understanding of 'development' as a harsh meritocracy in which markets alone arbitrate the realisation of the potential of people, communities and places.

A progressive framework for local and regional development is focused upon a set of foundational, even universal, principles and values. These trans-historical notions might include justice, fairness, equality, equity, democracy, unity, cohesion, solidarity and internationalism (Harvey 1996). Such values have often been forged and evolve over time through the progressive and broader political ambition of what Raymond Williams called 'militant particularism' – the ideas and principles that can connect local, particular, struggles together in a more general, geographically encompassing common and shared interest (Harvey 2000). The specific form and articulation of such principles and values are subject to a greater or lesser degree of local and regional social determination, shaping and struggle within their particular national and international contexts. As our Ontario case study demonstrated, for example, the particular Canadian national social settlement is bound up within the confederation of its provinces. With comparatively weak local and regional institutions, North East England has struggled to adapt in its marginal position with the United Kingdom national political economy. Ireland's innovative national social partnership contributed to sustaining its rapid economic growth in the 1990s.

Universalist values are not simply fixed in stone, however. Neither are they simply the products of relativist definitions of 'development' determined by particular places in specific time periods. Such isolated and parochial understandings may only fuel inter-territorial competition and zero-sum interpretations of the development of some places at the expense of others. Instead, the local and regional articulation and determination of principles and values are normative issues that are subject to democratic dialogue, evolution and political choices. They are questions for localities and regions of what *their* local and regional development should be about. Keating *et al.* (2003) argue that territorial identities can become instrumentalised and utilised by political and social agents and can provide a socially rooted framework for politics. In their view, formalised institutions of government and governance in the state and civil society adapt and mould such interests. Drawing upon their particular interpretation of the concepts and theories discussed in Chapter 3, localities and regions then find or reach their own 'syntheses' of distinctive models of local and regional development – for example growth, social solidarity, sustainable development – conditioned by cultural values, institutions and prevailing modes of social and political mobilisation. Yet, while reflecting the particular and specific aspirations, needs and traits of localities and regions, such locally and regionally determined models should not be developed independently of the more fundamental and universal values outlined above. Moreover, such local and regional resolutions or settlements are shaped by the balance, dialogue, power and relations of local and regional interests and their context and mediated through institutions of government and governance.

The notion of broadly based principles and values can help frame the extent and nature of local and regional development irrespective of the relative levels of development of specific countries, regions and localities (Standing 1999). This last point is crucial to the understanding of locally and regionally determined and appropriate development for localities and regions. A progressive view is that there are basic or universal principles that matter irrespective of the levels of wealth and income. Controversy has accompanied this issue, however, as developing countries have sometimes portrayed developed world concern with environmental awareness and labour standards as relative luxuries and closet protectionism they cannot yet consider due to pressing social needs and a desire to grow and enhance their own living standards (Cypher and Dietz 2004). As we saw in our discussion of post-development in Chapter 3, the idea of universal progressive values has been criticised as encouraging a Eurocentric and modernist 'one-best-way' of 'development' overly reliant on the national state and tried and tested in the industrialised North. We reject a monolithic, 'one-size-fits-all' local and regional development. We envisage localities and regions constructing their own context-sensitive models but with reference to more universal principles and values.

The sustainable dimension

Sustainability is the third guiding principle for our holistic and progressive version of local and regional development. Drawing upon the discussion in Chapter 3, our interpretation of sustainable development is holistic in understanding 'development' as a broader idea of health, well-being and quality of life in localities and regions. It incorporates an understanding of the relations between the economic, social, ecological, political and cultural dimensions of sustainability. Such an approach encourages a closer look at whether particular forms of economic growth might be socially and ecologically damaging, even though they may offer short-term jobs and investment. Sustainability is progressive in prioritising the values and principles of equity and long-term thinking in access to and use of resources within and between current and future generations.

As Chapter 1 illustrated, previous forms of local and regional development have often been criticised as too short-term in their focus, design and delivery. Local and regional development in particular places can often be fleeting and based upon the fortuitous combination of circumstances that work together in a virtuous way for specific periods of time. Yet, the socio-spatial inequalities in Silicon Valley and Ireland raise issues of sustainability as such configurations can just as quickly unravel as distinctive advantages are not renewed or eroded by competition or cultures, institutions and networks enabling adaptation, innovation and learning are outgrown. Quick fixes have sometimes been sought for problems with deep historical roots such as Seville's ambitious strategy to create a 'technocity'. Forms of development have often proved short-lived, public money has been spent and enduring problems typically return. A sustainable approach seeks lasting and more resilient forms of local and regional development. Evidence suggests long-term strategies promote continuity and stability, and may foster the conditions for sustainable local and regional development.

Finally, our approach to sustainable development is context-sensitive. Connecting with the indigenous approaches discussed in Chapter 5, this view seeks to recognise distinctive structural problems, dovetail with local assets and social aspirations to encourage the kinds of local and regional development that are more likely to take root and succeed as locally and regionally grown solutions (Hirschman 1958; Storper 1997). Developing our ideas from Chapter 4's analysis of the institutions of government and governance, this connects to the recognition of the role of the state as one – among others including the social partners of labour, capital and wider civil society – of the leading agents of development and more holistic, programmatic and systemic forms of local and regional policy:

> environmentally sustainable development implies a more important role for the public sector, because sustainability requires a long-term – intergenerational –

> and holistic perspective, taking into account the full benefits and costs to society and the environment, not only the possibility of private profitability.
>
> (Geddes and Newman 1999: 22; see also Aufhauser *et al.* 2003)

Depending upon the circumstances and aspirations of particular localities and regions, balances and compromises will no doubt emerge from considerations of sustainable development when connected to the holistic and progressive principles.

Holistic, progressive and sustainable local and regional development in practice

Far from an infeasible wish list, the principles and values of holistic, progressive and sustainable local and regional development are being explored and put into practice by international, national, regional and local interests. The International Labour Organisation's framework focuses upon local development as the focus for human development and their conception of 'decent work'. Figure 8.1 describes the relations in this particular model. Some examples – among many others – include sustaining local and regional economies (Example 8.1), international fair trade and local development (Example 8.2), localising the food chain through creative public procurement (Example 8.3) and encouraging local development through 'demanufacturing' and recycling (Example 8.4).

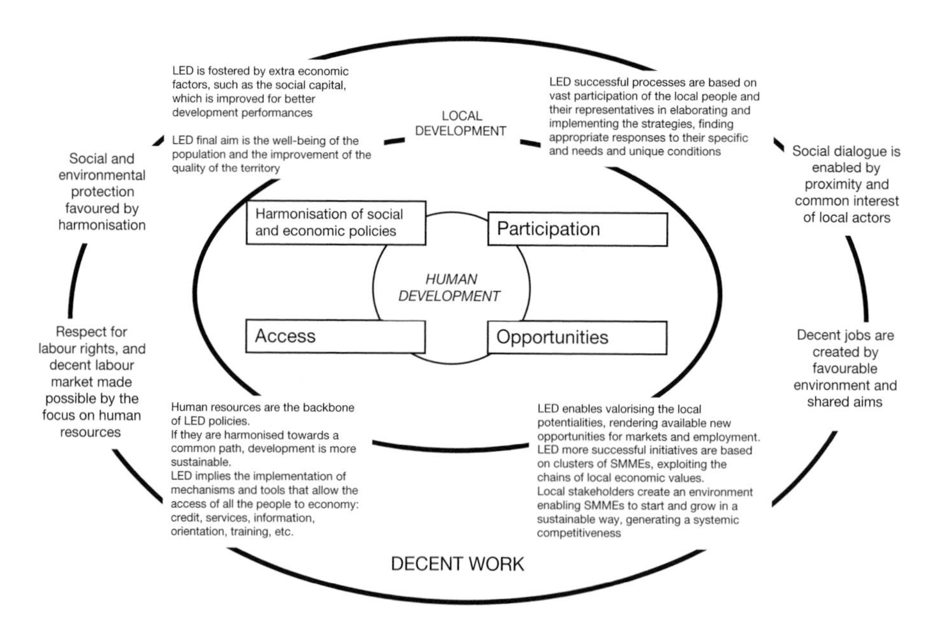

Figure 8.1 ***The ILO framework for 'Decent Work, Human Development and Local Development'***

Source: Adapted from Canzanelli (2001)

Example 8.1 Sustaining local and regional economies

The viability and sustainability of local and regional economies has received growing attention in recent years (Pike *et al.* 2006). In the context of the potentially 'delocalising' and damaging effects of globalisation, the preservation and promotion of local and regional economic linkages and circuits of value have come to the fore (Hines 2000). Drawing upon the concepts and theories in Chapter 3, Pike *et al.* (2006) have identified economic growth, economic income, investment and economic assets, people, employment structure, economic structure, economic roles and functions, innovation, learning and technological change, place-based factors and government and governance as the determinants of local economic viability and self-containment. Each can operate to support or undermine economic viability and self-containment across a variety of spatial scales and time periods.

In the United Kingdom, the localisation agenda has prompted the New Economics Foundation (2002) to develop a practical framework for use by local and regional institutions analysing and encouraging the recirculation of local spending within local economies. It argues that the leakage of local spending power outside the local economy, for example to supermarkets and food takeaways, undermines local economic vitality. Drawing on a simplified version of the ideas of Keynesian multipliers discussed in Chapter 3, their 'Plugging the Leaky Bucket' framework reveals that for every £100 that enters the local economy, if 80 per cent of each £1 spent stays in the local economy, the total amount of spending that the initial £100 will generate is about £500 – a multiplier of 5 (500/100). If only 20 per cent of each £1 spent stays in the local economy, the total spending is £125 – a multiplier of only 1.25 (125/100). Maximising the amount of local spending that circulates locally prevents financial leakage and promotes the economic viability and sustainability of local economies.

Similarly, in the United States, the Institute for Local Self-Reliance's Hometown Advantage initiative has sought to encourage the localisation of ownership and retail spending among local businesses due to their greater local linkages and impact on the local economy (Mitchell 2000). A comparative study of the economic impact of ten firms in Andersonville, Illinois, with their chain store competitors revealed that for every $100 spent in a local firm, $68 remains in the Chicago economy – only $43 is retained by $100 spent in a chain store (Civic Economics 2004). Stacy Mitchell argues:

> local ownership ensures that economic resources are broadly owned and locally controlled. It keeps decision-making local. While large corporations are required by law to maximise returns to shareholders, locally owned businesses can be guided by other values besides the bottom line. And because they are owned by people who live in the community, they tend to have a far greater concern about the community's welfare and long-term health and vitality.
>
> (Mitchell 2002: 2)

Such initiatives have encouraged supportive legislation on limiting retail unit floorspace to constrain 'big box' retail stores and preventing formula restaurants.

Practical initiatives such as 'Plugging the Leaky Bucket' and 'Hometown Advantage' have been criticised for their relatively simplistic understanding and application of Keynesian multipliers, particularly in their relations to the size and complexity of local and regional economies. Indeed, following the theory of comparative advantage and export base theory, some would advocate less localisation and more specialisation and trade in economic activities as a means of growing local and regional economies and income. Despite such criticisms, such practical and necessarily basic initiatives can represent a first step in promoting broader understanding and action in the formidable and ongoing challenge to localise economic activity in the context of globalisation and make it positive for local and regional development.

Sources: New Economics Foundation www.neweconomics.org/gen;
Hometown Advantage www.newrules.org/journal/hta.htm;
Civic Economics www.civiceconomics.com/index.html; Schuman (1998)

Example 8.2 International fair trade and local development

Fair trade promotes more equal economic relations and exchanges between producers and consumers in the North and 'developing' countries of the South globally. Acknowledging that globalisation has stretched commodity chains across the world and heightened price competition, especially in agricultural product markets, the terms of trade between advanced prosperous countries and less developed poorer countries are seen systematically to have disadvantaged producers in the South. This has historically contributed to negative implications for local and regional development and well-being as livelihoods and incomes are tied into volatile commodity markets and the politics of international trade regulation. Fair trade organisations attempt to connect producers from the South with consumers from the North, highlighting the impact of retail purchasing decisions in shops in the North on the income and living standards of people in the South. Encouraging progressive values such as ethical consumption patterns, respect for socially and ecologically responsible production and the purchase of fairly traded commodities among consumers in the North, the aim is to achieve a more equitable distribution of value between producers and consumers. In addition, retailers and distributors are encouraged to respect certain economic and social safeguards in their trading arrangements, for instance in terms and conditions of employment, health and safety and trade union membership.

Fair trade brands and products are developed with premium prices, identified by agreed and independent labelling protocols, for sales through wholesalers and retailers. The aim is to ensure price and income stability for producers and to increase the proportion of earnings that finds its way back along the commodity chain to producers in the South, raising living standards as a result. Self-help is embedded in the fair trade ethos, captured in the maxim 'Trade not aid'. Fair trade has been explicitly connected with sustainable forms of local and regional development among producers in the South. Economic, social and environmental objectives may simultaneously be attained. Fair trade in local development can encourage more ecologically balanced agriculture and provide a stimulus to local initiative and indigenous entrepreneurship.

People Tree is an example of a fair trade organisation that has worked with organic cotton farmers in Maharashtra, India. People Tree interpret fair trade as sustainable because of its ability to empower people as producers and ensure ecologically sensitive production. Fair trade products can compete in markets with strong design and distinctive characteristics bearing the hallmarks of particular localities. People Tree have paid 40 per cent in addition to the market price to purchase cotton from 70 local organic farmers. The additional income boost has allowed the farmers to sustain organic production, free from pesticide and genetically modified seeds, reduce indebtedness and propagate indigenous plant varieties. Community well-being and cohesion have benefited as a result. People Tree's organic cotton manufacturing activities started in 1998 with 8 employees and have built up workshops, now employing 150 including disabled women, with design and new fabric development capability. Market outlets include UK retailer Selfridges, London, with organic T-shirts competitive in price against other established brands due to their lower overhead costs for management and brand advertising.

Challenges for fair trade include the need to balance market integration and the preservation of indigenous cultures, the reliance upon consumerism and continued retail spending for their export markets and the displacement effects of producing fair trade products for export markets rather than concentrating on indigenous self-sufficiency. Criticism of fair trade initiatives highlights the potential tensions between increased global trade and local development.

Source: Fair Trade Forum (2001) http://fairtrade.socioeco.org
World Social Forum (2004) Fair Trade and Local Development
http://allies.alliance21.org/fsm/article.2004.en.php3?id_article=283
People Tree www.ptree.co.uk; Morgan (2004)

In our case studies too, examples of the kinds of holistic, progressive and sustainable local and regional development we envisage are evident. Sustainable local and regional development has underpinned calls for more 'liveable communities' in Silicon Valley, stimulated the establishment of the 'Smart Growth' panels in Ontario and underpinned the growth of organic farming and eco-tourism in Jalisco and regionally balanced national spatial strategies in Ireland and South Korea.

The preceding examples are obviously not exhaustive – many more exist and are under development and experimentation (Scott 1998; Beer *et al.* 2003) – but they provide concrete cases of our approach. A holistic, progressive and sustainable local and regional development is not intended as an 'off-the-shelf' template or 'one-size-fits-all' universal design or model. Neither is it a plea for local and regional relativism and voluntarism in the definitions of development driven solely by perhaps parochial and regressive local and regional interests. Rather, our approach seeks to provide guiding principles informed by the kinds of universal values discussed above that may influence the social determination of definitions, geographies, varieties, principles and values for local and regional development that are geographically differentiated and change over time. This version can shape local and regional thinking and discussion about the fundamental questions of what kinds of local and regional development and for whom.

Plate 8.1 ***Indigenous fair trade: woman entrepreneur and local Tunari brand of jeans at a trade fair in Cochabamba, Bolivia***

Source: Photograph by Nina Laurie

Example 8.3 Localising the food chain through creative public procurement

The international stretching of commodity chains and growing concerns about food quality and safety have stimulated interest in 'relocalising' the food chain in the interests of local and regional development. Consumers tend to place greater trust in products that display a strong attachment to place, making the geographical provenance of food increasingly important. Recent research has focused on the role of school meals in debates about the integrity of the public realm and the broadened notions of health, well-being and sustainable development. A focus on school meals may provide a 'multiple dividend', first, with more nutritious food reducing diet-related health problems (e.g. obesity, heart disease, diabetes). Second, locally produced school meals could create new local markets for local farmers and producers. Third, a more localised food chain delivers potential environmental benefits through reducing the 'food miles' or distance travelled by products from field to fork.

Creative public procurement in EU countries, including Austria, Denmark, France, Germany, Italy and Sweden, has played a leading role in promoting more nutritious, some times organic, food in public facilities, including schools. While UK public procurement managers purchasing goods and services for public bodies cite EU directives that seek to prevent discrimination in favour of local suppliers within the Single European Market's competition rules, other EU Member States are creatively interpreting such regulations. They are actively specifying contracts using product qualities – fresh seasonal produce, regionally certified products and organic ingredients – to implement local purchasing policies in all but name. In Ferrara, Italy, for example, 80 per cent of all food served in the city's nursery schools is organic. Recognition of the strategic role and innovative deployment of public procurement practices together with increased resources could provide health, educational and welfare benefits with positive impacts upon local and regional development. Such initiatives are potentially holistic, progressive and sustainable. Challenges for such an approach include encouraging awareness and education among public procurement managers, changing the dietary aspirations and norms among users of public service institutions, and expanding the capability of the supply-side of local markets to fulfil the scale of public contracts.

Source: Morgan and Morley (2002); Morgan (2004)

Example 8.4 Demanufacturing, recycling and local development

A potentially holistic and sustainable approach capable of generating local and regional development benefits is based on 'demanufacturing' and recycling. In the context of globalisation and intensified competition, rapid technological change is reducing the duration of the life cycles of products and increasing the turnover of new product development, production, sale and consumption. Increased obsolescence has created a growing problem of what to do with unwanted but potentially usable products, such as cars, mobile phones and personal computers. Some products have well-developed secondary markets that can extend their usable lives. Others further intensify the pressures on local and regional waste-management systems.

Manufactures are often disposed of into waste systems for free yet provide the potential for the responsible treatment and disposal of hazardous waste, the recycling of secondary materials, the development of secondary materials markets (e.g. glass, copper) and the municipal management of solid waste. 'Demanufacturing' is production 'in reverse'. Discarded manufactures are disassembled into their component parts. For local and regional development, economic activity can be generated from such formerly neglected resources and assets that hitherto were considered only as waste that required disposal. New activities have been utilised to create employment in disadvantaged communities and, simultaneously, remediate degraded local environments. In particular, entry-level semi-skilled jobs with training can be provided that may form the basis of intermediate labour market paths back into mainstream employment. Criticisms of such initiatives point to their small scale relative to the magnitude of the problem and the need for manufacturers to do much more to support the responsible disposal of discarded products.

Source: Roberts (2004)

The limits of local and regional development

Despite the undoubted growth in importance and the tangible evidence of the emergence of holistic, progressive and sustainable forms, there are clearly limits to what local and regional development can achieve. An emphasis upon local and regional development is a necessary but not sufficient condition for the more even territorial development and distribution of wealth and well-being in economic, social and environmental terms across and between localities and regions. The macroeconomics of growth and the nature of the engagement of national states within the international political economy fundamentally shape the problems and prospects for local and regional development (Scott 1998; Hudson 2001). While the national state alone clearly cannot hope to answer all the questions of local and regional development – as if it or any other institution ever could in the context of a global capitalist economy – the problems that inhibit the state's role in the current neo-liberal era are undoubtedly challenging. Namely:

> how, in a prospective global mosaic of regional economies, individual regions can maximise their competitive advantages through intra-regional policy efforts while simultaneously working together collaboratively to create an effective world-wide inter-regional division of labour with appropriate built-in mechanisms of mutual aid, and especially with some modicum of collective assistance for failing or backward regions.
>
> (Scott 1998: 7)

Such challenges are not insurmountable. Public institutions, particularly at the national level, can be vital, longstanding and potentially progressive guardians in concert with civil society of the principles and values of local and regional development.

There are four key issues in reflecting upon the constraints on local and regional development. First, some commentators believe globalisation has emasculated the nation state, limiting its scope for policy action and denuding its potential to achieve fairer and more progressive social and spatial distributions of wealth and well-being (for a review see Radice 1999). Yet, as we discussed in Chapter 4, we believe this argument is overdone. Globalisation itself is the product of the decisions of political agents: international institutions, national governments, transnational corporations, civil society, consumers and so on (Hirst and Thompson 1999). Concerted intergovernmental cooperation at the international scale may provide part of a potential response. Michie and Grieve Smith (1995: xxvi) are alive to this, noting that 'Action at the local, regional, national or bloc level, far from being a utopian alternative to the real international stage, might in reality prove a prerequisite to co-operation'. This is a laudable and complementary aspiration, notwithstanding questions regarding whether, how and in what ways such initiatives can 'reach down' through geographical scales to localities and regions to shape their development in meaningful ways. Nation states are integral too in regulating inter-territorial competition at the international and national levels in the context of globalisation (Rodríguez-Pose and Arbix 2001). Whether and to what extent national policy frameworks pursue the 'new regionalist' agenda and seek to encourage localities and regions to become agents of their own development and compete against other regions are critical questions. Having endowed every English region – irrespective of their relative prosperity – with development agencies, for example, recent debate in the United Kingdom has questioned how the national government's approach of treating unequals equally can ever reduce local and regional disparities in economic and social conditions (House of Commons 2003). The plight of lagging localities and regions like our case study North East England appears relatively bleak as a consequence.

Second, the increasingly evident 'quasi-governance' of local and regional development touched upon in Chapter 4 has raised concerns about accountability, coordination and transparency (Skelcher *et al.* 2000; Pike 2002b). Local and regional development institutions may seek to operate in a wholly functional and technocratic manner – making decisions, giving policy advice, spending public money and implementing public policy. But this way of working risks 'de-politicising' many issues that involve normative questions of principles and values that should properly lie in the public and political sphere. Enhanced local and regional – as well as national – level accountability and transparency may be a prerequisite for the deliberation and design of effective and locally and regionally rooted development strategy and policies (Geddes and Newman 1999; Pike 2004). Our contrasting case studies of Ireland, Seville and Jalisco testify to the effectiveness of such an approach. While democratised systems may deliver democratically legitimate political outcomes that do not measure up favourably against our version of a holistic, progressive and sustainable local and regional development – as our Ontario case study demonstrated – we have to respect the accountable process by which they were achieved. The myriad interests involved in localities and regions need accountable and transparent institutional mechanisms of representation, dialogue and resolution to ensure their voice, involvement and participation is secured.

Third, as we discussed in Chapter 4, coordination and integration issues are becoming more difficult for the institutions of government and governance in the context of increasingly complex multi-agent and multilevel systems operating across and between a range of geographical scales. For some local and regional development issues, it is hard to see exactly who is responsible for what and at which level. Assessments of policy effectiveness have also raised doubts. Recent devolved government and governance to the subnational level has largely failed to reduce local and regional disparities and, under particular conditions, has even served to exacerbate them (Rodríguez-Pose and Gill 2005). A decentralisation of austerity has often accompanied devolution programmes alongside the retrenchment of commitments to national redistribution and equalisation of living standards across the constituent local and regional territories of nation states. Within multilayered systems, government and governance relationships between specialised institutions and interdependent levels may hold out the prospect of more inclusive, networked and horizontal government and governance for local and regional development. Equally, however, the evolving system may lead to the reinforcement of exclusive relations and vertical hierarchies. Significantly, within multilevel systems, the national level typically retains much authority, power and resources to make decisions and allocate responsibility and finance for local and regional development (Morgan 2002). Nation states therefore remain pivotal within the institutional mechanisms of coordination within a multilevel system capable of steering and guiding the future development trajectories of local and regional economies (Scott 1998).

Last, as we suggested in Chapter 4, the declining faith and trust in public institutions, public services and traditional representative democracy are said to have undermined the authority, legitimacy and capacity of the state at the national, regional and local levels (Wainwright 2003). Civil society and experiments with participatory forms of democracy might then be necessary complements to the democratic renewal of state forms and the deeper, more meaningful engagement with especially local and regional 'stakeholders' (Humphrey and Shaw 2004). As the national social partnership in Ireland and other evidence suggests, collectively organised and systematic social negotiation can help foster local and regional development responses during periods of unprecedented change (Amin and Thomas 1996). Weak leadership, elite projects and shifting priorities can, conversely, undermine development strategies as we saw in Seville. Supplementing as well as challenging rather than replacing and substituting the state and representative democracy through the autonomous and independent agency of civic society may provide institutional supports for local and regional development. State and civil society interests may not necessarily always align, however (Moulaert *et al.* 2005).

The political renewal of local and regional development

Our assessment of the apparent shortcomings of the national state and possible responses points towards the need for a renewal of the politics of local and regional development. Politics explicitly recognises the normative choices about what local and regional development should be about, where and for whom. Such choices are value laden – not simply objective and technical assessments – and require institutional mechanisms of articulation, deliberation, representation and resolution. They also require participation. Local and regional development need not be something that is 'done' to people and places. Achieving answers and solutions to the question of 'what kind of local and regional development and for whom?' – in the manner of Keating *et al.*'s (2003) distinctive locally and regionally determined syntheses – may involve compromise, conflict and struggle between sometimes opposing priorities. Like Thompson's (1963) understanding of social history, the cut and thrust of political practice will forge the functional and geographical shape of the institutions of coordination and collective order for local and regional development (see Scott 1998). What constitutes 'success', 'failure' and 'development' in localities and regions are framed and shaped by such processes and politics of government and governance. Deciding upon locally and regionally appropriate and rooted strategies and forms of local and regional development may need a renewed, democratised, participatory and progressive politics capable of addressing the fundamental questions of what kinds of local and regional development and for whom?

As we argued above, the deliberation and responses to the fundamental questions are shaped by the interaction between the universal principles, our version of holistic, progressive and sustainable local and regional development and the specific concerns of particular localities and regions. As Scott (1998: 117) suggests: 'Successful development programmes must inevitably be judicious combinations of general principle and localized compromise, reflecting the actual geography and history of each individual region'. Some places may aspire to a particular extent or nature of growth that may take

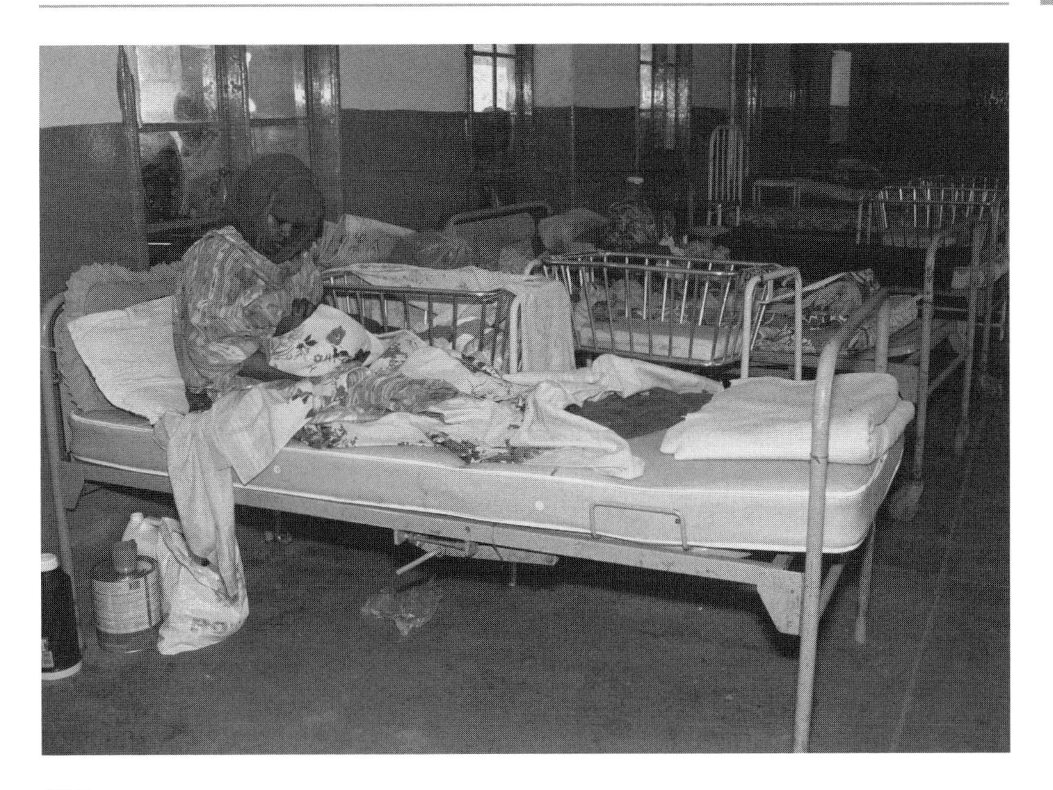

Plate 8.2 ***The state and public services: a maternity ward in Somalia***

Source: Photograph by Michele Allan

specific forms, perhaps less volatile, environmentally damaging, socially unequal or territorially uneven than hitherto. Others may seek to promote happiness as a common good more capable of encouraging well-being and community cohesion in preference to a narrow focus upon individual income and GDP (Layard 2005). Places may collectively decide to limit growth and focus upon a community-based approach to meeting local and regional social needs. Localities and regions may want better balances between economic, social and environmental development, for example in our case studies of Silicon Valley, Ontario and Jalisco.

Some places may work towards a local and regional development that is more territorially even, equal and socially just. Healing or ameliorating the socially and ethnically divided nature of the 'Two Valleys' in Silicon Valley connects to just such a progressive local and regional development. Other places may aspire to creating and encouraging a vibrant 'learning economy' and dynamic 'regional innovation system' – evident in Ontario and Seville – better able to adapt to the uncertainties and rapid changes in the international economy and capable of creating productive, high wage and high value-added and environmentally sustainable jobs central to the 'high road' variety of local and regional development. Some places may desire sustained public investment to renew the public infrastructure that propelled their earlier growth and development. Amidst the growth propelled by its laissez-faire privatism, even Silicon Valley has benefited positively from historical Federal defence, R&D and university public expenditure and North East England has historically been a state-managed region.

Localities and regions may demand the regulation of inter-territorial competition and wasteful subsidies for investment and jobs to promote inter-territorial cooperation and solidarity. Some may seek to regain their competitive standing vis-à-vis other places encouraged by national policy frameworks that encourage further growth and development in all regions – irrespective of their relative levels of prosperity and well-being. Places might seek to protect – or, conversely, dismantle – political settlements of often national state-provided social benefits, such as public health care, that provide wage advantages for comparable levels of skills and productivity with positive implications for particular forms of local and regional development. Ontario's experience revealed the future sustainability of the Canadian 'social envelope' is in question in the context of a productivity gap, intercontinental economic integration and the competition of its provinces with other North American region-states.

Localities and regions may question homogenous 'models' of 'one-size-fits-all' policy universalism. 'Off-the-shelf' local and regional development strategies may be deemed inappropriate, infeasible or undesirable. Not everywhere can – or even wants – to be Silicon Valley or the Third Italy. Even Silicon Valley itself is struggling with the sustainability of its hitherto highly dynamic development trajectory. Localities and

Plate 8.3 ***The politics of local and regional development: demonstrating against the abolition of the Inner London Education Authority (ILEA) during the 1990s***

Source: Photograph by Michele Allan

regions may want to find their own distinctive and particular forms and ways of development more fitting to their own specific aspirations and needs, albeit shaped by the combinations of internal and external assets, capabilities and constraints that people and places face. Yet each place will wrestle with the articulation and representation of its own politics of universal values, visions and particular concerns for local and regional development within a much more interdependent world. However uneven the reality, Ontario's recent emphasis upon a decentralised, regional innovation system based upon distinctive local and regional strengths, intersectoral cooperation, trustful relations and social learning situated within a North American and international context is evidence of this approach.

Ultimately, a renewed politics of local and regional development hinges upon the question of who governs? Who decides and how do they decide what forms, institutions and resources are available to frame, address and answer the questions of what kind of local and regional development and for whom? Our version of a holistic, progressive and sustainable approach and renewed politics of local and regional development is not a call for a parochial and introspective politics of place at the expense of other people, classes and places (Beynon and Hudson 1993). We explicitly reject any notion of a relativist 'free for all' of local and regional development models developed in splendid isolation. We believe in the potential of international intergovernmental coordination, the role of the national state in concert with local and regional institutions of the state and civil society, democratised institutions of government and governance with enhanced accountability and transparency capable of empowering people, communities and places, combining innovations in representative and participatory democracy, and decentralised decision-making structures coordinating and integrating their relationships within multilevel institutional structures operating across a range of scales. This agenda may be criticised as utopian or too reformist and insufficiently radical in its approach. Yet we see that even our aspiration for local and regional development is beset with potential barriers and problems – grandiose visions, regressive or insufficiently progressive political agendas, entrenched vested interest groups and collective views, weak governance and coordinating capacity and state traditions in multilevel systems, uninspiring leaders, disenchanted publics and so on – many of which we have addressed in this book. Building a holistic, progressive and sustainable local and regional development is by no means an easy or straightforward task. However, not having the vision of what we want local and regional development to do and to look like would make such a task even harder.

Further reading

On the holistic, progressive and sustainable nature of local and regional development, see Hudson, R. (2001) *Producing Places*. New York: Guilford; Morgan, K. (2004) 'Sustainable regions: governance, innovation and scale', *European Planning Studies* 12(6): 871–889; Perrons, D. (2004) *Globalisation and Social Change: People and Places in a Divided World*. London: Routledge; Scott, A.J. (1998) *Regions and the World Economy: The Coming Shape of Global Production, Competition and Political Order*. Oxford: Oxford University Press.

On the local and regional syntheses of appropriate models of development, see Keating, M., Loughlin, J. and Deschouwer, K. (2003) *Culture, Institutions and Economic Development: A Study of Eight European Regions*. Cheltenham: Edward Elgar.

On the political renewal of local and regional development, see Harvey, D. (2000) *Spaces of Hope*. Edinburgh: Edinburgh University Press.

SELECTED WEBSITES

Centre for International Earth Science Information Network, Columbia University – Decentralisation and Local Development
www.ciesin.org/decentralization/SB_entry.html

Centre for Urban and Regional Development Studies (CURDS), University of Newcastle, UK
www.ncl.ac.uk/curds

European Association for Information on Local Development
www.aeidl.be

European Association of Development Agencies
www.eurada.org/home.php

European Cities Network
www.eurocities.org

European Commission Regional Policy (DG Regio)
http://europa.eu.int/comm/dgs/regional_policy/index_en.htm

Eurostat – EU statistics division
http://epp.eurostat.cec.eu.int

German Federal Ministry Agency – Local and Regional Economic Development Toolkit
www.wiram.de/toolkit

ILO –Local Economic Development
http://learning.itcilo.org/entdev/led

International Network for Urban Research and Action
www.inura.org

OECD Local Economic and Employment Development Programme
www.oecd.org/department/0,2688,en_2649_34417_1_1_1_1_1,00.html

Regional Science Association International (RSAI)
www.regionalscience.org

Regional Studies Association
www.regional-studies-assoc.ac.uk

UN Centre for Regional Development
www.uncrd.or.jp

UNCTAD
www.unctad.org

World Bank – Local Economic Development
www.worldbank.org/urban/led

REFERENCES

Acioly, C. (2002) *Participatory Budgeting in the Municipality of Santo André, Brazil: The Challenge of Linking Short-term Problem-solving with Long-term Strategic Planning*. Rotterdam: Institute of Housing and Urban Development Studies.

Acs, Z. and Storey, D. (2004) 'Introduction: entrepreneurship and economic development', *Regional Studies* 38(8): 871–877.

Allen, J., Massey, D. and Cochrane, A. (1998) *Rethinking the Region*. London: Routledge.

Allen, K. (2000) *The Celtic Tiger: The Myth of Social Partnership in Ireland*. Manchester: Manchester University Press.

Amin, A. (1985) 'Restructuring and the decentralisation of production in Fiat', in J. Lewis and R. Hudson (eds) *Uneven Development in Southern Europe*. London: Methuen.

Amin, A. (ed.) (1994) *Post Fordism: A Reader*. Oxford: Blackwell.

Amin, A. (1999) 'An institutionalist perspective on regional economic development', *International Journal of Urban and Regional Research* 23(2): 365–378.

Amin, A. (2000) 'Industrial districts', in T.J. Barnes and E. Sheppard (eds) *A Companion to Economic Geography*. Oxford: Blackwell.

Amin, A. and Thomas, D. (1996) 'The negotiated economy: state and civic institutions in Denmark', *Economy and Society* 25: 255–281.

Amin, A. and Thrift, N. (1995) 'Globalization, institutional "thickness" and the local economy', in P. Healey, S. Cameron, S. Davoudi, S. Graham and A. Madanipour (eds) *Managing Cities: The New Urban Context*. Chichester: Wiley.

Amin, A. and Thrift, N. (1999) 'Neo-Marshallian nodes in global networks', in J. Bryson, N. Henry, D. Keeble and R. Martin (eds) *The Economic Geography Reader*. Chichester: Wiley.

Amin, A. and Tomaney, J. (eds) (1995a) *Behind the Myth of European Union: Prospects for Cohesion*. London: Routledge.

Amin, A. and Tomaney, J. (1995b) 'The regional dilemma in a neo-liberal Europe', *European Urban and Regional Studies* 2(2): 171–188.

Amin, A. and Tomaney, J. (1998) 'The regional development potential of inward investment', in M. Storper, S.B. Thomadakis and L.J. Tsipouri (eds) *Latecomers in the Global Economy*. London and New York: Routledge.

Amin, A., Bradley, D., Howells, J., Tomaney, J. and Gentle, C. (1994) 'Regional incentives and the quality of mobile investment in the less favoured regions of the EC', *Progress in Planning* 41(1): 1–122.

Amin, A., Cameron, A. and Hudson, R. (2002) *Placing the Social Economy*. London: Routledge.

Amsden, A.H. (1992) *Asia's Next Giant: South Korea and Late Industrialisation*. New York: Oxford University Press.

Amsden, A.H. (2001) *The Rise of 'The Rest': Challenges to the West from Late-Industrialising Countries*. Oxford: Oxford University Press.

Anderson, J. (1996) 'The shifting stage of politics: new medieval and postmodern territorialities?', *Environment and Planning D: Society and Space* 14(2): 133–153.

Angel, D.P. (2000) 'Environmental innovation and regulation', in T.J. Barnes and E. Sheppard (eds) *A Companion to Economic Geography*. Oxford: Blackwell.

Armstrong, H. (1997) 'Regional-level jurisdictions and economic regeneration initiatives', in M. Danson, G. Lloyd and S. Hill (eds) *Regional Governance and Economic Development*. London: Jessica Kingsley.

Armstrong, H. and Taylor, J. (2000) *Regional Economics and Policy* (3rd edn). Oxford: Blackwell.

Armstrong, H. and Vickerman, R. (1995) *Convergence and Divergence among European Regions*. London: Pion.

Arthur, B. (1996) *Increasing Returns and Path Dependence*. Ann Arbor, MI: University of Michigan Press.

Aschauer, D.A. (1989) 'Is public expenditure productive?', *Journal of Monetary Economics* 23(2): 177–200.

Asheim, B. and Clark, E. (2001) 'Creativity and cost in urban and regional development in the "New Economy"', *European Planning Studies* 9(7): 805–811.

Atkinson, M.M. and Coleman, W.D. (1989) *The State, Business and Industrial Change in Canada*. Toronto: University of Toronto Press.

Audretsch, D.B. and Feldman, M.P. (1996) 'R&D spillovers and the geography of innovation and production', *American Economic Review* 86: 630–640.

Audretsch, D. and Keilbach, M. (2004) 'Entrepreneurship capital and economic performance', *Regional Studies* 38(8): 949–959.

Aufhauser, E., Herzog, S., Hinterleitner, V., Oedl-Wieser, T. and Reisinger, E. (2003) *Principles for a 'Gender-Sensitive Regional Development': On Behalf of the Austrian Federal Chancellery, Division IV/4 for Co-ordination of Regional Planning and Regional Policies*. Vienna: Institut für Geographie & Regionalforschung, Universität Wien.

Azzoni, C.R. (2001) 'Economic growth and regional income inequality in Brazil', *Annals of Regional Science* 35(1): 133–152.

Bache, I. and Flinders, M. (2004) *Multi-level Governance*. Oxford: Oxford University Press.

Baddeley, M., Martin, R. and Tyler, P. (1998) 'European regional unemployment disparities: convergence or persistence?', *European Urban and Regional Studies* 5(3): 195–215.

Baker, S., Kousis, M., Richardson, D. and Young, S. (eds) (1997) *The Politics of Sustainable Development: Theory, Policy and Practice within the EU*. London: Routledge.

Bardhan, P. (2002) 'Decentralization of governance and development', *Journal of Economic Perspectives* 16(4): 185–205.

Barnes, T.J and Gertler, M. (eds) (1999) *The New Industrial Geography: Regions, Regulation and Institutions*. London and New York: Routledge.

Barnes, T.J. and Sheppard, E. (eds) (2000) *A Companion to Economic Geography*. Oxford: Blackwell.

Barnett, V. (1997) *Kondratiev and the Dynamics of Economic Development: Long Cycles and Industrial Growth in Historical Context*. London: Palgrave Macmillan.

Barratt Brown, M. (1995) *Models in Political Economy*. Harmondsworth: Penguin.

Barro, R.J. and Sala-i-Martin, X. (1991) 'Convergence across states and regions', *Brookings Papers on Economic Activity* 2: 107–158.

Barro, R.J. and Sala-i-Martin, X. (1995) *Economic Growth*. New York: McGraw Hill.

Barron, S., Field, J. and Schuller, T. (eds) (2000) *Social Capital: Critical Perspectives*. Oxford: Oxford University Press.

Bartlett, C.A. and Ghoshal, S. (1986) 'Tap your subsidiaries for global reach', *Harvard Business Review* 64: 87–94.

Bay Area Economic Forum (2004) *Downturn and Recovery: Restoring Prosperity*. San Francisco, CA: Bay Area Economic Forum.

Beauregard, R.A. (1993) 'Constituting economic development: a theoretical perspective', in R.D. Bingham and R. Mier (eds) *Theories of Local Economic Development*. Newbury Park, CA: Sage.

Becattini, G. (1990) 'The Marshallian industrial district as a socio-economic notion', in F. Pyke, G. Becattini and W. Sengenberger (eds) *Industrial Districts and Inter-firm Cooperation in Italy*. Geneva: International Institute for Labour Studies.

Becker, G.S. (1962) 'Investment in human capital: a theoretical analysis', *Journal of Political Economy* LXX: 9–49.

Beer, A., Haughton, G. and Maude, A. (2003) *Developing Locally: An International Comparison of Local and Regional Economic Development*. Bristol: Policy Press.

Bellini, N. and Pasquini, F. (1998) 'The case of ERVET in Emilia-Romagna: towards a second-generation Regional Development Agency', in H. Halkier, M. Danson and C. Damborg (eds) *Regional Development Agencies in Europe*. London: Jessica Kingsley.

Belt, V. (2003) 'Social labour, employability and social exclusion: pre-employment training for call centre work', Paper presented at the Employability and Labour Market Policy Seminar Series, Warwick: Institute for Employment Research, University of Warwick.

Benjamin, G. and Nathan, R. (2001) *Regionalism and Realism: A Study of Governments in the New York Metropolitan Area*. Washington, DC: Brookings Institution Press.

Benner, C. (2002) *Work in the New Economy: Flexible Labor Markets in Silicon Valley*. Malden, MA: Blackwell.

Benner, C. (2003) 'Labour flexibility and regional development: the role of labour market intermediaries', *Regional Studies* 37(6–7): 621–633.

Bennett, B., Krebs, G. and Zimmerman, H. (1990) *Local Economic Development in Britain and Germany*. London: Anglo-German Foundation.

Bennett, F. and Roberts, M. (2004) *From Input to Influence: Participatory Approaches to Research and Inquiry into Poverty*. York: York Publishing Services.

Bennett, R.J. (1998) 'Business associations and their potential to contribute to economic development: re-exploring an interface between the state and market', *Environment and Planning A* 30: 1367–1387.

Bennett, R.J. and McCoshan, A. (1993) *Enterprise and Human Resource Development: Local Capacity Building*. London: Paul Chapman.

Bennett, R.J. and Robson, P.J.A. (2000) 'The small business service: business support, use, fees and satisfaction', *Policy Studies* 21(3): 173–190.

Benz, A. and Burkhard, E. (1999) 'The Europeanization of regional policies: patterns of multi-level governance', *Journal of European Public Policy* 6(2): 329–348.

Berg, P-O., Linde-Laursen, A. and Lofgren, O. (2000) *Invoking a Transnational Metropolis: The Making of the Øresund Region*. Lund, Sweden: Studentlitteratur AB.

Bergman, E.M., den Hertog, P., Charles, D.R. and Remoe, S. (2001) *Innovative Clusters: Drivers of National Innovation Systems*. Paris: OECD.

Best, M. (1991) *The New Competition: Institutions of Industrial Restructuring*. Cambridge: Polity.

Beynon, H. and Hudson, R. (1993) 'Place and space in contemporary Europe: some lessons and reflections', *Antipode* 25(3): 177–190.

Birch, D.L. (1981) 'Who creates jobs?', *Public Interest* 65: 3–14.

Birkinshaw, J. and Fry, N. (1998) 'Subsidiary initiatives to develop new markets', *Sloan Management Review* Spring: 51–61.

Blair, J.P. (1995) *Local Economic Development: Analysis and Practice*. Thousand Oaks, CA: Sage.

Blomstrom, M., Lipsey, R. and Zejan, M. (1996) 'Is fixed investment the key to economic growth?', *Quarterly Journal of Economics* 111: 269–276.

Bloom, D.E. and Sachs, J. (1998) 'Geography, demography, and economic growth in Africa', *Brookings Papers on Economic Activity* 2: 207–295.

Bluestone, B. and Harrison, B. (1982) *The Deindustrialisation of America: Plant Closing, Community Abandonment and the Dismantling of Basic Industry*. New York: Basic Books.

Bluestone, B. and Harrison, B. (2000) *Growing Prosperity: The Battle for Growth with Equity in the 21st Century*. Berkeley, CA: University of California Press.

Bogason, P. (2004a) 'Postmodern public administration', in E.W. Ferlie, L. Lynne and C. Pollitt (eds) *Handbook of Public Management*. Oxford: Oxford University Press.

Bogason, P. (2004b) 'Local democratic governance: allocative, integrative or deliberative?', in P. Bogason, H. Miller and S. Kensen (eds) *Tampering with Tradition: The Unrealized Authority of Democratic Agency*. Lanham, MD: Lexington Books.

Boldrin, M. and Canova, F. (2001) 'Inequality and convergence: reconsidering European regional policies', *Economic Policy* 16: 207–253.

Borts, G.H. and Stein, J.L. (1964) *Economic Growth in a Free Market*. New York: Columbia University Press.

Bosch, G. (1992) *Retraining Not Redundancy: Innovative Approaches to Industrial Restructuring in Germany and France*. Geneva: ILO.

Bourdieu, P. (1986) 'The forms of capital', in J.G Richardson (ed.) *Handbook of Theory and Research for the Sociology of Education*. Westport, CT: Greenwood Press.

Boyer, R. and Drache, D. (1996) *States against Markets: The Limits of Globalization*. London: Routledge.

Braczyk, H-J., Cooke, P. and Heidenreich, H. (eds) (1998) *Regional Innovation Systems*. London: UCL Press.

Bradley, J. (2005) 'Committing to growth: experiences in small European countries', in D. Coyle, W. Alexander and B. Ashcroft (eds) *New Wealth for Old Nations: Scotland's Economic Prospects*. Princeton, NJ: Princeton University Press.

Brakman, S. and Garretsen, H. (2003) 'Rethinking the "new" geographical economics', *Regional Studies* 37(6–7): 637–648.

Braun, D. (1991) *The Rich Get Richer: The Rise of Income Inequality in the United States and the World*. Chicago, IL: Nelson Hall.

Braunstein, E. (2003) 'Gender and foreign direct investment', in J. Michie (ed.) *The Globalization Handbook*. Cheltenham: Edward Elgar.

Brenner, N. (2002) 'Decoding the newest "metropolitan regionalism" in the USA: a critical overview', *Cities: International Journal of Policy and Planning* 19(1): 3–21.

Brenner, N. (2003) 'Metropolitan institutional reform and the rescaling of state space in contemporary Western Europe', *European Urban and Regional Studies* 10(4): 297–324.

Bristow, G. (2005) 'Everyone's a "winner": problematising the discourse of regional competitiveness', *Journal of Economic Geography* 5(3): 285–304.

Brusco, S. (1982) 'The Emilian model', *Cambridge Journal of Economics* 6(2): 167–184.

Buck, N., Gordon, I.R., Hall, P., Harloe, M. and Kleinman, M. (2002) *Working Capital: Life and Labour in Contemporary London*. London: Routledge.

Burton-Jones, A. (1999) *Knowledge Capitalism*. Oxford: Oxford University Press.

Byrne, D. and Wharton, C. (2004) 'Loft living – Bombay calling: culture, work and everyday life on post-industrial Tyneside (a joint polemic)', *Capital and Class* 84: 191–198.

Byrne, K. (2004) 'Action needed to boost south-east's economy', *Irish Times* 4 May.

Cain, P.J. and Hopkins, A.G. (1993a) *British Imperialism: Innovation and Expansion, 1688–1914* (Volume I). Harlow: Longman.

Cain, P.J. and Hopkins, A.G. (1993b) *British Imperialism: Crisis and Destruction, 1914–1990* (Volume II). Harlow: Longman.

Camagni, R. (1996) *Regional Strategies for an Innovative Economy: The Relevance of the Innovative Milieu Concept*. Østersund, Sweden: SIR.

Camagni, R. (2002) 'On the concept of territorial competitiveness: sound or misleading?', *Urban Studies* 39(13): 2396–2411.

Campbell, M. (1990) *Local Economic Policy*. London: Cassells.

Campbell, M., Sanderson, I. and Walton, F. (1998) *Local Responses to Long Term Unemployment*. York: Joseph Rowntree Foundation.

Cano, W. (1993) *Reflexoes sobre o Brasil e a nova (des)ordem internacional*. Campinas, Brazil: Editora da Unicamp.

Canzanelli, G. (2001) *Overview and Learned Lessons on Local Economic Development, Human Development, and Decent Work*. Geneva: ILO/Universitas Working Paper, www.ilo.org/public/english/universitas/publi.htm.

Cárdenas, E. (1996) *La Política Económica en México, 1950–1994*. Mexico: Fondo de Cultura Económica-El Colegio de México.

Casson, M., Cooke, P., Merfyn Jones, R. and Williams, C.H. (1994) *Quiet Revolution? Language, Culture and the Economy in the Nineties*. Aberystwyth: Menter a Busnes.

Castells, M. (1983) *The City and the Grassroots: A Cross-Cultural Theory of Urban Social Movements*. Berkeley, CA: University of California Press.

Castells, M. (1989) *The Informational City: Information Technology, Economic Restructuring and Urban-Regional Process*. Oxford: Blackwell.

Castells, M. and Hall, P. (1994) *Technopoles of the World: The Making of 21st Century Industrial Complexes*. London: Routledge.

Castree, N. (1999) 'Envisioning capitalism: geography and the renewal of Marxian political economy', *Transactions of the Institute of British Geographers* 24(2): 324–340.

Cato, M.S. (2004) *The Pit and the Pendulum: A Cooperative Future for Work in the Welsh Valleys*. Cardiff: University of Wales Press.

Central Statistics Office (CSO) (2005) *Measuring Ireland's Progress, 2004*. Dublin: CSO.

Chatterton, P. (2000) 'Will the real creative city please stand up', *City* 4(3): 390–397.

Chatterton, P. (2002) 'Be realistic. Demand the impossible. Moving towards "strong" sustainable development in an old industrial region?', *Regional Studies* 35(5): 552–561.

Checkland, S.G. (1976) *The Upas Tree: Glasgow 1875–1975*. Glasgow: Glasgow University Press.

Cheshire, P. and Gordon, I. (1998) 'Territorial competition: some lessons for policy', *Annals of Regional Science* 32: 321–346.

Chibber, V. (2003) *Locked in Place: State-building and Late Industrialisation in India*. Princeton, NJ: Princeton University Press.

Chinitz, B. (1961) 'Contrasts in agglomeration: New York and Pittsburgh', *American Economic Review Papers and Proceedings* 51: 279–298.

Chisholm, M. (1990) *Regions in Recession and Resurgence*. London: Unwin Hyman.

Civic Economics (2004) *The Andersonville Study of Retail Economics*. Chicago, IL: Civic Economics.

Clark, C. (1939) *The Conditions of Economic Growth*. London: Macmillan.

Clark, G.L. (1990) 'Piercing the corporate veil: the closure of Wisconsin Steel in South Chicago', *Regional Studies* 24: 405–420.

Clark, G.L., Gertler, M. and Whiteman, J. (1986) *Regional Dynamics: Studies in Adjustment Theory*. Boston, MA: Allen & Unwin.

Clark, G.L., Feldman, M. and Gertler, M. (2000) *The Oxford Handbook of Economic Geography*. Oxford: Oxford University Press.

Clark, J. and Guy, K. (1997) *Innovation and Competitiveness*. Brighton: Technopolis.

Coe, D.T. and Helpman, E. (1995) 'International R&D spillovers', *European Economic Review* 39: 859–887.

Coe, D.T., Helpman, E. and Hoffmeister, A.W. (1997) 'North–South R&D spillovers', *Economic Journal* 107: 134–149.

Coe, N. (1997) 'US transnationals and the Irish software industry: assessing the nature, quality and stability of a new wave of foreign direct investment', *European Urban and Regional Studies* 4: 211–230.

Coe, N., Hess, M., Yeung, H.W-C., Dicken, P. and Henderson, J. (2004) '"Globalizing" regional development: a global production networks perspective', *Transactions of the Institute of British Geographers* 29: 468–484.

Coffield, F. (1999) 'Breaking the consensus: lifelong learning as social control', Paper delivered at the Second European Conference on Lifelong Learning, University of Bremen, Germany.

Cohen, J. and Rogers, J. (2003) 'Power and reason', in A. Fung and E.O. Wright (eds) *Deepening Democracy: Institutional Innovations in Empowered Participatory Governance*. London: Verso.

Collaborative Economics (1998) *Linking the New Economy to the Livable Community*. Mountain View, CA: Collaborative Economics.

Cooke, P. (1995) 'Keeping to the high road: learning, reflexivity and associative governance in regional economic development', in P. Cooke (ed.) *The Rise of the Rustbelt*. London: UCL Press.

Cooke, P. (1997) 'Institutional reflexivity and the rise of the region state', in G. Benko and U. Strohmayer (eds) *Space and Social Theory: Interpreting Modernity and Post-Modernity*. Oxford: Blackwell.

Cooke, P. and Morgan, K. (1998) *The Associational Economy: Firms, Regions and Innovation*. Oxford: Oxford University Press.

Cooke, P., Uranga, M. and Etxebarria, G. (1998) 'Regional systems of innovation: an evolutionary perspective', *Environment and Planning A* 30: 1563–1584.

Cornford, J., Naylor, R. and Driver, S. (2000) 'New media and regional development: the case of the UK computer and video games industry', in A. Guinta, A. Lagendijk and A. Pike (eds) *Restructuring Industry and Territory: The Experience of Europe's Regions*. Norwich: The Stationery Office.

Courchene, T.J. (2001) 'Ontario as a North American region-state, Toronto as a global city-region: responding to the NAFTA challenge', in A.J. Scott (ed.) *Global City-Regions: Trends, Theory, Policy*. Oxford: Oxford University Press.

Crafts, N. (1996) *Endogenous Growth: Lessons for and from Economic History*, Discussion Paper 1333. London: Centre for Economic Policy Research.

Crotty, R. (1986) *Ireland in Crisis: A Study of Capitalist Colonial Underdevelopment*. Dingle: Brandon Books.

Crouch, C. (2004) *Post-democracy*. Cambridge: Polity.

Crouch, C. and Marquand, D. (1989) *The New Centralism: Britain Out of Step in Europe?* Oxford: Basil Blackwell.

Crouch, C., le Galès, P., Trigilia, C. and Voelzkow, H. (eds) (2001) *Local Production Systems in Europe: Rise or Demise?* Oxford: Oxford University Press.

Cuadrado Roura, J.R. (1994) 'Regional disparities and territorial competition in the EC', in J.R. Cuadrado Roura, P. Nijkamp and P. Salvá (eds) *Moving Frontiers: Economic Restructuring, Regional Development and Emerging Networks*. Aldershot: Avebury.

Culliton, J. (1992) *A Time for Change: Industrial Policy for the 1990s* (Culliton Report). Dublin: Industrial Policy Review Group and The Stationery Office.

Cumings, B. (2005) *Korea's Place in the Sun: A Modern History* (2nd edn). New York: Norton.

Cypher, J.M. and Dietz, J.L. (2004) *The Process of Economic Development*. London: Routledge.

Dahl, R. (2000) *On Democracy*. New Haven, CT: Yale University Press.

Danson, M., Halkier, H. and Cameron, G. (2000) *Governance, Institutional Change and Regional Development*. London: Ashgate.

Danson, M., Halkier, H. and Damborg, C. (1998) 'Regional development agencies in Europe: an introduction and framework for analysis', in H. Halkier, M. Danson and C. Damborg (eds) *Regional Development Agencies in Europe*. London: Jessica Kingsley.

Dawkins, C.J. (2003) 'Regional development theory: conceptual foundations, classic works and recent developments', *Journal of Planning Literature* 18(2): 131–172.

Dawley, S. (2003) 'High-tech industries and peripheral region development: the case of the semiconductor industry in the North East Region of England', unpublished PhD thesis, Newcastle upon Tyne: Centre for Urban and Regional Development Studies (CURDS), University of Newcastle.

Dejonckheere, J., Ramioul, M. and Van Hootegem, G. (2003) *Is Small Finally Becoming Beautiful? Small and Medium Enterprises in the New Economy*. Brighton: Institute for Employment Studies.

Department of Trade and Industry (DTI) (2004) *The UK Contact Centre Industry: A Study*. London: DTI.

Díaz-Cayeros, A., Magaloni, B. and Weingast, B. (2001) 'Democratization and the economy in Mexico: equilibrium (PRI) hegemony and its demise', typescript, Stanford, CA: Stanford University.

Dicken, P. (2003) *Global Shift: Reshaping the Global Economic Map in the 21st Century* (4th edn). London: Sage.

Dixon, R.J. and Thirlwall, A.P. (1975) 'A model of regional growth rate differentials along Kaldorian lines', *Oxford Economic Papers* 27: 201–214.

Donahue, J. (1997) *Disunited States: What's at Stake as Washington Fades and the States Take the Lead*. New York: Basic Books.

Dowling, M. (1999) 'Social exclusion, inequality and social work', *Social Policy and Administration* 33(3): 245–261.

Drache, D. and Gertler, M. (eds) (1991) *The New Era of Global Competition*. Toronto: McGill-Queens University Press.

Dunford, M. (1988) *Capital, the State and Regional Development*. London: Pion.

Dunford, M. (1990) 'Theories of regulation', *Environment and Planning D: Society and Space* 8: 297–321.

Dunford, M. (1993) 'Regional disparities in the European Community: evidence from the REGIO databank', *Regional Studies* 27(8): 727–743.

Dunford, M. and Perrons, D. (1994) 'Regional inequality, regimes of accumulation and economic integration in contemporary Europe', *Transactions of the Institute of British Geographers* 19: 163–182.

Dunning, J.H. (1988) *Explaining International Production*. London: Unwin Hyman.

Dunning, J.H. (2000) 'Regions, globalization and the knowledge economy: the issues stated', in J.H. Dunning (ed.) *Regions, Globalization and the Knowledge-based Economy*. Oxford: Oxford University Press.

Dussel Peters, E. (2000) *Polarizing Mexico: The Impact of the Liberalization Strategy*. Boulder, CO: Lynne Rienner.

Dymski, G. (1996) 'On Krugman's model of economic geography', *Geoforum* 27: 439–452.

Edwards, M. (1989) 'The irrelevance of development studies', *Third World Quarterly* 11: 116–135.

Enright, M. (1993) 'The geographic scope of competitive advantage', in E. Dirven, J. Groenewegen and S. van Hoof (eds) *Stuck in the Region? Changing Scales of Regional Identity*. Utrecht: Geographical Studies.

Enright, M.J. and Ffowcs-Williams, I. (2001) *Local Partnership, Clusters and SME Globalisation: C.* Background Paper Prepared for Workshop 2 in Enhancing SME Competitiveness, OECD Bologna Ministerial Conference. Paris: OECD.

Ernst, D. and Kim, L. (2002) 'Global production networks, knowledge diffusion and local capability formation', *Research Policy* 31: 1417–1429.

ERVET (Emilia Romagna Development Agency) www.insme.org/documenti/Ervet%20presentation1.pdf, accessed 17 March 2006.

Escobar, A. (1995) *Encountering Development: The Making and Unmaking of the Third World*. Princeton, NJ: Princeton University Press.

Esping-Andersen, G. (1999) *Social Foundations of Postindustrial Economies*. Oxford: Oxford University Press.
European Commission (1999) *Achieving the Balanced and Sustainable Development of the Territory of the EU: The Contribution of the Spatial Development Policy*. Luxembourg: Office for Official Publications of the European Communities.
European Commission (2004) A *New Partnership for Cohesion: Convergence Competitiveness Cooperation*. Luxembourg: Office for Official Publications of the European Communities.
Evans, P.B. (1989) *Embedded Autonomy: States and Industrial Transformation*. Princeton, NJ: Princeton University Press.
Fainstein, S. (2001) 'Inequality in global city-regions', in A.J. Scott (ed.) *Global City-Regions: Trends, Theory, Policy*. Oxford: Oxford University Press.
Farr, J. (2004) 'Social capital: a conceptual history', *Political Theory* 32(1): 6–33.
Feldman, M.P. (2000) 'Location and innovation: the new economic geography of innovation, spillovers and agglomeration', in G.L. Clark, M. Feldman and M. Gertler (eds) *The Oxford Handbook of Economic Geography*. Oxford: Oxford University Press.
Field, D. (2002) *Social Capital*. London: Routledge.
Fingleton, B. and McCombie, J. (1997) 'Increasing returns and economic growth: some evidence for manufacturing from the European Union regions', *Oxford Economic Papers* 50: 89–105.
Firn, J. (1975) 'External control and regional development: the case of Scotland', *Environment and Planning A* 7: 393–414.
Fischer, S. (2003) 'Globalization and its challenges', *American Economic Review* 93(2): 1–30.
Fisher, A. (1939) 'Primary, secondary, tertiary production', *Economic Record* June: 24–38.
Fisher, E. and Reuber, R. (2000) *Industrial Clusters and SME Promotion in Developing Countries*. London: Commonwealth Secretariat.
Florida, R. (2002a) 'The learning region', in M. Gertler and D. Wolfe (eds) *Innovation and Social Learning*. Basingstoke: Palgrave Macmillan.
Florida, R. (2002b) *The Rise of the Creative Class: And How It is Transforming Work, Leisure, Community and Everyday Life*. New York: Basic Books.
Florida, R. and Kenney, M. (1990) 'Silicon Valley and Route 128 won't save us', *California Management Review* 33(1): 68–88.
Fothergill, S. (2005) 'A new regional policy for Britain', *Regional Studies* 39(5): 659–667.
Frank, A.G. (1978) *Dependent Accumulation and Underdevelopment*. London: Macmillan.
Frankel, J.A. and Romer, D. (1999) 'Does trade cause growth?', *American Economic Review* 89(3): 379–399.
Freeman, C. and Perez, C. (1988) 'Structural crises of adjustment: business cycles and investment behaviour', in G. Dosi, C. Freeman, R. Nelson, G. Silverberg and L. Soete (eds) *Technical Change and Economic Theory*. London: Pinter.
Friedmann, J. (1972) 'A general theory of polarized development', in N.M. Hansen (ed.) *Growth Centres in Regional Economic Development*. New York: Free Press.
Fröbel, F., Heinrichs, J. and Kreye, O. (1980) *The New International Division of Labour: Structural Unemployment in Industrialised Countries and Industrialisation in Developing Countries (Studies in Modern Capitalism)*. Cambridge: Cambridge University Press.
Fu, X.L. (2004) 'Limited linkages from growth engines and regional disparities in China', *Journal of Comparative Economics* 32(1): 148–164.
Fuller, C. and Phelps, N.A. (2004) 'Multinational enterprises, repeat investment and the role of aftercare services in Wales and Ireland', *Regional Studies* 38(7): 783–801.
Fuller, D. and Jonas, A.E.G. (2002) 'Institutionalising future geographies of financial inclusion: national legitimacy versus local autonomy in the British credit union movement', *Antipode* 34(1): 85–110.
Fung, A. and Wright, E.O. (2003) 'Thinking about empowered participatory governance', in A. Fung and E.O. Wright (eds) *Deepening Democracy: Institutional Innovations in Empowered Participatory Governance*. London: Verso.
Gamble, A. (1994) *The Free Economy and the Strong State* (2nd edn). London: Macmillan.
Garcia, B. (2004) 'Urban regeneration, arts programming and major events: Glasgow 1990, Sydney 2000 and Barcelona 2004', *International Journal of Cultural Policy* 10(1): 103–118.
Gardiner, B., Martin, R. and Tyler, P. (2004) 'Competitiveness, productivity and economic growth across the European regions', *Regional Studies* 38(9): 1045–1067.
Garmise, S. (1994) 'Economic development strategies in Emilia Romagna', in M. Burch (ed.) *The Regions and the New Europe*. Manchester: Manchester University Press.

Ge, W. (1999) *The Dynamics of Export-Processing Zones. United Nations Conference on Trade and Development Discussion Paper no. 144*. Geneva: United Nations.

Geddes, M. (2000) 'Tackling social exclusion in the European Union', *International Journal of Urban and Regional Research* 24(4): 782–800.

Geddes, M. and Newman, I. (1999) 'Evolution and conflict in local economic development', *Local Economy* 13(5): 12–25.

Gentle, C. and Marshall, J.N. (1992) 'The deregulation of the financial services industry and the polarisation of regional economic prosperity', *Regional Studies* 26: 581–592.

Geroski, P.A. (1989) 'The choice between scale and diversity', in E. Davis (ed.) *1992: Myths and Realities*. London: Centre for Business Strategy, London Business School.

Gerschenkron, A. (1962) *Economic Backwardness in Historical Perspective*. Cambridge, MA: Harvard University Press.

Gertler, M. (1984) 'Regional capital theory', *Progress in Human Geography* 8(1): 50–81.

Gertler, M. (1992) 'Flexibility revisited: districts, nation states, and the forces of production', *Transactions of the Institute of British Geographers* 17: 259–278.

Gertler, M.S. (1995) 'Groping towards reflexivity: responding to industrial change in Ontario', in P. Cooke (ed.) *The Rise of the Rustbelt*. London: UCL Press.

Gertler, M. (2004) *Manufacturing Culture: The Institutional Geography of Industrial Practice*. Oxford: Oxford University Press.

Gertler, M.S., Florida, R., Gates, G. and Vinodrai, T. (2002) *Competing on Creativity: Placing Ontario's Cities in North American Context*. Toronto: Program on Globalization and Regional Innovation Systems, Centre for International Studies, University of Toronto.

Gibbs, D. (2000) 'Ecological modernisation, regional economic development and regional development agencies', *Geoforum* 31(1): 9–19.

Gibbs, D. (2002) *Local Economic Development and the Environment*. London: Routledge.

Gibson-Graham, J.K. (2000) 'Poststructural interventions', in T.J. Barnes and E. Sheppard (eds) *A Companion to Economic Geography*. Oxford: Blackwell.

Gibson-Graham, J.K. (2004) 'The violence of development: two political imaginaries', *Development* 47(1): 27–34.

Gibson-Graham, J.K. and Ruccio, D. (2001) '"After" development: negotiating the place of class', in J.K. Gibson-Graham, S. Resnick and R. Wolff (eds) *Re-presenting Class: Essays in Postmodern Political Economy*. Durham, NC: Duke University Press.

Giddens, A. (1998) *The Third Way: Prospects for Social Democracy*. Cambridge: Polity.

Giloth, R.P. (2003) *Workforce Intermediaries for the 21st Century*. Philadelphia, PA: Temple University Press.

Giordano, B. and Roller, E. (2004) '"Té para todos"? A comparison of the processes of devolution in Spain and the UK', *Environment and Planning A* 36(12): 2163–2181.

Giuliani, E., Pietrobelli, C. and Rabellotti, R. (2005) 'Upgrading in global value chains: lessons from Latin American clusters', *World Development* 33(4): 549–573.

Glasmeier, A. (2000) 'Economic geography in practice: local economic development policy', in G.L. Clark, M. Feldman and M. Gertler (eds) *The Oxford Handbook of Economic Geography*. Oxford: Oxford University Press.

Goddard, J.B., Thwaites, A.T., Gillespie, A.E., James, V., Nash, P., Oakey, R.P. and Smith, I.J. (1979) *The Mobilisation of Indigenous Potential in the UK* (Final Report for the Regional Policy Directorate, EEC). Newcastle upon Tyne: CURDS, University of Newcastle.

Goetz, E. and Clarke, S. (1993) *The New Localism: Comparative Politics in a Global Era*. Newbury Park, CA: Sage.

Gordon, I. and McCann, P. (2000) 'Industrial clusters: complexes, agglomeration and/or social networks?', *Urban Studies* 37(3): 513–532.

Government of Canada (1972) *Foreign Direct Investment in Canada*. Ottawa: Information Canada.

Government of Ireland (2002) *National Spatial Strategy for Ireland, 2002–2020: People, Places and Potential*. Dublin: The Stationery Office.

Government of the State of Jalisco (2001) *Jalisco crece: Una visión de su presente y su futuro*. Guadalajara, Mexico: Secretaría de Promoción Económica.

Grabher, G. (1993) 'The weakness of strong ties: the lock-in of regional development in the Ruhr Area', in G. Grabher (ed.) *The Embedded Firm: On the Socio-Economics of Industrial Networks*. London: Routledge.

Grabher, G. (1994) 'The disembedded regional economy: the transformation of East German industrial complexes into western enclaves', in A. Amin and N. Thrift (eds) *Globalization, Institutions and Regional Development in Europe*. Oxford: Oxford University Press.

Granovetter, M. and Swedberg, R. (1992) *The Sociology of Economic Life*. Boulder, CO: Westview Press.

Grass, G. (2005) 'The high price of freedom', *Guardian* 7 May.

Greene, F.J. (2002) 'An investigation into enterprise support for younger people, 1975–2000', *International Small Business Journal* 20: 315–336.

Grossman, G. and Helpman, E. (1991) *Innovation and Growth in the Global Economy*. Cambridge, MA: MIT Press.

Håkanson, H. (1990) 'International decentralization of R&D: the organizational challenges', in C.A. Bartlett, Y. Doz and G. Hedlund (eds) *Managing the Global Firm*. London: Routledge.

Halkier, H., Danson, M. and Damborg, C. (1998) *Regional Development Agencies in Europe*. London: Jessica Kingsley.

Hall, P. and Markusen, A. (1985) *Silicon Landscapes*. Boston, MA: Allen & Unwin.

Hall, P. and Preston, P. (1988) *The Carrier Wave: New Information Technology and the Geography of Innovation*. Boston, MA: Unwin Hyman.

Hall, P.A. and Soskice, D. (eds) (2001) *Varieties of Capitalism: The Institutional Foundations of Comparative Advantage*. Oxford: Oxford University Press.

Halpern, D. (2004) *Social Capital*. Cambridge: Polity.

Halpern, D. (2005) 'A matter of respect', *Prospect* July: 40–43.

Hamnett, C. (2003) *Unequal City: London in the Global Arena*. London: Routledge.

Harrison, B. (1994) *Lean and Mean: Corporate Power in the Age of Flexibility*. New York: Basic Books.

Harvey, D. (1982) *Limits to Capital*. Oxford: Basil Blackwell.

Harvey, D. (1989a) 'From managerialism to entrepreneurialism: the transformation in urban governance in late capitalism', *Geografiska Annaler* 71B(1): 3–17.

Harvey, D. (1989b) *The Condition of Postmodernity*. Oxford: Blackwell.

Harvey, D. (1996) *Justice, Nature and the Geography of Difference*. Oxford: Blackwell.

Harvey, D. (2000) *Spaces of Hope*. Edinburgh: Edinburgh University Press.

Harvie, C. (1994) *The Rise of Regional Europe*. London: Routledge.

Haskel, J. and Jackman, R. (1987) *Long-term Unemployment and Special Employment Measures in Britain*. London: Centre for Labour Economics, London School of Economics.

Haughton, G. (ed.) (1999) *Community Economic Development*. London: The Stationery Office and Regional Studies Association.

Haughton, G. and Counsell, D. (2004) *Regions, Spatial Strategies and Sustainable Development*. London: Routledge and Regional Studies Association.

Hausner, J. (1995) 'Imperative versus interactive strategy of systematic change in Central and Eastern Europe', *Review of International Political Economy* 2(1): 246–266.

Hauswirth, I., Herrschel, T. and Newman, P. (2003) 'Incentives and disincentives to city-regional cooperation in the Berlin-Brandenburg conurbation', *European Urban and Regional Studies* 10(2): 119–134.

Hayek, F.A. (1944) *The Road to Serfdom*. London: Routledge.

Hayter, R. (1997) *The Dynamics of Industrial Location: The Factory, the Firm and the Production System*. London: Wiley.

Hechter, M. (1999) *Internal Colonialism: The Celtic Fringe in British National Development* (2nd edn). New Brunswick, NJ: Transaction.

Held, D. (ed.) (1993) *Prospects for Democracy*. Cambridge: Polity.

Held, D. (1995) *Democracy and the Global Order: From the Modern State to Cosmopolitan Governance*. Cambridge: Polity.

Held, D., McGrew, A., Goldblatt, D. and Perraton, J. (1999) *Global Transformations: Politics, Economics and Culture*. Cambridge: Polity.

Henderson, J., Dicken, O., Hess, M., Coe, N. and Yeung, H.W-C. (2002) 'Global production networks and the analysis of economic development', *Review of International Political Economy* 9(3): 436–464.

Henson, S.J. and Loader, R.J. (2001) 'Barriers to agricultural exports from developing countries: the role of sanitary and phytosanitary requirements', *World Development* 29(1): 85–102.

Henton, D. (2001) 'Lessons from Silicon Valley: governance in a global city-region', in A.J. Scott (ed.) *Global City-Regions*. Oxford: Oxford University Press.

Hernández Laos, E. (1985) *La Productividad y el Desarrollo Industrial en México*. Mexico, DF: Fondo de Cultura Económica.

Hess, M. (2004) '"Spatial" relationships? Re-conceptualising embeddedness', *Progress in Human Geography* 28(2): 165–186.

Hines, C. (2000) *Localization: A Global Manifesto*. London: Earthscan.

Hirschman, A.O. (1958) *The Strategy of Economic Development*. New Haven, CT: Yale University Press.

Hirst, P. and Thompson, G. (1999) *Globalization in Question* (2nd edn). Cambridge: Polity.

Hirst, P. and Zeitlin, J. (1991) 'Flexible specialization versus post Fordism: theory, evidence and policy implications', *Economy and Society* 20(1): 1–56.

HM Treasury and DTI (2001) *Productivity in the UK: 3 – The Regional Dimension*. Norwich: The Stationery Office.

Holl, A. (2004) 'The role of transport in firms' spatial organization: evidence from the Spanish food processing industry', *European Planning Studies* 12(4): 537–550.

Holland, S. (1976) *Capital Versus the Regions*. London: Macmillan.

Hooghe, L. and Marks, G. (2001) *Multi-level Governance and European Integration*. Lanham, MD: Rowman & Littlefield.

House of Commons (2003) *Reducing Regional Disparities in Prosperity*. Housing, Planning, Local Government and the Regions Select Committee Report (HC 492-I). Norwich: The Stationery Office.

Howells, J. and Wood, M. (1993) *The Globalisation of Production and Technology*. London: Belhaven Press.

Howes, C. and Markusen, A. (1993) 'Trade, industry and economic development', in H. Naponen, J. Graham and A. Markusen (eds) *Trading Industries, Trading Regions*. New York: Guilford.

Hudson, R. (1989) *Wrecking a Region*. London: Pion.

Hudson, R. (1999) 'The learning economy, the learning firm and the learning region: a sympathetic critique of the limits to learning', *European Urban and Regional Studies* 6(1): 59–72.

Hudson, R. (2001) *Producing Places*. New York: Guilford.

Hudson, R. (2003) 'Fuzzy concepts and sloppy thinking: reflections on recent developments in critical regional studies', *Regional Studies* 37(6–7): 741–746.

Hudson, R. and Weaver, P. (1997) 'In search of employment creation via environmental valorization: exploring a possible eco-Keynesian future for Europe', *Environment and Planning A* 29: 1647–1661.

Hudson, R. and Williams, A. (1994) *Divided Britain* (2nd edn). Chichester: Wiley.

Hudson, R., Dunford, M., Hamilton, D. and Kotter, R. (1997) 'Developing regional strategies for economic success: lessons from Europe's economically successful regions', *European Urban and Regional Studies* 4: 365–373.

Humphrey, L. and Shaw, K. (2004) 'Regional devolution and democratic renewal: developing a radical approach to stakeholder involvement in the English regions', *Environment and Planning A* 36(12): 2183–2202.

Hymer, S. (1972) 'The multinational corporation and the law of uneven development', in J. Bagwathi (ed.) *Economics and World Order*. New York: Macmillan.

Hymer, S. (1979) *The Multinational Corporation: A Radical Approach*. Cambridge: Cambridge University Press.

IDA Ireland (2004) *Ireland: Vital Statistics*. Dublin: IDA Ireland.

Ietto-Gilles, G. (2003) 'The role of transnational corporations in the globalisation process', in J. Michie (ed.) *The Handbook of Globalisation*. Cheltenham: Edward Elgar.

Indergaard, M. (2004) *Silicon Alley: The Rise and Fall of a New Media District*. London: Routledge.

Innis, H. (1920) *The Fur Trade in Canada*. New Haven, CT: Yale University Press.

Institute for Competitiveness and Prosperity (2005) *Realizing Canada's Prosperity Potential*. Toronto: Institute for Competitiveness and Prosperity.

International Monetary Fund (1995) 'Gender issues in economic adjustment discussed at UN Conference on Women', *IMF Survey* (25 September): 286–288.

International Monetary Fund (2000) *World Investment Report*. Washington, DC: IMF.

Jacobs, J., Berridge, J., Broadbent, A., Crombie, D., Gertler, M.S., Gilbert, R., Mendelson, M., Nowlan, D.M., Sewell, J. and Slack, E. (2000) *Toronto: Considering Self-government*. Owen Sound, Ontario: Ginger Press.

Jaffe, A.B., Trajtenberg, M. and Henderson, R. (1993) 'Geographic localization of knowledge spillovers as evidenced by patent citations', *Quarterly Journal of Economics* 108: 577–598.

Jauch, H. (2002 'Export processing zones and the quest for sustainable development: a southern African perspective', *Environment and Urbanization* 14(1): 101–114.

Javed Burki, S., Perry, G. and Dillinger, W. (1999) *Beyond the Center: Decentralising the State*. Washington, DC: Latin American and Caribbean Studies, World Bank.

Jessop, B. (1994) 'Post-Fordism and the state', in A. Amin (ed.) *Post-Fordism: A Reader*. Oxford: Blackwell.

Jessop, B. (1997) 'Capitalism and its future: remarks on regulation, government and governance', *Review of International Political Economy* 4(3): 561–581.

Jessop, B. (2002) *The Future of the Capitalist State*. Cambridge: Polity.

Johnson, C. (1999) 'The developmental state: odyssey of a concept', in M. Woo-Cumings (ed.) *The Developmental State*. Ithaca, NY: Cornell University Press.

Joint Venture: Silicon Valley Network (2005) *2005 Index of Silicon Valley*. San Jose, CA: Joint Venture: Silicon Valley Network.

Jones, J. and Wren, C. (2004) 'Inward foreign direct investment and employment: a project-based analysis in north-east England', *Journal of Economic Geography* 4(5): 517–543.

Jones, M., Jones, R. and Woods, M. (2004) *An Introduction to Political Geography: Space, Place and Politics*. London: Routledge.

Kaldor, N. (1970) 'The case for regional policies', *Scottish Journal of Political Economy* 18: 337–348.

Kaldor, N. (1981) 'The role of increasing returns, technical progress and cumulative causation in the theory of international trade and economic growth', in F. Targetti and A. Thirlwall (eds) *The Essential Kaldor*. London: Duckworth.

Kang M-G. (2003) 'Decentralization and restructuring of regionalism in Korea', *Korea Journal* 43(2): 5–31.

Kang, N.H. and Johansson, S. (2000) 'Cross-border mergers and acquisitions: their role in industrial globalisation', STI Working Paper. Paris: Directorate for Science, Technology and Industry, OECD.

Keating, M. (1998) *The New Regionalism in Western Europe: Territorial Restructuring and Political Change*. Cheltenham: Edward Elgar.

Keating, M. (2005) 'From functional to political regionalism: England in comparative perspective', in R. Hazell (ed.) *The English Question*. Manchester: Manchester University Press.

Keating, M., Loughlin, J. and Deschouwer, K. (2003) *Culture, Institutions and Economic Development: A Study of Eight European Regions*. Cheltenham: Edward Elgar.

Keep, E. and Mayhew, K. (1999) 'Knowledge, skills and competitiveness', *Oxford Review of Economic Policy* 15(1): 1–15.

Kennedy, K. (1992) 'The context of economic development', in J. Goldthorpe and C. Whelan (eds) *The Development of Industrial Society in Ireland* (Proceedings of the British Academy, 79). Oxford: Oxford University Press.

Keynes, J.M. (1936) *The General Theory of Employment, Interest and Money*. London: Macmillan.

Kim B-W. and Kim P-S. (1992) *Korean Public Administration: Managing Uneven Development* (2nd edn). Seoul: Hollym.

Kim D-J. (1996) *Mass Participatory Economy: Korea's Road to World Economic Power*. Lanham, MD: University Press of America.

Kim W-B. (2003) 'Regionalism: its origins and substance with competition and exclusion', *Korea Journal* summer: 5–31.

Kimenyi, M.S., Wieland, R.C. and Von Pischke, J.D. (eds) (1998) *Strategic Issues in Microfinance*. Aldershot: Ashgate.

Kirby, P. (2002) *The Celtic Tiger in Distress: Growth with Inequality in Ireland*. Basingstoke: Palgrave.

Kirby, P. (2004) 'Globalization, the Celtic Tiger and social outcomes: is Ireland a model or a mirage?', *Globalizations* 1(2): 205–222.

Kitson, M., Martin, R. and Tyler, P. (2004) 'Regional competitiveness: an elusive yet key concept', *Regional Studies* 38(9): 991–999.

Kjær, A.M. (2004) *Governance*. Cambridge: Polity.

Klagge, B. and Martin, R. (2005) 'Decentralized versus centralized financial systems: is there a case for local capital markets?', *Journal of Economic Geography* 5: 387–421.

Kominos, N. (2002) *Intelligent Cities: Innovation, Knowledge Systems and Digital Spaces*. London: Spon Press.

Korea National Commission for UNESCO (2001) *The Korean Economy: Reflections at the New Millennium*. Seoul: Hollym.

Kozul-Wright, R. (1995) 'Transnational corporations and the nation state', in J. Michie and J. Grieve Smith (eds) *Managing the Global Economy*. Oxford: Oxford University Press.

Kozul-Wright, R. and Rowthorn, R. (1998) 'Spoilt for choice: multinational corporations and the geography of international production', *Oxford Review of Economic Policy* 14(2): 74–92.

Kroszner, R.S. (2003) 'Currency competition in the digital age', in D. Altig and B.D. Smith (eds) *Evolution and Procedures in Central Banking*. Cambridge: Cambridge University Press.

Krugman, P. (1986) *Strategic Trade Policy and the New International Economics*. Cambridge, MA: MIT Press.

Krugman, P. (1990) *Rethinking International Trade*. Cambridge, MA: MIT Press.

Krugman, P. (1991) *Geography and Trade*. Leuven, Belgium: Leuven University Press.

Krugman, P. (1993) 'On the relationship between trade theory and location theory', *Review of International Economics* 1: 110–122.

Krugman, P. (1994) 'Competitiveness: a dangerous obsession', *Foreign Affairs* March–April, 73(2): 28–44.

Krugman, P. (1995) *Development, Geography and Economic Theory*. Cambridge, MA: MIT Press.

Krugman, P. (1996) 'Making sense of competitiveness', *Oxford Review of Economic Policy* 12(3): 17–25.

Krugman, P. (1997) 'Good news from Ireland: a geographical perspective', in A. Gray (ed.) *International Perspectives on the Irish Economy*. Dublin: Inderon.

Kuznets, S. (1960) *Population Change and Aggregate Output*. Princeton, NJ: Princeton University Press.

Kuznets, S. (1966) *Modern Economic Growth: Rate, Structure and Spread*. New Haven, CT: Yale University Press.

Lagendijk, A. (2003) 'Towards conceptual quality in regional studies: the need for a subtle critique – a response to Markusen', *Regional Studies* 37(6–7): 719–727.

Lagendijk, A. and Cornford, J. (2000) 'Regional institutions and knowledge: tracking new forms of regional development policy', *Geoforum* 31: 209–218.

Lambooy, J.G. (1980) *Economy and Space: Introduction into the Economic Geography and Regional Economics*. Assen: van Gorcum.

Layard, R. (2005) *Happiness: Lessons from a New Science*. London: Allen Lane.

Lee, J.J. (1989) *Ireland 1912–1985*. Cambridge: Cambridge University Press.

Lee M-H., Choi N-H. and Park, M. (2005) 'A systems thinking approach to the new administrative capital: balanced development or not?', *Systems Dynamic Review* 21(1): 69–85.

Lee, R., Leyshon, A., Aldridge, T., Tooke, T., Williams, C. and Thrift, N. (2004) 'Making geographies and histories? Constructing local circuits of value', *Environment and Planning D: Society and Space* 22: 595–617.

Lee, S. (2004) 'Economic change and regional development disparities in the 1990s', *Korea Journal* spring: 75–102.

Le Galès, P. and Lequesne, C. (1998) 'Introduction', in P. Le Galès and C. Lequesne (eds) *Regions in Europe*. London: Routledge

Leibovitz, J. (2003) 'Institutional barriers to associative city-region governance: the politics of institution-building and economic governance in "Canada's technology triangle"', *Urban Studies* 40(13): 2613–2642.

Leslie, S.W. (1993) *The Cold War and American Science: The Military-Industrial Complex at MIT and Stanford*. New York: Columbia University Press.

Leyshon, A. and Thrift, N. (1995) 'Geographies of financial exclusion: financial abandonment in Britain and the United States', *Transactions of the Institute of British Geographers* 20: 312–341.

Leyshon, A., Lee, R. and Williams, C.C. (eds) (2003) *Alternative Economic Spaces*. London: Sage.

Lipietz, A. (1980) 'The structuration of space, the problem of land, and spatial policy', in J. Carney, R. Hudson and J. Lewis (eds) *Regions in Crisis*. London: Croom Helm.

Local Government Commission (2004) *Local Economies*, Smart Growth: Economic Development for the 21st Century, www.lgc.org/economic/localecon.html.

Loewendahl, H.B. (2001) *Bargaining with Multinationals: The Investment of Siemens and Nissan in North East England*. Basingstoke: Palgrave.

Logan, J. and Molotch, H. (1997) *Urban Fortunes: The Political Economy of Place*. Berkeley, CA: University of California Press.

Loughlin, J. (ed.) (2001) *Subnational Democracy in the European Union: Challenges and Opportunities*. Oxford: Oxford University Press.

Love, J. (1994) 'Economic ideas and ideologies in Latin America since 1930', in L. Bethell (ed.) *The Cambridge History of Latin America*, Vol. VI, Cambridge: Cambridge University Press.
Lovering, J. (1989) 'The restructuring debate', in R. Peet and N. Thrift (eds) *New Models in Geography*. London: Unwin Hyman.
Lovering, J. (1991) 'The changing geography of the military industry in Britain', *Regional Studies* 25(4): 279–293.
Lovering, J. (1999) 'Theory led by policy: the inadequacy of the new regionalism (as illustrated from the case of Wales)', *International Journal of Urban and Regional Research* 23(2): 379–398.
Lovering, J. (2001) 'The coming regional crisis (and how to avoid it)', *Regional Studies* 35(4): 349–354.
Lundvall, B-A. (ed.) (1992) *National Innovation Systems: Towards a Theory of Innovation and Interactive Learning*. London: Pinter.
Lundvall, B-A. and Maskell, P. (2000) 'Nation states and economic development: from national systems of production to national systems of knowledge creation and learning', in G.L. Clark, M. Feldman and M. Gertler (eds) *The Oxford Handbook of Economic Geography*. Oxford: Oxford University Press.
Luria, D. (1997) 'Toward lean or rich? What performance benchmarking tells us about SME performance and some implications for extension centers' services and mission', in P. Shapira and J. Youtie (eds) *Manufacturing Modernization: Learning from Evaluation Practices and Results*. Washington, DC: National Institute of Standards and Technology.
McCall, C. and Williamson, A. (2001) 'Governance and democracy in Northern Ireland: the role of the voluntary and community sector after the Agreement', *Governance* 14(2): 363–383.
McCann, P. and Mudambi, R. (2004) 'The location behaviour of the multinational enterprise', *Growth and Change* 35(4): 491–524.
McCombie, J. and Thirlwall, A. (1997) 'The dynamic Harrod foreign trade multiplier and the demand-orientated approach to economic growth: an evaluation', *International Review of Applied Economics* 11: 5–26.
McCrone, G. (1969) *Regional Policy in Britain*. London: Allen & Unwin.
McDonald, F. (2005) 'Dublin's growth way off balance', *Irish Times* 26 June.
MacGillavray, A. and Walker, P. (2002) 'Local social capital', in S. Barron, J. Field and T. Schuller (eds) *Social Capital: Critical Perspectives*. Oxford: Oxford University Press.
MacKay, R. (2001) 'Regional taxing and spending: the search for balance', *Regional Studies* 35(6): 563–575.
MacKay, R. (2003) 'Twenty five years of regional development', *Regional Studies* 37(3): 303–317.
MacKinnon, D. and Cumbers, A. (2007) *An Introduction to Economic Geography: Globalisation, Uneven Development and Place*. Harlow: Pearson.
MacKinnon, D. and Phelps, N.A. (2001) 'Devolution and the territorial politics of foreign direct investment', *Political Geography* 20(3): 353–379.
MacKinnon, D., Cumbers, A. and Chapman, K. (2002) 'Learning innovation and regional development: a critical appraisal of recent debates', *Progress in Human Geography* 26 (3): 293–311.
McLean, I. (2004) 'Fiscal federalism in Australia', *Public Administration* 82(1): 21–38.
McLean, I. (2005) *Fiscal Crisis of the United Kingdom*. London: Palgrave.
Macleod, G. (1997) 'Globalising Parisian thought-waves: recent advances in the study of social regulation, politics, discourse, and space', *Progress in Human Geography* 21: 530–554.
Macleod, G. (2001) 'Beyond soft institutionalism: accumulation, regulation and their geographical fixes', *Environment and Planning A* 33: 1145–1167.
McMichael, P. (1996) *Development and Social Change: A Global Perspective*. Thousand Oaks, CA: Pine Forge Press.
McQuaid, R.W. (1996) 'Social networks, entrepreneurship and regional development', in M.W. Danson *Small Firm Formation and Regional Economic Development*. London: Routledge.
McQuaid, R.W., Green, A.E. and Danson, M. (2005) 'Introducing employability', *Urban Studies* 42(2): 191–195.
MacSharry, R. and White, P. (2000) *The Making of the Celtic Tiger*. Dublin: Mercier Press.
Malecki, E. (1997) *Technology and Economic Development: The Dynamics of Local, Regional and National Competitiveness* (2nd edn). London: Addison Wesley Longman.
Malecki, E. (2004) 'Jockeying for position: what it means and why it matters to regional development policy when places compete', *Regional Studies* 38(9): 1101–1120.
Marks, G. and Hooghe, M. (2004) 'Contrasting visions of multi-level governance', in I. Bache and M. Flinders (eds) *Multi-level Governance*. Oxford: Oxford University Press.

Markusen, A. (1985) *Profit Cycles, Oligopoly and Regional Development*. Cambridge, MA: MIT Press.

Markusen, A. (1991) 'The military industrial divide: Cold War transformation of the economy and the rise of new industrial complexes', *Environment and Planning D: Space and Society* 9(4): 391–416.

Markusen, A. (1996) 'Sticky places in slippery space: a typology of industrial districts', *Economic Geography* 72(3): 293–313.

Markusen, A. (2004) 'Targeting occupations in regional and community economic development', *Journal of the American Planning Association* 70 (3): 253–268.

Markusen, A. (2003) 'Fuzzy concepts, scanty evidence, policy distance: the case for rigor and policy relevance in critical regional studies', *Regional Studies* 33(9): 869–884.

Marquand, D. (2004) *Decline of the Public*. Cambridge: Polity.

Marshall, J.N. (1978) 'Ownership, organisation and industrial linkage: a case study in the Northern region of England', *Regional Studies* 13: 531–557.

Marshall, J.N. (1982) 'Linkages between manufacturing industry and business services', *Environment and Planning A* 14: 1523–1540.

Marshall, J.N., Bradley, D., Hodgson, N., Alderman, N. and Richardson, R. (2005) 'Relocation, relocation, relocation: assessing the case for public sector dispersal', *Regional Studies* 39(6): 769–789.

Marshall, M. (1987) *Long Waves of Regional Development*. New York: St Martin's Press.

Marshall, N. and Wood, P. (1995) *Services and Space: Key Aspects of Urban and Regional Development*. London: Prentice Hall

Martin, P. (1999) 'Public policies, regional inequalities and growth', *Journal of Public Economics* 73(1): 85–105.

Martin, R. (1989) 'The new economics and politics of regional restructuring: the British experience', in L. Albrechts, F. Moulaert, P. Roberts and E. Swyngendouw (eds) *Regional Policy at the Crossroads*. London: Jessica Kingsley.

Martin, R. (1999) 'Institutional approaches in economic geography', in T.J. Barnes and E. Sheppard (eds) *A Companion to Economic Geography*. Oxford: Blackwell.

Martin, R. and Morrison, P. (2003) *Geographies of Labour Market Inequality*. London: Routledge.

Martin, R. and Sunley, P. (1996) 'Paul Krugman's geographical economics and its implications for regional development theory: a critical assessment', *Economic Geography* 72: 259–292.

Martin, R. and Sunley, P. (1997) 'The Post-Keynesian state and the space-economy', in R. Lee and J. Wills (eds) *Geographies of Economies*. London: Edward Arnold.

Martin, R. and Sunley, P. (1998) 'Slow convergence? Post neo-classical endogenous growth theory and regional development', *Economic Geography* 74(3): 201–227.

Martin, R. and Sunley, P. (2003) 'Deconstructing clusters: chaotic concept or policy panacea?', *Journal of Economic Geography* 3(1): 5–35.

Maskell, P. (2002) 'Social capital, innovation and competitiveness', in S. Barron, J. Field and T. Schuller (eds) *Social Capital: Critical Perspectives*. Oxford: Oxford University Press.

Maskell, P. and Törnqvist, G. (1999) *Building a Cross-border Learning Region. Emergence of the North European Øresund Region*. Copenhagen: Handelshøjskolens Forlag.

Maskell, P., Eskelinen, H., Hannibalson, I., Malmburg, A. and Vatne, E. (1998) *Competitiveness, Localized Learning and Regional Development*. London: Routledge.

Mason, C. and Harrison, R. (1999) 'Financing entrepreneurship: venture capital and regional development', in R. Martin (ed.) *Money and the Space Economy*. Chichester: Wiley.

Mason, C.M. and Harrison, R.T. (2002) 'Barriers to investment in the informal venture capital sector', *Entrepreneurship and Regional Development* 14(3): 271–287.

Massey, D. (1993) 'Power-geometry and a progressive sense of place', in J. Bird, B. Curtis, T. Putnam and G. Robertson (eds) *Mapping the Futures: Local Cultures, Global Change*. London and New York: Routledge.

Massey, D. (1995) *Spatial Divisions of Labour: Social Structures and the Geography of Production* (2nd edn). London: Macmillan.

Massey, D. and Allen, J. (eds) (1984) *Geography Matters!* Cambridge: Cambridge University Press.

Massey, D., Quintas, P. and Wield, D. (1992) *High-tech Fantasies: Science Parks in Society, Science and Space*. London: Routledge.

Matthews, A. (1994) *Managing the EU Structural Funds*. Cork: Cork University Press.

Matzner, E. and Streeck, W. (1991) 'Towards a socioeconomics of employment in a post-Keynesian society', in E. Matzner and W. Streeck (eds) *Beyond Keynesianism: The Socioeconomics of Production and Full Employment*, Aldershot: Edward Elgar.

Mayer, T. (2002) *Media Democracy: How the Media Colonize Politics*. Cambridge: Polity.

Mayes, D. (1995) 'Introduction: conflict and cohesion in the Single European Market', in A. Amin and J. Tomaney (eds) *Behind the Myth of European Union: Prospects for Cohesion*. London: Routledge.
Meadows, P. (2001) *Lessons for Employment Policy*. York: Joseph Rowntree Foundation.
Meardon, S.J. (2000) 'Eclecticism, inconsistency, and innovation in the history of geographical economics', in R.E. Backhouse and J. Biddle (eds) *Toward a History of Applied Economics*. Durham, NC: Duke University Press.
Meheroo, J. and Taylor, R. (eds) (2003) *Information Technology Parks of the Asia Pacific: Lessons for the Regional Digital Divide*. New York: M.E. Sharp.
Metcalf, H. (ed.) (1995) *Future Skill Demand and Supply: Trends, Shortages and Gluts*. London: Policy Studies Institute.
Michie, J. and Grieve Smith, J. (eds) (1995) *Managing the Global Economy*. Oxford: Oxford University Press.
Miller, D. (2002) *The Regional Governing of Metropolitan America*. Boulder, CO: Westview Press.
Milne, K. (2005) *Manufacturing Dissent: Single-issue Protest, the Public and the Press*. London: Demos.
Mitchell, S. (2000) *The Hometown Advantage*. Minneapolis, MN: Institute for Local Self-Reliance.
Mitchell, S. (2002) 'New rules for the new localism: favoring communities, deterring corporate chains', *Multinational Monitor* 23(10–11): 1–10.
Mjøset, L. (1992) 'The Irish economy in a comparative institutional perspective', *NESC Report (93)*. Dublin: The Stationery Office.
Moggridge, D. (1973) *Collected Writings of John Maynard Keynes, Vol. 7 – The General Theory*. London: Macmillan for the Royal Economic Society.
Mohan, G. and Mohan, J. (2002) 'Placing social capital', *Progress in Human Geography* 26(2): 191–210.
Moore, B. and Rhodes, J. (1986) *The Effects of Regional Economic Policy*. London: HMSO.
Morgan, K. (1997) 'The learning region: institutions, innovation and regional renewal', *Regional Studies* 31: 491–504.
Morgan, K. (1998) 'Regional renewal: the development agency as animateur', in H. Halkier, M. Danson and C. Damborg (eds) *Regional Development Agencies in Europe*. London: Jessica Kingsley.
Morgan, K. (2001) 'The new territorial politics: rivalry and justice in post-devolution Britain', *Regional Studies* 35(4): 343–348.
Morgan, K. (2002) 'The new regeneration narrative: local development in the multi-level polity', *Local Economy* 17(3): 191–199.
Morgan, K. (2004) 'Sustainable regions: governance, innovation and scale', *European Planning Studies* 12(6): 871–889.
Morgan, K. and Henderson, D. (2002) 'Regions as laboratories: the rise of regional experimentalism in Europe', in M. Gertler and D. Wolfe (eds) *Innovation and Social Learning*. Basingstoke: Palgrave Macmillan.
Morgan, K. and Morley, A. (2002) *Re-localising the Food Chain: The Role of Creative Public Procurement*. Cardiff: The Regeneration Institute, Cardiff University.
Morgan, K. and Mungham, G. (2000) *Redesigning Democracy: The Making of the Welsh Assembly*. Bridgend: Seren.
Morgan, K. and Nauwelaers, C. (eds) (1999) *Regional Innovation Strategies: The Challenge for Less Favoured Regions*. London: The Stationery Office.
Morgan, K. and Sayer, A. (1988) *Microcircuits of Capital*. Cambridge: Polity.
Morgenroth, E. (2000) 'Regionalisation and the functions of regional and local government', in A. Barrett (ed.) *Budget Perspectives*. Dublin: Economic and Social Research Institute.
Moulaert, F., Martinelli, F., Swyngedouw, E. and González, S. (2005) 'Towards alternative model(s) of local innovation', *Urban Studies* 42(11): 1969–1990.
Mulgan, G. (2005) 'Lessons of power', *Prospect* May: 24–29.
Mulgan, G. (1994) *Politics in an Antipolitical Age*. Cambridge: Polity.
Mullins, R. (2005) 'Silicon Valley jobs continue to vanish', *Silicon Valley/San Jose Business Journal* 27 January.
Munck, R. (1993) *The Irish Economy: Results and Prospects*. London: Pluto.
Mur, J. (1996) 'A future for Europe? Results with a regional prediction model', *Regional Studies* 30(6): 549–565.
Murphy, A.E. (2000b) 'The "Celtic Tiger": an analysis of Ireland's economic growth performance', Robert Schuman Centre for Advanced Studies, Paper No. 2000/16, European University Institute: Florence.
Murphy, J (2000a) 'Ecological modernisation', *Geoforum* 31(1): 1–8.

Myrdal, G. (1957) *Economic Theory and Underdeveloped Regions*. London: Duckworth.
Mytelka, L.K. (2000) 'Location tournaments for FDI: inward investment into Europe in a global world', in S. Hood (ed.) *The Globalization of Multinational Enterprise Activity and Economic Development*. London: Macmillan.
National Assembly for Wales (NAW) (2000) *Learning to Live Differently*. Cardiff: NAW.
National Economic and Social Council (NESC) (1982) *A Review of Industrial Policy*. Dublin: NESC.
National Software Directorate (2004) *Software Industry Statistics, 1991–2003*, www.nsd.ie, accessed 14 February 2005.
Nelson, R. and Winter, S. (1985) *An Evolutionary Theory of Economic Change*. Cambridge, MA: Harvard University Press.
New Economics Foundation (NEF) (2002) *Plugging the Leaks*. London: NEF.
Newman, C. (2004) 'Creation of greater Dublin authority urged', *Irish Times* 18 October.
Newman, P. (2000) 'Changing patterns of regional governance in the EU', *Urban Studies* 37(5–6): 895–908.
Noponen, H., Graham, J. and Markusen, A. (eds) (1993) *Trading Industries, Trading Regions*. New York: Guilford.
North, D.C. (1955) 'Location theory and regional economic growth', *Journal of Political Economy* 63: 243–258.
North, D.C. (1990) *Institutions, Institutional Change, and Economic Performance*. Cambridge: Cambridge University Press.
North, P. (2005) 'Scaling alternative economic practices? Some lessons from alternative currencies', *Transactions of the Institute of British Geographers* 30(2): 233–235.
Norton, R.D. and Rees, J. (1979) 'The product cycle and the spatial decentralisation of American manufacturing', *Regional Studies* 13: 141–151.
Nurske, R. (1961) *Equilibrium and Growth in the World Economy*. Cambridge, MA: Harvard University Press.
Nussbaum, M. and Sen, A. (1993) *The Quality of Life*. Oxford: Oxford University Press.
O'Brien, P. (2004) 'Trade unions and regional governance', unpublished PhD thesis, Newcastle upon Tyne: Centre for Urban and Regional Development Studies (CURDS), University of Newcastle.
O'Brien, P., Pike, A. and Tomaney, J. (2004) 'Devolution, the governance of regional development and the Trades Union Congress in the North East region of England', *Geoforum* 35(1): 59–68.
O'Donnell, R. (1993) *Ireland and Europe*. Dublin: Economic and Social Research Institute.
O'Donnell, R. (1997) 'The competitive advantage of peripheral regions: conceptual issues and research approaches', in B. Fynes and S. Ennis (eds) *Competing from the Periphery*. London: Dryden Press.
O'Donnell, R. (2004) 'Ireland: social partnership and the "Celtic Tiger" economy', in J. Perraton and B. Clift (eds) *Where are National Capitalisms Now?* London: Palgrave Macmillan.
O'Donnell, R. and Thomas, D. (1998) 'Partnership and policy-making', in S. Healy and B. Reynolds (eds) *Social Policy in Ireland*. Dublin: Oak Tree Press.
O'Donnell, R. and Walsh, J. (1995) 'Ireland: region and state in the European Union', in M. Rhodes (ed.) *The Regions and the New Europe*. Manchester: Manchester University Press.
OECD (1994) *Review of Foreign Direct Investment: Ireland*. Paris: OECD.
OECD (1996) *Networks of Enterprises and Local Development: Competing and Co-operating in Local Productive Systems*. Paris: OECD.
OECD (2001) *Devolution and Globalisation: Implications for Local Decision-makers*. Paris: OECD.
OECD (2003a) *Entrepreneurship and Local Economic Development: Programme and Policy Recommendations*. Paris: OECD.
OECD (2003b) *OECD Territorial Reviews: Øresund, Denmark/Sweden*. Paris: OECD.
OECD (2004a) *New Forms of Governance for Economic Development*. Paris: OECD.
OECD (2004b) *Territorial Reviews: Busan, Korea*. Paris: OECD.
OECD (2005a) *Economic Survey: Korea*. Paris: OECD.
OECD (2005b) *OECD Regions at a Glance*. Paris: OECD.
OECD (2005c) *Strengthening Entrepreneurship and Economic Development at Local Level in Eastern Germany*. Paris: OECD.
OECD (2005d) *Territorial Reviews: Seoul, Korea*. Paris: OECD.
Office for National Statistics (2004) *Regional Trends*. Norwich: The Stationery Office.
O'Hearn, D. (1998) *Inside the Celtic Tiger: The Irish Economy and the Asian Model*. London: Pluto.
O'Hearn, D. (2000) 'Globalisation, "New Tigers", and the end of the development state', *Politics and Society* 28(1): 67–92.

Ohmae, K. (1990) *Borderless World*. London: Collins.

Olsson, J. (1998) 'Regional development and political democracy', in H. Halkier, M. Danson and C. Damborg (eds) *Regional Development Agencies in Europe*. London: Jessica Kingsley.

O'Malley, E. (1992) 'Problems of industrialisation in Ireland', in J. Goldthorpe and C. Whelan (eds) *The Development of Industrial Society in Ireland* (Proceedings of the British Academy, 79). Oxford: Oxford University Press.

O'Malley, E. (1998) 'Industrial policy in Ireland', in M. Storper, S. Thomadakis and L. Tsipouri (eds) *Latecomers in the Global Economy*. London: Routledge.

O'Malley, E. and O'Gorman, C. (2001) 'Competitive advantage in the Irish indigenous software industry and the role of inward FDI', *European Planning Studies* 9(3): 303–321.

Ontario Jobs and Investment Board (1999) *A Roadmap to Prosperity*. Toronto: Ontario Jobs and Investment Board.

Ó Riain, S. (2000) 'The flexible developmental state: globalization, information technology and the Celtic Tiger', *Politics and Society* 28(2): 157–193.

Ó Riain, S. (2004) *The Politics of High Tech Growth: Developmental Network States in the Global Economy*. Cambridge: Cambridge University Press.

O'Toole, F. (2003) *After the Ball: What is the Legacy of the Celtic Tiger?* Dublin: New Island.

O'Toole, T. (2004) 'Regions need to work for themselves', *Irish Times* 29 October.

Owen, K.A. (2002) 'The Sydney Olympics and urban entrepreneurialism: local variations in urban governance', *Australian Geographical Studies* 40(3): 323–336.

Owen Smith, E. (1994) *The German Economy*. London: Routledge.

Paasi, A. (1991) 'Deconstructing regions: notes on the scales of spatial life', *Environment and Planning A* 23: 239–256.

Park, S.O. (2000) 'Innovation systems, networks and the knowledge-based economy in Korea', in J.H. Dunning (ed.) *Regions, Globalization and the Knowledge-based Economy*. Oxford: Oxford University Press.

Parthasarathy, B. (2004) 'India's Silicon Valley or Silicon Valley's India? Socially embedding the computer software industry in Bangalore', *International Journal of Urban and Regional Research* 28(3): 664–685.

Pavlinek, P. (2004) 'Regional development implications of foreign direct investment in central Europe', *European Urban and Regional Studies* 11(1): 47–70.

Peck, J. (1999) 'Neoliberalizing states: thin policies/hard outcomes', *Progress in Human Geography* 25(3): 445–455.

Peck, J. (2000) 'Doing regulation', in G.L. Clark, M. Feldman and M. Gertler (eds) *The Oxford Handbook of Economic Geography*. Oxford: Oxford University Press.

Peck, J. (2001) *Workfare States*. New York: Guilford.

Peck, J. (2003) 'Fuzzy old world: a response to Markusen', *Regional Studies* 37(6–7): 729–740.

Peck, J. (2005) 'Struggling with the creative class', *International Journal of Urban and Regional Research* 29(4): 740–770.

Peck, J. and Tickell, A. (1995) 'The social regulation of uneven development: "regulatory deficit", England's South East and the collapse of Thatcherism', *Environment and Planning A* 27: 15–40.

Peck, J. and Tickell, A. (2002) 'Neoliberalizing space', *Antipode* 34(3): 380–404.

Peck, J. and Yeung, H.W-C. (eds) (2003) *Remaking the Global Economy*. London: Sage.

Peet, R. (1998) *Modern Geographical Thought*. Oxford: Blackwell.

Peet, R. (2002) 'Ideology, discourse, and the geography of hegemony: from socialist to neoliberal development in post-apartheid South Africa', *Antipode* 34(1): 54–84.

Pellow, D.N. and Park, L.S. (2002) *The Silicon Valley of Dreams: Environmental Justice, Immigrant Workers and the High-Tech Global Economy*. New York: New York University Press.

Perloff, H.S., Edgar, S., Dunn Jr, E.S., Lampard, E.E. and Muth, R.F. (1960) *Regions, Resources and Economic Growth*. Baltimore, MD: Johns Hopkins University Press.

Perrons, D. (2004) *Globalisation and Social Change: People and Places in a Divided World*. London: Routledge.

Perroux, F. (1950) 'Economic space: theory and applications', *Quarterly Journal of Economics* 64(1): 89–104.

Perroux, F. (1957) *Théorie générale du progrès économique*. Paris: ISEA.

Peters, B.G and Pierre, J. (2001) 'Developments in intergovernmental relations: towards multi-level governance', *Policy and Politics* 29(2): 131–135.

Pharr, S. and Putnam, R. (eds) (2000) *Disaffected Democracies: What's Troubling the Trilateral Countries?* Princeton, NJ: Princeton University Press.
Phelps, N.A. (1993) 'Branch plants and the evolving spatial division of labour: a study of material linkage change in the Northern Region of England', *Regional Studies* 27(2): 87–101.
Phelps, N.A. and Fuller, C. (2000) 'Multinationals, intracorporate competition and regional development', *Economic Geography* 76(3): 224–243.
Phelps, N.A. and Tewdwr-Jones, M. (1998) 'Institutional capacity building in a strategic policy vacuum: the case of the Korean firm LG in South Wales', *Environment and Planning C: Government and Policy* 16(6): 735–755.
Phelps, N.A. and Waley, P. (2004) 'Capital versus the districts: the story of one multinational company's attempts to disembed itself', *Economic Geography* 80(2): 191–215.
Phelps, N.A., MacKinnon, D., Stone, I. and Braidford, P. (2003) 'Embedding the multinationals? Institutions and the development of overseas manufacturing affiliates in Wales and North East England', *Regional Studies* 37(1): 27–40.
Pike, A. (1998) 'Making performance plants from branch plants? *In-situ* restructuring in the automobile industry in UK Region', *Environment and Planning A* 30: 881–900.
Pike, A. (2002a) 'Post-devolution blues? Economic development in the Anglo-Scottish Borders', *Regional Studies* 36(9): 1067–1082.
Pike, A. (2002b) 'Task forces and the organisation of economic development: the case of the North East region of England', *Environment and Planning C* 20: 717–739.
Pike, A. (2004) 'Heterodoxy and the governance of economic development', *Environment and Planning A* 36: 2141–2161.
Pike, A. (2005) 'Building a geographical political economy of closure: the closure of *R&DCo* in North East England', *Antipode* 37(1): 93–115.
Pike, A. (2006) '"Shareholder value" versus the regions: the close of the Vaux Brewery in Sunderland', *Journal of Economic Geography* 6: 201–222.
Pike, A. and Tomaney, J. (1999) 'The limits to localization in declining industrial regions? Transnational corporations and economic development in Sedgefield Borough', *European Planning Studies* 7(4): 407–428.
Pike, A. and Tomaney, J. (2004) 'Sub-national governance and economic and social development', *Environment and Planning A* 36(12): 2091–2096.
Pike, A., Lagendijk, A. and Vale, M. (2000) 'Critical reflections on "embeddedness" in economic geography: the case of labour market governance and training in the automotive industry in the North-East region of England', in A. Giunta, A. Lagendijk and A. Pike (eds) *Restructuring Industry and Territory: The Experience of Europe's Regions*. London: The Stationery Office.
Pike, A., Champion, T., Coombes, M., Humphrey, L. and Tomaney, J. (2006) *The Economic Viability and Self-Containment of Geographical Economies* (Report for ODPM). Newcastle upon Tyne: CURDS, University of Newcastle.
Piore, M.J. and Sabel, C.F. (1984) *The Second Industrial Divide: Possibilities for Prosperity*. New York: Basic Books.
Polanyi, K. (1944) *The Great Transformation: The Political and Economic Origins of our Time*. New York: Farrar & Rinehart.
Pollard, S. (1981) *Peaceful Conquest: The Industrialization of Europe, 1760–1970*. Oxford: Oxford University Press.
Pollard, S. (1999) *Labour History and the Labour Movement in Britain*. Aldershot: Ashgate.
Porter, M. (1985) *Competitive Advantage: Creating and Sustaining Superior Performance*. New York: Free Press.
Porter, M. (1990) *The Competitive Advantage of Nations*. New York: Free Press.
Porter, M. (1995) 'The competitive advantage of the inner city', *Harvard Business Review* 73: 55–71.
Porter, M. (1998) *On Competition*. Boston, MA: Harvard Business School Press.
Porter, M. (2000) 'Locations, clusters and company strategy', in G.L. Clark, M. Feldman and M. Gertler (eds) *The Oxford Handbook of Economic Geography*. Oxford: Oxford University Press.
Porter, M. (2003) 'The economic performance of regions', *Regional Studies* 37(6–7): 549–578.
Prime Minister's Strategy Unit (2004) *Strategy Survival Guide*, www.strategy.gov.uk/downloads/survivalguide/downloads/alternative_policy_instruments.pdf, accessed 16 March 2006.
Puga, D. (2002) 'European regional policies in light of recent location theories', *Journal of Economic Geography* 2(4): 373–406.

Putnam, R. (1993) *Making Democracy Work: Civic Traditions in Modern Italy*. Princeton, NJ: Princeton University Press.

Putnam, R. (1995) 'Bowling alone: America's declining social capital', *Journal of Democracy* 6(1): 65–78.

Putnam, R. (2000) *Bowling Alone: The Collapse and Revival of American Community*. New York: Simon & Schuster.

Putnam, R. (2002) *Democracies in Flux: The Evolution of Social Capital in Contemporary Society*. Oxford: Oxford University Press.

Pyke, F. and Sengenberger, W. (1992) *Industrial Districts and Local Economic Regeneration*. Geneva: International Institute for Labour Studies.

Quinn, D.P. and Woolley, J.T. (2001) 'Democracy and national economic performance: the preference for stability', *American Journal of Political Science* 45: 634–657.

Rabellotti, R. (1999) 'Recovery of a Mexican cluster: devaluation bonanza or collective efficiency?', *World Development* 27(9): 1571–1585.

Radice, H. (1999) 'Taking globalisation seriously', in L. Panitch and C. Leys (eds) *Global Capitalism Versus Democracy* (*Socialist Register 1999*). Rendlesham: Merlin Press.

Rahnema, M. and Bawtree, V. (eds) (1997) *The Post-Development Reader*. London: Zed Books.

Rainnie, A. and Grobelaar, M. (2004) *New Regionalism in Australia*. Aldershot: Ashgate.

Rainnie, A. and Hardy, J. (1996) *Restructuring Krakow, Desperately Seeking Capitalism*. London: Cassell Mansell.

Ram, M. and Smallbone, D. (2003) 'Policies to support ethnic minority enterprise: the English experience', *Entrepreneurship and Regional Development* 15(2): 151–166.

Ramey, G. and Ramey, V.A. (1995) 'Cross-country evidence on the link between volatility and growth', *American Economic Review* 85: 1138–1151.

Rees, T. (2000) 'The learning region! Integrating gender equality into regional economic development', *Policy and Politics* 28(2): 179–191.

Reese, L. (1992) 'Explaining the extent of local economic development activity: evidence from Canadian cities', *Environment and Planning C: Government and Policy* 10: 105–120.

Reese, L. (1997) *Local Economic Development Policy: The United States and Canada*. New York: Garland.

Reese, L.A. and Fasenfest, D. (1999) 'Critical perspectives on local development policy evaluation', *Economic Development Quarterly* 13(1): 3–7.

Reich, R.B. (1991) *The Work of Nations: Preparing Ourselves for 21st Century Capitalism*. London: Simon & Schuster.

Reynolds, P., Storey, D.J. and Westhead, P. (1994) 'Cross-national comparison of the variation in new firm formation rates', *Regional Studies* 28: 443–456.

Rhodes, R. (1996) 'The new governance: governing without government', *Political Studies* 44(4): 652–667.

Richardson, H.W. (1979) 'Aggregate efficiency and interregional equity', in H. Folmer and J. Oosterhaven (eds) *Spatial Inequalities and Regional Development*. Boston, MA: Martinus Nijhoff.

Richardson, H.W. (1980) 'Polarization reversal in developing countries', *Papers of the Regional Science Association* 45: 67–85.

Richardson, R., Belt, V. and Marshall, J.N. (2000) 'Taking calls to Newcastle: the regional implications of the growth in call centres', *Regional Studies* 34(4): 357–369.

Ringen, S. (2004) 'Wealth and decay: Norway funds a massive political self-examination – and finds trouble for all', *Times Literary Supplement* 13 February: 3–5.

Roberts, P. (2004) 'Wealth from waste: local and regional economic development and the environment', *The Geographical Journal* 170(2): 126–134.

Robinson, F. (ed.) (1989) *Post-Industrial Tyneside*. Newcastle upon Tyne: Newcastle upon Tyne City Libraries and Arts.

Robinson, F. (2002) 'The North East: a journey through time', *City* 6(3): 317–334.

Robinson, J. (1964) *The Economics of Imperfect Competition*. London: Macmillan.

Rodríguez, V.E. and Ward, P.M. (1995) *Opposition Government in Mexico*. Albuquerque, NM: University of New Mexico Press.

Rodríguez-Pose, A. (1994) 'Socioeconomic restructuring and regional change: rethinking growth in the European Community', *Economic Geography* 70(4): 325–343.

Rodríguez-Pose, A. (1996) 'Growth and institutional change: the influence of the Spanish regionalization process on economic performance', *Environment and Planning C: Government and Policy* 14(1): 71–87.

Rodríguez-Pose, A. (1998) *Dynamics of Regional Growth in Europe*. Oxford: Clarendon Press.
Rodríguez-Pose, A. (1999) 'Instituciones y desarrollo económico', *Ciudad y Territorio Estudios Territoriales* 31(122): 775–784.
Rodríguez-Pose, A. (2001) 'Is R&D investment in lagging areas of Europe worthwhile? Theory and empirical evidence', *Papers in Regional Science* 80(3): 275–295.
Rodríguez-Pose, A. (2002a) *The European Union: Economy, Society, and Polity*. Oxford: Oxford University Press.
Rodríguez-Pose, A. (2002b) *The Role of the ILO in Implementing Local Economic Development Strategies in a Globalised World*. Geneva: ILO.
Rodríguez-Pose, A. and Arbix, G. (2001) 'Strategies of waste: bidding wars in the Brazilian automobile sector', *International Journal of Urban and Regional Research* 25(1): 134–154.
Rodríguez-Pose, A. and Fratesi, U. (2004) 'Between development and social policies: the impact of European Structural Funds in Objective 1 regions', *Regional Studies* 38(1): 97–113.
Rodríguez-Pose, A. and Gill, N. (2003) 'The global trend towards devolution and its implications', *Environment and Planning C* 21(3): 333–351.
Rodríguez-Pose, A. and Gill, N. (2004) 'Is there a global link between regional disparities and devolution?', *Environment and Planning A* 36(12): 2097–2117.
Rodríguez-Pose, A. and Gill, N. (2005) 'On the "economic dividend" of devolution', *Regional Studies* 39(4): 405–420.
Rodríguez-Pose, A. and Refolo, M.C. (2003) 'The link between local production systems and public and university research in Italy', *Environment and Planning A* 35(8): 1477–1492.
Rodríguez-Pose, A. and Tomaney, J. (1999) 'Industrial crisis in the centre of the periphery: stabilization, economic restructuring and policy responses in São Paulo', *Urban Studies* 36(3): 497–498.
Rodríguez-Pose, A., Tomaney, J. and Klink, J. (2001) 'Local empowerment through economic restructuring in Brazil: the case of the Greater ABC region', *Geoforum* 32(4): 459–469.
Rodrik, D. (2000) 'How far will international economic integration go?', *Journal of Economic Perspectives* 14(1): 177–186.
Rogers, J. (2004) 'Devolve this!', *The Nation* 30 August: 20–28.
Romer, P. (1990) 'Endogenous technological change', *Journal of Political Economy* 98: 71–102.
Romer, P. (1994) 'The origins of endogenous growth', *Journal of Economic Perspectives* 8(1): 3–22.
Rosenstein-Rodan, P.N. (1943) 'Problems of industrialization of Eastern and South-Eastern Europe', *Economic Journal* 33: 202–211.
Rostow, W.W. (1971) *The Stages of Economic Growth: A Non-Communist Manifesto* (2nd edn). Cambridge: Cambridge University Press.
Royal Commission on the Economic Union and Development Prospects for Canada (MacDonald Commission) (1985) *Report*, vol. 1. Ottawa: Queen's Printer.
Rozman, G. (2002) 'Decentralization in East Asia: a reassessment of its background and potential', *Development and Society* 31(1): 1–22.
Ruigrok, W. and van Tulder, R (1995) *The Logic of International Restructuring*. London: Routledge.
Sachs, J.D. and Warner, A (1995) 'Economic reform and the process of global integration', *Brookings Papers on Economic Activity* 1: 1–95.
Sachs, J.D. and Warner, A. (1997) 'Fundamental sources of long-run growth', *American Economic Review* 87: 184–188.
Sánchez-Reaza, J. and Rodríguez-Pose, A. (2002) 'The impact of trade liberalization on regional disparities in Mexico', *Growth and Change* 33: 72–90.
Sawers, L. and Tabb, W.K. (1984) *Sunbelt/Snowbelt: Urban Development and Regional Restructuring*. New York: Oxford University Press.
Saxenian, A. (1989) 'The Cheshire Cat's grin: innovation, regional development and the Cambridge case', *Economy and Society* 18(4): 448–477.
Saxenian, A. (1994) *Regional Advantage: Culture and Competition in Silicon Valley and Route 128*. Cambridge, MA: Harvard University Press.
Saxenian, A. (1995) 'Creating a twentieth century technical community: Frederick Terman's Silicon Valley', Paper for the Inventor and the Innovative Society Symposium, Lemelson Center for the Study of Invention and Innovation, November.
Saxenian, A. (1999) *Silicon Valley's New Immigrant Entrepreneurs*. San Francisco, CA: Public Policy Institute of California.
Sayer, A. (1985) 'Industry and space: a sympathetic critique of radical research', *Environment and Planning D: Society and Space* 3: 3–29.

Sayer, A. (1989) 'Dualistic thinking and rhetoric in geography', *Area* 21: 301–305.
Scharpf, F.W. (1991) *Crisis and Choice in European Social Democracy*. Ithaca, NY: Cornell University Press.
Schneider, F. and Enste, D.H. (2000) 'Shadow economies: size, causes, and consequences', *Journal of Economic Literature* 38: 77–114.
Schoenberger, E. (1989) 'Thinking about flexibility: a response to Gertler', *Transactions of the Institute of British Geographers* 14: 98–108.
Schoenberger, E. (2000) 'The management of time and space', in G.L. Clark, M.P. Feldman and M.S. Gertler (eds) *The Oxford Handbook of Economic Geography*. Oxford: Oxford University Press.
Schuman, M. (1998) *Going Local: Creating Self-Reliant Communities in a Global Age*. New York: Free Press.
Schumpeter, J.L. (1994) *Capitalism, Socialism and Democracy*. London: Routledge.
Scott, A.J. (1986) 'High technology industry and territorial development: the rise of the Orange County complex, 1955–1984', *Urban Geography* 7: 3–45.
Scott, A.J. (1988) *New Industrial Spaces*. London: Pion.
Scott, A.J. (1998) *Regions and the World Economy: The Coming Shape of Global Production, Competition and Political Order*. Oxford: Oxford University Press.
Scott, A.J. (ed.) (2003) *Global City-Regions: Trends, Theory, Policy*. Oxford: Oxford University Press.
Scott, A.J. (2004) 'A perspective of economic geography', *Journal of Economic Geography* 4: 479–499.
Scott, A.J. and Storper, M. (2003) 'Regions, globalization, development', *Regional Studies* 37(6–7): 579–593.
Seers, D. (1967) 'The limitations of the special case', in K. Martin and J. Knapp (eds) *The Teaching of Development Economics*. London: Cass.
Sen, A. (1999) *Development as Freedom*. Oxford: Oxford University Press.
Sennett, R. (1998) *The Corrosion of Character*. New York: Norton.
Shaiken, H. (1993) 'Going south: Mexican wages and US jobs after NAFTA', *American Prospect* 15: 58–64.
Sharpe, L.J. (1993) *The Rise of Meso Government in Europe*. Newbury Park, CA: Sage.
Shin D-H. (2000) 'Governing interregional conflicts: the planning approach to managing spillovers of extended metropolitan Pusan, Korea', *Environment and Planning A* 32: 507–518.
Shin D-H. (2004) 'Restructuring the footwear cluster in Busan, Korea', *Journal of the Korean Regional Science Association* 20(1): 79–101.
Shin D-H. (2006) 'Regional innovation and reform policies of the new government in Korea', Paper presented at the Forty-fifth Annual Meeting of the Western Regional Science Association, Sante Fe, New Mexico, 22–25 February.
Shirk, D.A. (2000) 'Mexico's victory: Vicente Fox and the rise of the PAN', *Journal of Democracy* 11(4): 25–32.
Shirlow, P. (1995) 'Transnational corporations in the Republic of Ireland and the illusion of economic well-being', *Regional Studies* 29: 687–705.
Shoval, N. (2002) 'A new phase in the competition for Olympic Gold: the London and New York bids for the 2012 games', *Journal of Urban Affairs* 24(5): 583–600.
Siegel, L. (1998) 'New chips in old skins: work and labour in Silicon Valley', in G. Sussman and J.A. Lent (eds) *Global Productions: Labor in the Making of the 'Information Society'*. Cresskills, NJ: Hampton Press.
Siggins, L. (2005) 'Bishops call for more investment in the west', *Irish Times* 13 June.
Singer, H.W. (1961) 'Trends in economic thought or underdevelopment', *Social Research* 4: 387–414.
Singer, H.W. (1975) *The Strategy of International Development*. London: Macmillan.
Skelcher, C., Weir, S. and Wilson, L. (2000) *Advance of the Quango State*. London: Local Government Information Unit.
Smith, A., Rainnie, A., Dunford, M., Hardy, J., Hudson, R. and Sadler, D. (2002) 'Networks of value, commodities and regions: reworking divisions of labour in macro-regional economies', *Progress in Human Geography* 26(1): 41–63.
Smith, E.O. (1994) *The German Economy*. London: Routledge.
Smith, I. (1985) 'Takeovers and rationalisation: impacts on the Northern economy', *Northern Economic Review* 12: 30–38.
Smith, N. (2004) 'Deconstructing "globalisation" in Ireland', *Policy and Politics* 32(4): 503–519.

Sokol, M. (2001) 'Central and Eastern Europe a decade after the fall of state-socialism: regional dimensions of transition processes', *Regional Studies* 35(7): 645–655.

Soto, H. de (1989) 'Structural adjustment and the informal sector', in J. Levitsky (ed.) *Microenterprises in Developing Countries*. London: Intermediate Technology Publications.

Standing, G. (1999) *Global Labour Flexibility: Seeking Distributive Justice*. London: Macmillan.

Sternberg, R. (1996) 'Regional growth theories and high-tech regions', *International Journal of Urban and Regional Research* 20(3): 518–538.

Stiglitz, J. (2002) *Globalization and its Discontents*. New York: Norton.

Stöhr, W.B. (ed.) (1990) *Global Challenge and Local Response: Initiatives for Economic Regeneration in Contemporary Europe*. London: The United Nations University, Mansell.

Stoker, G. (1995) 'Governance as theory: five propositions', *International Social Science Journal* 50(155): 17–28.

Stoker, G. (1997) 'Public–private partnerships and urban governance', in J. Pierre (ed.) *Partnerships in Urban Governance: European and American Experience*. London: Macmillan.

Storey, D.J. (1994) *Understanding the Small Business Sector*. London: Routledge.

Storper, M. (1985) 'Oligopoly and the product cycle: essentialism in economic geography', *Economic Geography* 61: 260–282.

Storper, M. (1995) 'The resurgence of regional economies, ten years later: the region as a nexus of untraded interdependencies', *European Urban and Regional Studies* 2(3): 191–221.

Storper, M. (1997) *The Regional World: Territorial Development in a Global Economy*. London: Guilford.

Storper, M. and Scott, A.J. (1988) 'The geographical foundations and social regulation of flexible production complexes', in J. Wolch and M. Dear (eds) *The Power of Geography*. Boston, MA: Allen & Unwin.

Storper, M. and Scott, A.J. (eds) (1992) *Pathways to Industrialization and Regional Development*. London: Routledge.

Storper, M. and Walker, R. (1989) *The Capitalist Imperative: Territory, Technology and Industrial Growth*. Oxford: Blackwell.

Storper, M., Thomadakis, S.B. and Tsipouri, L.J. (eds) (1998) *Latecomers in the Global Economy*. London: Routledge.

Stough, R.R. (1998) 'Endogenous growth in a regional context', *Annals of Regional Science* 32(1): 1–5.

Strange, S. (1994) *States and Markets* (2nd edn). London: Pinter.

Stutz, F.P. and Warf, B. (2005) *The World Economy* (4th edn). Englewood Cliffs, NJ: Pearson/Prentice Hall.

Sunley, P. (1996) 'Context in economic geography: the relevance of pragmatism', *Progress in Human Geography* 20(3): 338–355.

Sunley, P. (2000) 'Urban and regional growth', in T.J. Barnes and E. Sheppard (eds) *A Companion to Economic Geography*. Oxford: Blackwell.

Sweeney, P. (1997) *The Celtic Tiger: Ireland's Economic Miracle Explained*. Dublin: Oak Tree Press.

Sweeney, P. (2004) *Tax Cuts Did Not Create the Celtic Tiger*. Dublin: Irish Congress of Trade Unions (ICTU).

Swyngedouw, E. (1997) 'Neither global nor local: "glocalisation" and the politics of scale', in K.R. Cox (ed.) *Spaces of Globalisation: Reasserting the Power of the Local*. New York: Guilford.

Szreter, S. (2002) 'The state of social capital: bringing back in power, politics and history', *Theory and Society* 31(2): 573–621.

Taylor, J. and Wren, C. (1997) 'UK regional policy: an evaluation', *Regional Studies* 31(9): 835–848.

Taylor, M. (1986) 'The product-cycle model: a critique', *Environment and Planning A* 18: 751–761.

Taylor, P. and Flint, C. (2000) *Political Geography: World-Economy, Nation-State and Locality* (4th edn). Harlow: Prentice Hall.

Taylor, P.J. and Walker, D.R.F. (2001) 'World cities: a first multivariate analysis of their service complexes', *Urban Studies* 38: 23–47.

Telesis (1982) *A Review of Industrial Policy* (NESC Report no. 62). Dublin: National Economic and Social Council.

Therborn, G. (1995) *European Modernity and Beyond*. London: Sage.

Thomas, A. (2000) 'Development as practice in a liberal capitalist world', *Journal of International Development* 12: 773–787.

Thompson, E.P. (1963) *The Making of the English Working Class*. London: Victor Gollancz.

Thompson, W.R. (1968) *A Preface to Urban Economics*. Baltimore, MD: Johns Hopkins University Press.

Tomaney, J. (1995) 'Recent developments in Irish industrial policy', *European Planning Studies* 3(1): 99–113.

Tomaney, J. (1996) 'Regional government and economic development: possibilities and limits', *Local Economy* 11(1): 27–38.

Tomaney, J. (2000) 'End of the Empire State? New Labour and devolution in the United Kingdom', *International Journal of Urban and Regional Research* 24(3): 677–690.

Tomaney, J. (2002) 'The evolution of English regional governance', *Regional Studies* 36(7): 721–731.

Tomaney, J. (2003) 'Politics, institutions and the decline of coalmining in North East England', *Transactions of the Institute of Mining and Metallurgy A: Mining Technology* 112(1): 40–47.

Tomaney, J. (2005) 'Anglo-Scottish relations: a borderland perspective', in W.L. Miller (ed.) *Anglo-Scottish Relations since 1900 to Devolution and Beyond* (Proceedings of the British Academy 128). Oxford: Oxford University Press.

Tomaney, J., Pike, A. and Cornford, J. (1999) 'Plant closure and the local economy: the case of Swan Hunter on Tyneside', *Regional Studies* 33(5): 401–411.

Townroe, P.M. and Keen, D. (1984) 'Polarization reversed in the State of São Paulo', *Regional Studies* 18: 45–54.

Toye, J. (1987) *Dilemmas of Development: Reflections on the Counter-Revolution in Development Theory and Policy*. Oxford: Basil Blackwell.

Trends Business Research (2003) *Business Clusters in the UK – A First Assessment* (Final Report to the DTI). Newcastle upon Tyne: Trends Business Research.

Trigilia, C. (1992) *Sviluppo senza Autonomia: Effetti Perversi delle Politiche nel Mezzogiorno*. Bologna: Il Mulino.

Trigilia, C. (2001) 'Social capital and local development', *European Journal of Social Theory* 4: 427–442.

Turner, R. and Gregory, M. (1996) 'Life after the pit: the post-redundancy experiences of mineworkers', *Local Economy* 10: 149–162.

United Nations Conference on Trade and Development (UNCTAD) (2004) *World Investment Report 2004: The Shift Towards Services*. Geneva: UNCTAD.

United Nations Development Programme (UNDP) (2001) *Human Development Report 2001*. Oxford: Oxford University Press.

United Nations Economic Commission for Europe (2000) *Best Practice in Business Incubation*. Geneva: United Nations.

United Nations Industrial Development Organization (UNIDO) (1995) *Export Processing Zones: Principles and Practice*. Vienna: UNIDO.

Vale, M. (2004) 'Innovation and knowledge driven by a focal corporation: the case of the Autoeuropa supply chain', *European Urban and Regional Studies* 11(2): 124–140.

Vanhoudt, P., Mathä, T. and Smid, B. (2000) 'How productive are capital investments in Europe?', *EIB Papers* 5: 81–105.

van Stel, A.J. and Storey, D.J. (2004) 'The link between firm births and job creation: is there an Upas Tree effect?', *Regional Studies* 38(8): 893–909.

Vázquez Barquero, A. (1999) *Desarrollo, redes e innovación: lecciones sobre desarrollo endógeno*. Madrid: Pirámide.

Vázquez Barquero, A. (2003) *Endogenous Development: Networking, Innovation, Institutions and Cities*. London and New York: Routledge.

Vázquez Barquero, A. and Carrillo, E. (2004) 'Cartuja 98: a technological park located at the site of Sevilla's World's Fair', Paper presented to the Forty-fourth European Regional Science Association (ERSA) 2004 Congress, University of Porto, Portugal, 25–29 August.

Vernon, R. (1966) 'International investment and international trade in the product cycle', *Quarterly Journal of Economics* 80(2): 190–207.

Vernon, R. (1979) 'The product cycle in a new international environment', *Oxford Bulletin of Economics and Statistics* 41: 255–267.

Viesti, G. (2000) *Come nascono i distretti industriali*. Rome: Laterza.

Vigor, A. (2002) 'Cows: the Center on Wisconsin Strategy', *Local Economy* 17(4): 273–288.

Wade, R. (2003) *Governing the Market: Economic Theory and the Role of Government in East Asian Industrialization* (2nd edn). Princeton, NJ: Princeton University Press.

Wade, R. (2004) 'Is globalization reducing poverty and inequality?', *World Development* 32(4): 567–589.

Wainwright, H. (2003) *Reclaim the State: Experiments in Popular Democracy*. London and New York: Verso.

Wales TUC (n.d.) *Economic Development and Regeneration*. Cardiff: Wales TUC.

Walker, R. (1995) 'California rages against the dying of the light', *New Left Review* 209: 42–74.

Walsh, J. (1995) 'EC Structural Funds and economic development in the Republic of Ireland', in P. Shirlow (ed.) *Development Ireland: Contemporary Issues*. London: Pluto.

Warde, A. (1985) 'Spatial change, politics and the division of labour', in D. Gregory and J. Urry (eds) *Social Relations and Spatial Structures*. London: Macmillan.

Weinstein, B.L., Gross, H.T. and Rees, J. (1985) *Regional Growth and Decline in the United States*. New York: Praeger.

White, M. (2004) 'Inward investment, firm embeddedness and place: an assessment of Ireland's multinational software sector', *European Urban and Regional Studies* 11(3): 243–260.

White, M. (2005) 'Assessing the role of Dublin's International Financial Services Centre in Irish regional development', *European Planning Studies* 13(3): 387–405.

White, S. and Gasser, M. (2001) *Local Economic Development: A Tool For Supporting Locally Owned and Managed Development Processes that Foster the Global Promotion of Decent Work*. Geneva: Job Creation and Enterprise Development Department, ILO.

Williams, C.C. (1996) 'Local purchasing schemes and rural development: an evaluation of local exchange trading systems (LETS)', *Journal of Rural Studies* 12(3): 231–244.

Williams, C.C. and Millington, A.C. (2004) 'The diverse and contested meanings of sustainable development', *The Geographical Journal* 170(2): 99–104.

Williams, R. (1983) *Keywords*. London: HarperCollins.

Williamson, J.G. (1965) 'Regional inequalities and the process of national development', *Economic Development and Cultural Change* 13: 1–84.

Williamson, J.G. (1997) 'Globalization and inequality, past and present', *World Bank Research Observer* 12: 117–135.

Wills, J. (1998). 'A stake in place? The geography of employee ownership and its implications for a stakeholding society', *Transactions of the Institute of British Geographers* 23(1): 79–94.

Wolfe, D.A. (2002) 'Negotiating order: sectoral policies and social learning in Ontario', in M.S. Gertler and D.A. Wolfe (eds) *Innovation and Social Learning: Institutional Adaptation in an Era of Technological Change*. Basingstoke: Palgrave Macmillan.

Wolfe, D.A. and Creutzberg, T. (2003) *Community Participation and Multilevel Governance in Economic Development Policy*. Toronto: Program on Globalization and Regional Innovation Systems, Centre for International Studies, University of Toronto.

Wolfe, D.A. and Gertler, M. (2001) 'Globalization and economic restructuring in Ontario: from industrial heartland to learning region?', *European Planning Studies* 9(5): 575–592.

Wolfe, D.A. and Gertler, M. (2002) 'Innovation and social learning: an introduction', in M. Gertler and D.A. Wolfe (eds) *Innovation and Social Learning: Institutional Adaptation in an Era of Technological Change*. Basingstoke: Palgrave Macmillan.

Wood, A. and Valler, D. (2001) 'Turn again? Rethinking institutions and the governance of local and regional economies', *Environment and Planning A* 33: 1139–1144.

Woods, A. (1994) *North–South Trade, Employment and Inequality: Changing Fortunes in a Skill-Driven World*. Oxford: Clarendon Press.

Woo Gómez, G. (2002) *La regionalización: Nuevos horizontes para la gestión pública*. Guadalajara, Mexico: Universidad de Guadalajara and Centro Lindavista.

Woolcock, M. (1998) 'Social capital and economic development: toward a theoretical synthesis and policy framework', *Theory and Society* 27: 151–208.

World Bank (1992) *Export Processing Zones*, Policy and Research Series 20. Washington, DC: World Bank.

World Bank (1995) *The East Asian Miracle: Economic Growth and Public Policy*. Oxford: Oxford University Press.

World Commission on Environment and Development (1987) *Our Common Future*. Oxford: Oxford University Press.

Young, S. and Hood, N. (1995) 'Attracting, managing and developing inward investment in the single market', in A. Amin and J. Tomaney (eds) *Behind the Myth of European Union*. London: Routledge.

Young, S., Hood, N. and Peters, E. (1994) 'Multinational enterprises and regional development', *Regional Studies* 28(7): 657–677.

Zysman, J. (1996) 'The myth of a "global" economy: enduring national foundations and emerging regional realities', *New Political Economy* 1(2): 157–184.

INDEX

Note: page numbers in *italics* denote references to Figures, Tables and Plates.